# BUILD-IT BOOK OF DIGITAL CLOCKS

2ND EDITION

ROBERT HAVILAND

TAB BOOKS Inc.
Blue Ridge Summit, PA 17214

SECOND EDITION

FIRST PRINTING


Printed in the United States of America



Library of Congress Cataloging in Publication Data

Haviland, Robert P.
Build-it book of digital clocks.

Bibliography: p.
Includes index.
1. Digital electronic clocks. I. Title.
TS544.H32 1986 681.1'13 85-27879
ISBN 0-8306-0683-1
ISBN 0-8306-1683-7 (pbk.)

# Contents

# Introduction

In revising this volume for a second edition, my primary goal was to cover the developments in electronic clock and clock chip design that occurred in the five years since the original work was completed. In doing this, it became apparent to me that the range of chips specifically designed for clocks had decreased greatly, and that they were somewhat more difficult to obtain. In part, this always occurs as a field matures—the "solid" design features are kept and the "marginal" ones, those that are less popular, are dropped. The reduction in availability is partly a consequence of drop in demand but also results from more careful buying. When business is poor, scrap at model changeover is just too expensive to allow.

While clock chips per se are less available, there has been a great expansion in time-keeping chips for special purposes. These include clocks for computers, for TV screen presentation, and general and special-purpose timers in simple and complex form.

Rather than replacing original material by new to accommodate these changes, the approach has been to add material. Partly, the old material is useful in following the course of development. In addition, the older design chips are sometimes found in surplus, and in radio amateur and hobby "flea market" sales.

In addition, there is another source of these earlier chips. Production clocks based on these are approaching the time of high failure rate, largely due to drying out of electrolytic capacitors, but also due to displays reaching end of life. These failed units, as always, are a good source of parts for experimenters.

No change has been made in the construction projects. Several use general purpose components to develop timekeeping functions, and the original components are readily available, as are several families of identical function types. In designs using specific clock chips, the chip used is not critical, and data on use of other chips was already included.

One of the newly added chapters is termed "Clocks for the Handicapped." It is a new field to me; in fact, it seems to have been neglected generally. The material included covers what I have learned. There is a real opportunity here for creative experimentation.

# Chapter 1

# Time and Timekeeping

The idea of time seems to have arisen from man's observation of natural events. The Sun was observed to rise and set, to rise and set again. This obviously led to the concept of day and of night, and then to the more abstract concept of a day. Associated with these observations was the idea that these were unique, identifiable events which could be counted. The concept of the flow of time followed as a consequence of the successive counts.

Another event which was observed was the lunar cycle, the phenomenon of the changing phases of the moon with successive days. This apparently led to the concept of a month, because many primitive societies developed the idea of a unit of time, a moon. In the same way the stars were observed to return to the same evening positions, especially observable for the three bright stars which make up the belt of Orion. This, coupled with the pattern of the seasons, seems to have led to the concept of a longer period, the year.

It is easy to underestimate the observational ability of these ancients and their ability to draw conclusions. For example, there is a relation between the motion of the moon with respect to Earth and Sun which causes the same lunar phenomena to repeat with a cycle of just over eighteen years. If an eclipse happens in a particular area, another eclipse will occur in the same area just eighteen years later. This cycle, called the Saros, was known to the Babylonians and is the basis of the many stories of magical predic-

Table 1-1. Brief History of Timekeeping.

| | |
|---|---|
| 3500 BC | Sundial, using Vertical Gnomon |
| 1400 | Water Clock, Rate-Controlled Flow |
| 800 | Egypt, Sundial, Horizontal Gnomon |
| 300 | Equal division, Hemispherical Sundial |
| 100 AD | Sand Glass |
| 1335 | Striking Clock, Italy |
| 1500 | Spring Driven Clock |
| 1582 | Galileo developed Pendulum |
| 1600 | Candle Clocks |
| 1840 | Battery driven Pendulum Clock |
| 1906 | Electric Drive Clock |
| 1918 | Synchronous Motor developed |
| 1929 | Quartz Crystal Clocks |
| 1952 | Caesium Beam Clocks |

tions of eclipses. One wonders how many lifetimes it took to observe this repeated cycle and also to deduce that it was a recurrent natural event. One also wonders at the scale of record keeping needed to accomplish this. And it is interesting to speculate on the reaction of the astronomical community if such a cycle were first proposed today. Table 1-1 shows a look at timekeeping throughout the ages.

It seems that man is a counting and measuring creature. Following the discovery of time and of the month and year, a calendar was invented to organize this counting. Similarly, to measure time, subdivisions of a day were worked out. The concept of the hour, the minute, and the second seem to have come from consideration of the subdivisions of a circle. The fact that the units are the same must have been started by dividing the day into twelve parts. The relationship of time and angle also implies that there was an understanding of the circular motion of bodies, and also of rotation of the earth. The parallelism between time and angle can hardly be accidental. Incidentally, it may be noted that astronomers still often measure angles in units of time: a particular item is the quantity known as right ascension.

## TIME MEASURING INSTRUMENTS

It appears definite that the first instrument developed to measure time was the sundial. This device functions by measuring the angular position of the sun on a calibrated dial face marked with hours. The commonly seen type is shown in Fig. 1-1. The flat plate scale, mounted horizontally with respect to the earth's surface, has a nonuniform measuring scale, with the hour marks bunched together around noon, and well separated around sunrise and sunset. However, the ancients also worked out other types of sundials, including ones having linear time calibrations. The most common of these is a semicircular strip mounted with the concave side up and with the axis of the strip carrying the index, which points to the North Pole.

All of these simple instruments measure sun time. This has disadvantages for some purposes, since the duration of time required for the shadow to pass a given mark varies throughout the year, due to the eccentricity of the earth's orbit. The error from an average calibration, or "mean time," can be as great as 16.4 minutes, this occurring in the month of November. The fashion in which this error varies with date is known as the "equation of time," and can be found on many globes as the "Analemma," resembling a figure eight lying on the equator, the figure carrying date markings which gives the angular error as a function of time. Some sundials provide means for correcting from sun time to average time, possibly by changeable dial faces, or possibly by use of shadow producing elements which are curved to give the correct compensation.

An interesting feature of the sundial is that it is possible to construct one which will give the correct sun time for many places on earth simultaneously. To see how this can happen, imagine that

Fig. 1-1. A classical sundial. Made for use in the USA, the gnomen or shadow arm is fixed at an angle of 43°, about the average for the 48 states. The correct value is equal to 90° minus the latitude of the location, or 51.1° for Washington, D.C.

a large sphere is resting on earth, say at Washington, D.C. A miniature sundial is mounted on this sphere at its very top, so that it is horizontal and so that its shadow arm, the gnomen, points correctly to the elevated North Pole, and is exactly parallel to another sundial on the ground. (The elevated North Pole is near the North Star.) Since the sun's light rays move essentially in parallel lines, the sundial on the sphere and the sundial on the ground will have the same reading.

Now suppose the sundial is moved along the sphere, say to the northeast, always keeping tangent to the sphere. Suppose that it is moved by such a distance and angle that it comes to rest on the sphere at a point which corresponds to the latitude and longitude of London. Now let the arm of the sundial be oriented to due north and adjusted till its arm points exactly at the elevated North Pole. Again this sundial will be exactly parallel to another sundial, this being one mounted in London. Also again, because of the essential parallelism of the sun's rays, both sundials will read the same—the small one on the sphere will indicate the sun time in London. Sundials of this type have been made with as many as a dozen of the small individual dials, giving time in as many as a dozen cities.

Many other means of measuring time have been invented and constructed. A simple one was a candle, intended to burn at a constant rate and carrying markings along the side to indicate the passage of hours. If carefully made, and shielded from drafts, the accuracy of this candle was not too bad. Another "clock" instrument measured the time required to fill a bucket of water from a constant-flow stream—some models of this even had an automatic device for emptying the bucket when it became full, and an early form of digital counter to keep track of how many times the bucket had been emptied. Somewhat similar in principal is the sand glass, in the larger sizes called the hour glass. Nowadays these devices are relegated to timing the three minute egg, or to indicate that a telephone call is running into extra charges.

Over the centuries, search for improved methods of measuring time and keeping track of the measurements went on. Improvements in subdividing the units of measurement, to get minutes and seconds were made, as were improvements in presentation. The pendulum was invented and improved to give temperature correction. It was discovered that accurate timekeeping by a pendulum requires constant angular motion, and means for assuring this were worked out. But it was not until 1867, when Harrison invented the chronometer, that good portable timekeeping devices became available. Watches, of course, were known earlier, and clocks of the large fixed location type and of good accuracy were quite common. However, prior to a period just following Harrison's invention, accurate setting of the clock depended on astronomical observations which had to be solved for both time and position to get the local time. The problem of time measurements was responsible for the distortions of longitude in early charts of the earth.

After Harrison's development, improvement in timekeeping continued; in fact, accelerated. Clocks and watches became smaller and more accurate, reaching a high state of mechanical perfection. As this occurred, a new method of timekeeping also developed—the field of electronic timekeeping. This was made possible by the development of the vacuum tube, and developed with considerable rapidity. The quartz crystal oscillator driven by vacuum tubes, plus the counters and frequency dividers used to bring the frequency of the quartz oscillator to a reasonable value, and finally electromechanical displays to indicate and keep track of the time were developed. These rapidly surpassed older methods of time measurement in accuracy. Atomic frequency oscillators were invented and found to be even more stable than quartz crystals, replacing these in the most accurate measurements. As a result of these developments, it was possible to confirm that the earth's rotational period varied in both cyclic and unpredictable ways. This led, after many centuries, to a divorcement of the definition of time from earth's motion. To indicate this, new terms appeared—universal time and ephemeris time.

Development in precision measurements of time and in the concepts associated with these measurements still continues. In fact, change in the field of timekeeping has been a long, continuing process. For example, the ancients knew and recognized the differences between Babylonian, Egyptian and Greek time, depending on whether the day started at sunrise, at midnight, or at noon. Even as late as 1925, the navigator's day was considered to start at 12 noon, sun time. The appearance of the sun at maximum altitude is relatively easy to measure, even from a ship at sea. It is also easy to calculate latitude from this measurement. The change to the standard day became desirable when precomputed solutions were developed and made available to navigators as bound volumes. With these, any navigational problem is easily worked out, and there was no longer a reason for the special emphasis on the noon sun.

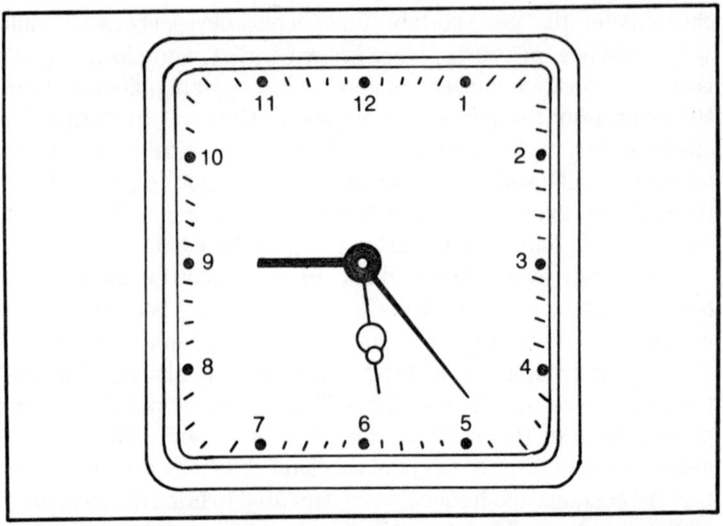


Fig. 1-2. A standard clock face. Many variations exist, the most common not having seconds markings even though a second hand is provided. The seconds data is sometimes shown by a small dial located above the six.

## HARDWARE DEVELOPMENT

Improvements in accuracy were by no means the only development. Of equal importance were reductions in size, and the associated increase in portability. Readability also increased, through development of a standard presentation, the familiar clock face shown in Fig. 1-2. Possibly even more important was the development of mass production techniques, which made these new, more accurate, and more useful devices available at low cost. One major advance in this combination of accuracy and low cost was the advent of electric power-line frequency synchronization, which made possible the low-cost electric clock, two types being shown in Fig. 1-3. These have timekeeping ability, on the average, just as accurate as the master time standard of the country. This was a major advance, since prior to that time, all clocks and watches had depended on their internal "elements" for maintenance of accuracy.

There were also important developments in portable devices, such as watches, first by using a tuning fork mechanically coupled to the gearing, and then by the introduction of miniature electronics, which made quartz crystal watches practical.

While this was going on there was also a revolution in the method of presenting time in the display. This revolution started with a mechanical device, the flipping chip clock face, with chips numbered to read hours, minutes and seconds. Movement is synchronized so that the chips flip at the correct time point. Electronic devices, in particular light emitting diodes in arrays, to give numerical displays, followed. A "kit" version is shown in Fig. 1-4.

Today, electronic clocks have converged toward a standardized design. The basic element is a "large-scale" integrated circuit (an LSI clock chip) several being shown in Fig. 1-5. These are driven from the 60 Hz power line (Table 1-2), and contain all of the electronics required to keep track of time accumulation and to drive the display. The displays are equally standardized, in the form of numerical presentation of hours, minutes and seconds.

Of course, there are a goodly number of variations in the detail around this standardized design. Some of the chips, and the clocks constructed from them, provide an electronic alarm. Some have radio control provisions, to turn a radio off after going to sleep and to turn it on in the morning as a wakeup signal. Some chips have calendar provisions, with the display alternating between time

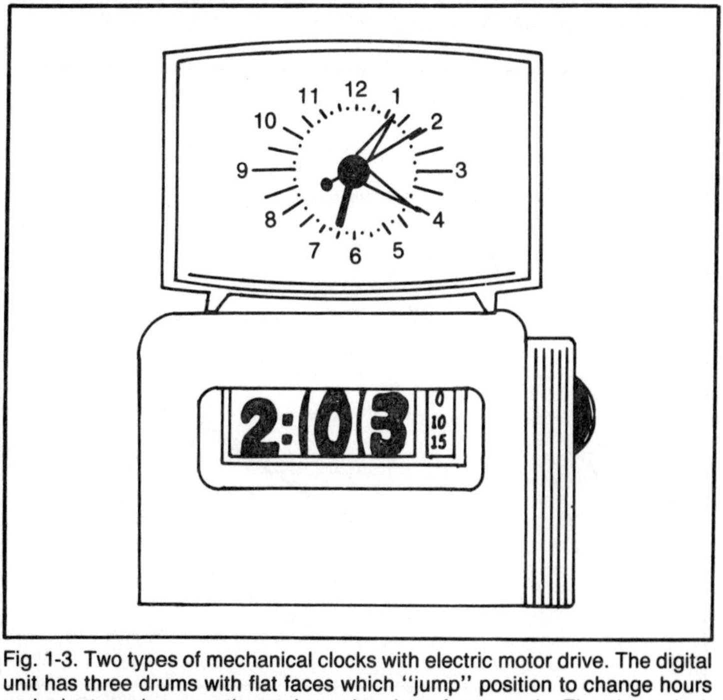


Fig. 1-3. Two types of mechanical clocks with electric motor drive. The digital unit has three drums with flat faces which "jump" position to change hours and minutes, plus a continuously turning drum for seconds. The alarm clock has no seconds markings, but instead has marks to indicate the alarm setting.

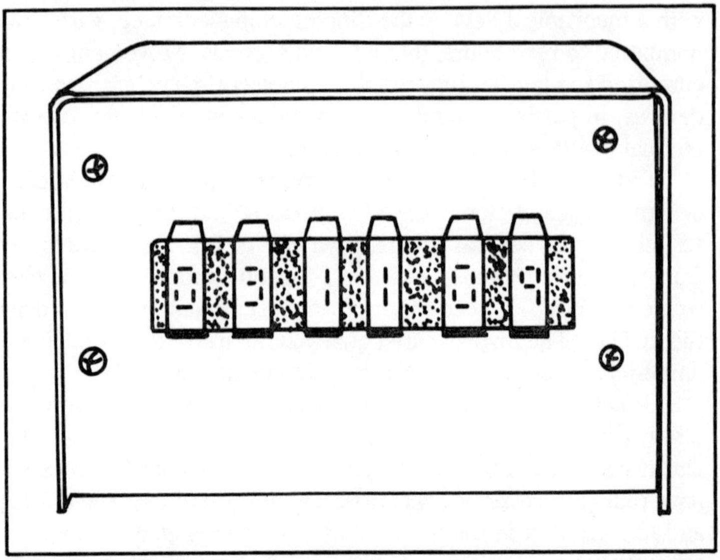

Fig. 1-4. A six-digit electronic clock built from a kit and fitted with a home-built case. The red filter which serves as a front panel has been removed to show the digit arrangement. Many types of these clock kits are available.

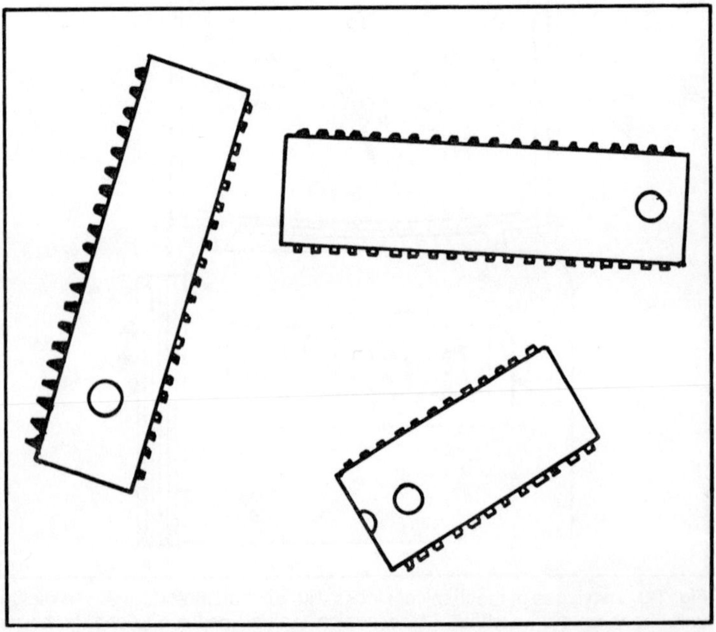

Fig. 1-5. Several clock chip integrated circuits. The 24 and 28 pin types are the most common.

Table 1-2. Conversion Factors (for 60 Hertz).

| | Cycles | Seconds | Minutes | Hours | Days |
|---|---|---|---|---|---|
| Seconds | 60 | 1 | | | |
| 10 Seconds | 600 | 10 | | | |
| Minutes | 3600 | 60 | 1 | | |
| 10 Minutes | $3.6 \times 10^4$ | 600 | 10 | | |
| Hours | $2.16 \times 10^5$ | 3600 | 60 | 1 | |
| 10 Hours | $2.16 \times 10^6$ | $3.6 \times 10^4$ | 600 | 10 | |
| Days | $5.184 \times 10^6$ | $8.64 \times 10^4$ | 1440 | 24 | 1 |
| Month (30 D) | $1.5552 \times 10^8$ | $2.592 \times 10^6$ | $4.32 \times 10^4$ | 720 | 30 |
| Year, Mean | $1.893456 \times 10^9$ | $3.15576 \times 10^7$ | $5.2596 \times 10^5$ | 8766 | 365.25 |

and date. There may be other features, for example, provisions for crystal drive, or for a standby oscillator in case of power failure. But all modern clocks of the electronic type are basically the same—a clock chip and a display.

The following chapters describe the way the clock chips work, and give data on chips readily available to the amateur experimenter. The principles of operation and use requirements for two types of displays follow, these being the type most likely to be encountered by the amateur experimenter. These chapters are followed by a series of How to Build chapters. Some of these use clock chips as intended strictly for timekeeping. Others show ways of duplicating the results obtained with a clock chip with other types of experimenter's devices. A new type of clock combining digital techniques with the standard clock face is described. There are also descriptions of a device which uses the capability of clock chips for other than timekeeping. Finally, there are some suggestions for additional experiments and designs which the experimenter might want to undertake. It is hoped that the data presented in this book will be sufficient for an experimenter to proceed far beyond the simple projects described, even if he is a relative newcomer to electronic construction.

# Chapter 2

# Elements of Electronic Timekeeping

In this chapter the elements of timekeeping as accomplished by electronic means are reviewed first. This is followed by a discussion of the major elements of an electronic clock, their principles of operation, and some of the details of their uses. These elements may be found in combined form, as in the commercial clock chip, or they may be used as building blocks for the design of one's own clock.

## THE CLOCK AS A WHOLE

All clocks, mechanical, electrical or electronic are basically the same. As indicated in Fig. 2-1, there are three main functions—oscillation, accumulation and display. In the mechanical clock, the oscillator may be a pendulum, as in the grandfather clock, or a balance wheel, a form of torsion pendulum. The accumulator is invariably gearing, and almost without exception, the indicator is an analog position display, the hour, minute and second hands.

Electric clocks use the generator at the power station as the oscillator, with the frequency of this being utilized to control clock rate. This electrical power drives a motor and gearing, and in the older clocks, a display just like the mechanical clock displays. Some electric clocks have a digital display, the older ones a multi-sided wheel containing numbers on each face, the later ones small chips which flip, changing the number reading.

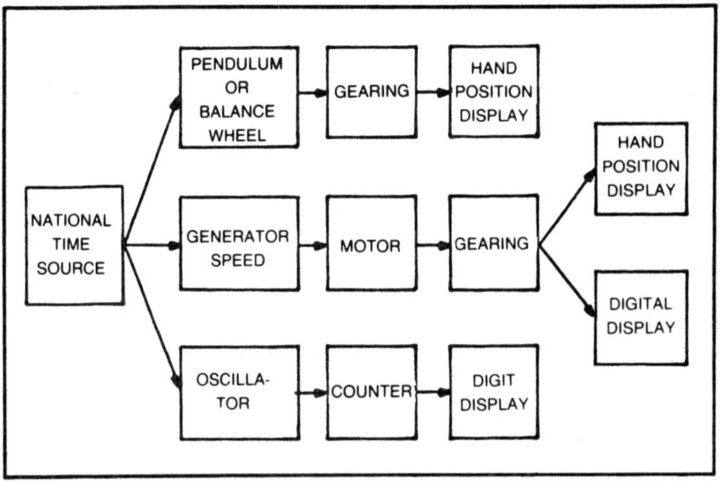


Fig. 2-1. Comparison of timekeeping principles for the commonly used clocks. Today, the most common is the electric clock (center). Most watches use the balance wheel principle, but large numbers of electronic watches are appearing (bottom).

The newest member of the clock family, the electronic clock, has no mechanical elements. The oscillator may be a frequency generator in the clock, but most likely it is the power line. The accumulators are electronic, based on the binary counter. The display is invariably electronic, some form of electronic-to-light converter or coupler.

All of these clocks may have extras—calendars, or perhaps indicators of the phase of the moon or the tide level. Many also contain timer provisions, to serve as an alarm. Some are also intentionally noise producers, striking the hours or marking them with a cuckoo's call.

In the electronic clock the basic elements, and most additional elements in the form of control and auxiliary circuits, have been integrated into a single package. Figure 2-2 shows the main elements, plus some of the additional ones which may be found. In most electronic clocks, this single integrated circuit includes all elements except the display and the setting circuits. This single package is commonly called a "clock chip." (Figure 1-5 showed several of these chips.) Externally they are no different from other large scale integrated circuits.

The following material is devoted entirely to the details of these electronic circuits. Most of the material relates to the basic clock, the part shown in solid lines in Fig. 2-2. Additional material covers the other circuits which may be found, those indicated by dotted lines in Fig. 2-2. The matter of displays and display drivers is discussed in the following chapter.

## THE ELECTRONIC CLOCK OSCILLATOR

For all practical purposes, there are only two sources of frequency to consider in clock design. One is the power line and the second is the family of crystal oscillators. Virtually all clock chips

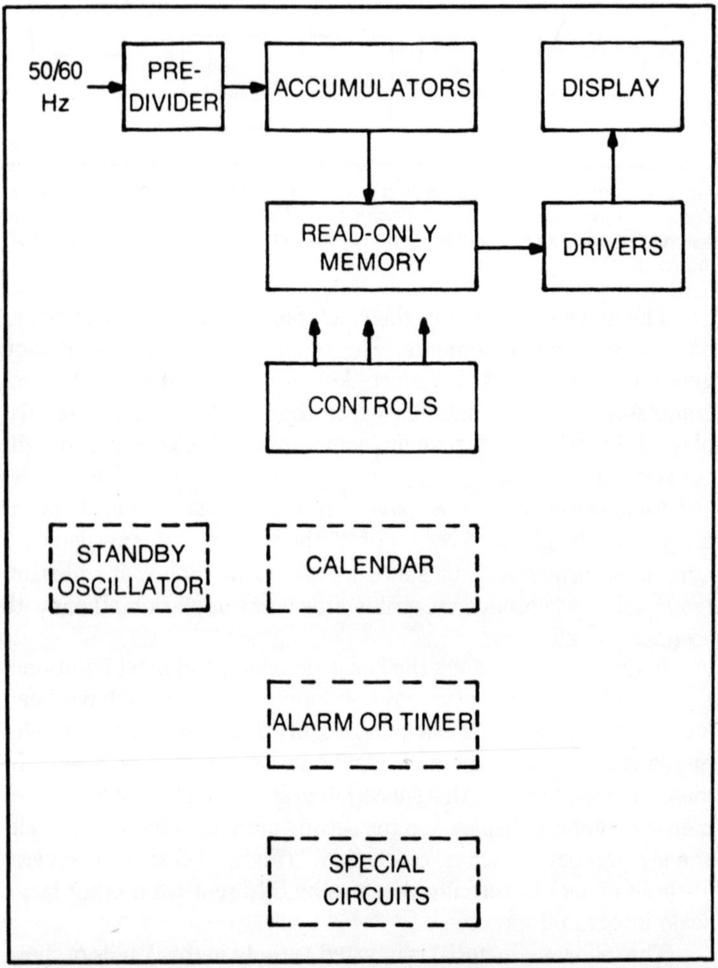


Fig. 2-2. Basic elements of a power line-operated electronic clock. The blocks with solid lines are present in all types; complex designs add one or more of the dotted block elements.

are designed to work with the power line. The crystal oscillator is common in portable timekeepers, watches in particular. A crystal oscillator has been used in a few mass produced clocks, but this is rare. There are a few other types of oscillators which have been used in the past, and a few used yet today. The tuning fork once was common for some purposes. Currently a few clock chips have provision for a standby oscillator, possibly crystal, but more likely a relaxation type, to use if the power line fails.

## POWER LINE FREQUENCY FACTORS

The power line is 60 Hz in U.S. practice, but 50 Hz in many other countries, and most clock chips have built-in capability of changing divider ratios to accommodate this difference. Derivation of the signal required to drive a clock chip from the power line has been essentially standardized, and has been combined with the power supply circuit providing dc power to the chip. The most common circuit is shown in Fig. 2-3. A bridge rectifier is used, the plus voltage going to the chip by means of a filter. The frequency drive for the chip is taken from one of the ac terminals. The reason for this arrangement is that the voltage at this point swings from the common or ground level to the maximum positive level, rather than going from positive to negative voltage as true ac. The pulsating current simplifies the design of the clock input circuits.

In Fig. 2-3, this voltage is shown as being a sine-wave, offset in the positive direction; practically, some wave shaping occurs so that the waveform is distorted in the direction of being a square-wave. The chip input circuits, however, always contain further squaring elements so that the following stages have square-waves of good rise time on which to operate.

Use of the line frequency oscillator does introduce one major factor and several problems. The major factor is that the frequency of the power line is not exactly 60 cycles but instead varies. As the amount of power demand changes, and as generators are put on the line or removed, the speed of the operating generators changes, thereby changing the frequency and also clock rate. The magnitude of this change is not enormous, but for timekeeping purposes it is appreciable. Figure 2-4 shows a graph of the major changes which occurred during the course of one evening. There were many small changes during this period, and several periods of major change. Probably these larger changes were due to the shutting down of a generator completely as the local load decreased—this would be expected shortly after 10 P.M., as most

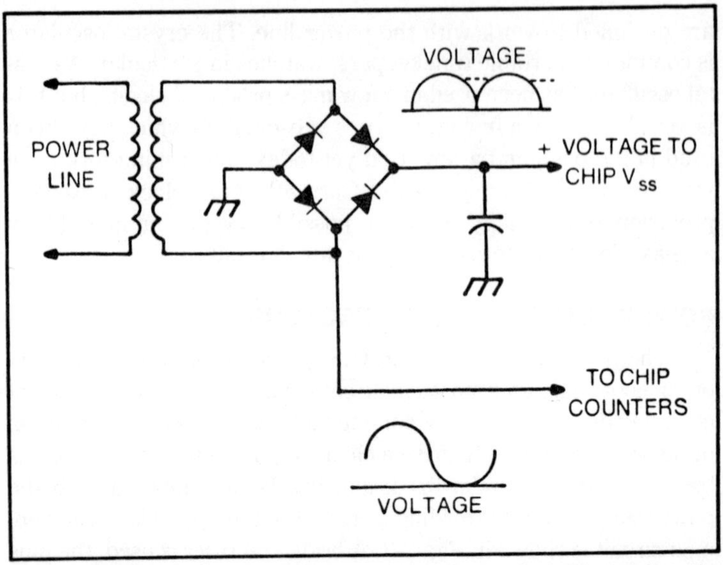


Fig. 2-3. Typical power supply circuit for an electronic clock. The tap from the bridge supplies pulsating dc to the counter chain.

people turn off television sets and lights. Over the period of time covered, the clock rate would be off appreciably. In the graph it reaches errors of nearly 500 parts per million, or an error of 0.02% in time rate. Also, these changes last for an appreciable period—the one near 11 P.M. on the graph shows a steady decrease in rate over a period of ten minutes.

The solution adopted by power companies to the problems of changing load and changing clock rate is to keep the average time

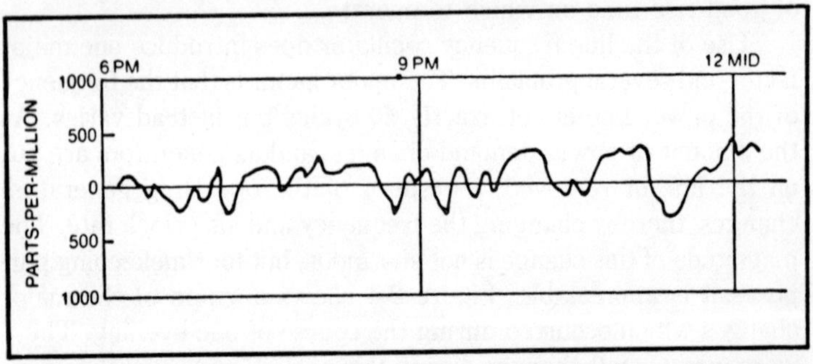


Fig. 2-4. Typical power line frequency variation during the evening hours. The variation is due to generator speed changes as the load on the electrical system varies, and causes error in clock reading. These errors may amount to several seconds before being corrected.

correct as closely as possible, letting the instantaneous time vary if necessary, but making efforts to keep the change low. The technique is to give two clocks to the load dispatcher, the man who controls the generator speed. One is run from the power system and the second, in essence, from the national time source. The dispatcher does have a frequency meter, which he attempts to keep at 60 Hz. However, from time to time, he also compares the two clocks. If the systems clock happens to be a little slow, he will increase the generator speed for a time, until the system clock catches up, and then will return to the speed which gives 60 Hz. His control is not perfect, but the systems clock is usually within a few seconds of true time. This degree of error is usually not important, but if precision work is to be undertaken, it should be remembered that the time can be off from true time, and the instantaneous time or frequency can be off by considerable amounts. As a result of the dispatcher's actions, any major error in clock reading is almost certainly associated with a power failure. (With most types of clocks, operation immediately stops, and restarts when power resumes.) There is a rather high probability that a power failure will be relatively short—a few tenths of a second or a few seconds, the time needed for the automatic switching circuits in the power system to cut out a defective component or line section, or just to permit a short circuit to clear itself. Some clocks provide special circuits for this type of event, so they will continue operating with essentially correct time.

A circuit for this is shown in Fig. 2-5. The 60 Hz line does not drive the clock chip counters directly, but instead synchronizes an internal oscillator which drives the accumulator. If power failure occurs, the clock oscillator continues to run, with possibly a small change in frequency. However, if the power failure is not too long, the error will be negligible. In some chips this oscillator may serve dual functions, as an example, for scanning, and may not be operated at 60 Hz, but the principle remains the same.

As the duration of the power failure lengthens out, another problem arises—that of continuing the dc power supplied to the chip. A typical way of handling this is shown in Fig. 2-5. Very simply, a very high value capacitor is connected across the positive bus. The size is chosen to provide power to the chip for at least several seconds, the actual size usually being determined by physical size and cost considerations. This simple method has a particular advantage with most chips—if the power failure occurs, the clock's reading is not lost, providing the power comes back on in

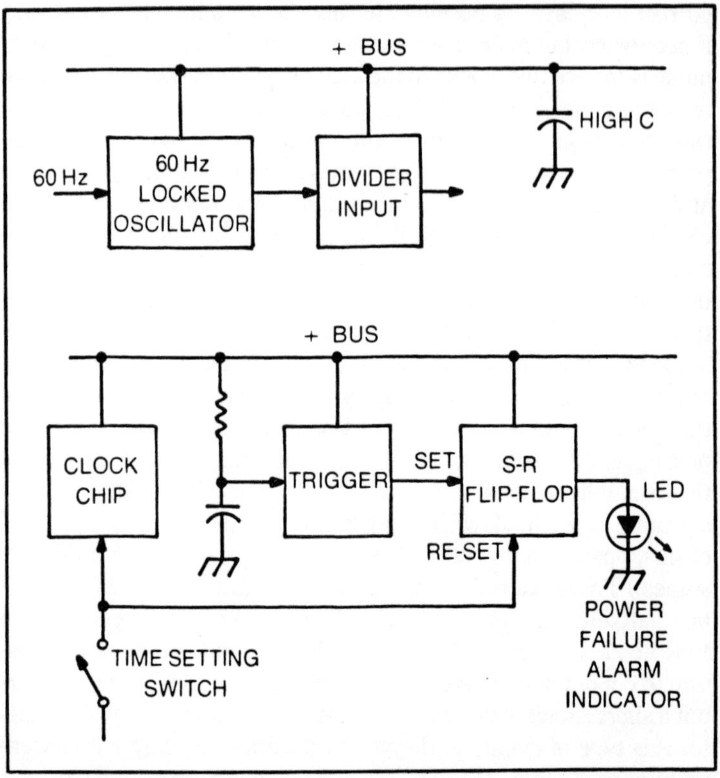


Fig. 2-5. Anti-power failure circuits. The upper one provides an oscillator of approximately 60 Hz, which can give nearly correct readings even though the line voltage fails; the protection lasts only as long as the supply to the chip functions, typically for a few seconds. The lower circuit, shown schematically, gives a visible warning that a power failure has occurred since the last setting.

a few seconds. The longer failures do cause a marked change in clock reading, but this in itself is a warning that a failure has occurred.

Some clocks do incorporate a separate method of warning that a failure has occurred and that the time readings are not reliable. The principle of a circuit for this is also shown in Fig. 2-5. The clock, in essence, contains a set-reset flip-flop, so constructed that its reset is lost if the power goes off. The flip-flop then causes a light or other warning to come on; this may be done by flashing the display. Usually the flip-flop is automatically reset when any of the setting buttons of the clock chip are operated, it being assumed that the clock will be set correctly. This failure indication will not be found in inexpensive clocks, but could be added.

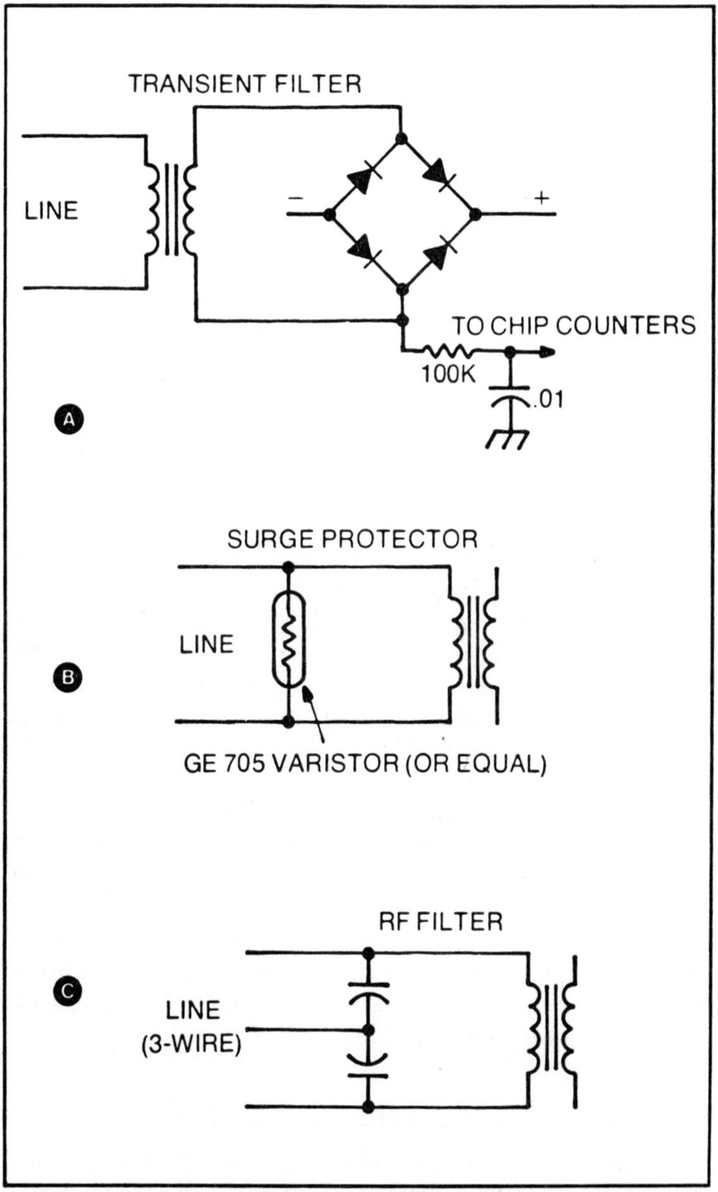


Fig. 2-6. Power line protective circuits. The transient filter is commonly used, and is intended to keep fast signals (transients or RF) from the chip input, where they could cause error, or even chip damage. The surge protector is seldom provided, but is recommended, especially in thunderstorm areas. The RF filter is seldom provided, but would be necessary if the clock is to be used in or near a radio transmitter.

Another problem associated with power lines is the matter of transient fluctuations in voltage. These sudden changes may be caused by the switching of loads, by failures of equipment, or especially by near lightning strokes. Many of these events are so rapid that the regulation controllers of the system do not have an opportunity to correct for the voltage change.

The usual clock has partial protection against these, since the dc voltage supply is heavily filtered, and voltage on the positive bus cannot change rapidly. Most clocks (and clock chip circuitry) show additional protection for the ac source of the chip counters. This takes the form of a low-pass filter, an RC combination, connected as shown in Fig. 2-6(A). The two protective elements shown in this figure are the only ones usually found in low cost clocks.

These protections are a definite help but do not give adequate protection against large events such as a close-by lightning stroke. To protect against these, a step such as shown in Fig. 2-6(B) can be taken. Here, a surge protector has been connected across the power line, in this case a General Electric metal-oxide varistor. These or equivalent type protective units are available at reasonably low cost, and experimenters might wish to add these elements to their clocks, and as well to their television receivers and other equipment. A complete set of protection for all equipment in the house would amount to an appreciable sum, but it would still be far less than the cost of replacing a set of rectifiers, for example, in a television receiver.

Radio hams, CBers, and others who generate radio frequencies (RF), or even people who live near a broadcast radio station, may find that their clock refuses to operate, or runs very fast, whenever the transmitter is on. This is almost always associated with RF pickup by the power line. Usually this can be eliminated by an RF filter of the type shown in Fig. 2-6(C). It may not be necessary to build this into the clock—the small type which plugs into the power socket may be good enough.

## CRYSTAL OSCILLATORS FOR CLOCK CHIP DRIVE

A basic oscillator consists of two devices. One, as sketched in Fig. 2-7, is a resonant element, and the second is a circuit to supply the power to this element and to provide useful output. In a mechanical clock, the resonant element is composed of the hairspring-balance wheel combination and the power supplying device is the mainspring-escapement combination.

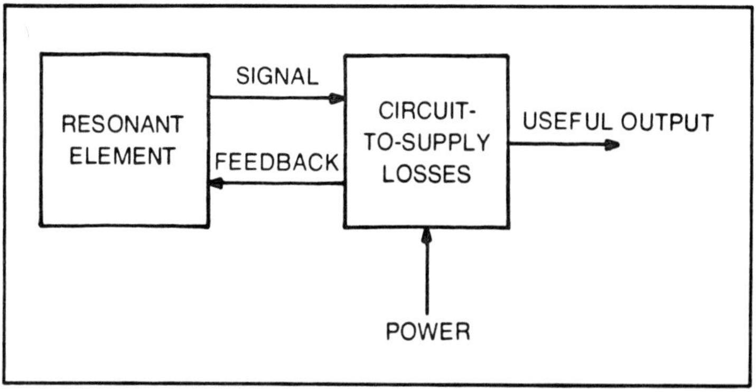


Fig. 2-7. Elements of an oscillator. These elements are necessary in all types of oscillators, whether electronic, mechanical, acoustic or other.

Basically a crystal oscillator is no different. The resonant element is just a hunk of material, but one which does have the property of changing shape when an electric potential is applied to it. This class of material is always crystalline in nature, and is called a piezoelectric crystal, to show that it has electrical and mechanical properties. Of course, the material also changes shape when a force is applied, and in this respect it behaves like a spring, specifically, like the hairspring. Further, the material has mass, and so behaves like a balance wheel. A piece of this material, properly shaped with respect to the rules which govern piezoelectric characteristic, is a complete resonator. Usually the shaping rules require relatively simple shapes, a plate, perhaps one-half inch square by a few thousandths of an inch thick, or a bar, perhaps one-eighth inch square and one-half inch long.

To function in an electric circuit, some method of applying electrical energy to the crystal is needed. In the modern crystal this is done by plating metal electrodes on two faces of the crystal, and supporting it from posts by means of small springs, shown schematically in Fig. 2-8. The entire assembly is usually protected by a metal case, which may be filled with inert gas. Incidentally, a crystal which refuses to operate probably has a failure between a lead and the plated electrode, the connection usually being a very thin layer of solder.

Since the piezoelectric material has properties of mass and of spring and is electrically coupled, its characteristics can be represented by a tuned LC circuit. Since the material is not ideal, there is also a resistance which produces a loss. Further, the plated electrodes act as the plates of a capacitor, having the crystalline

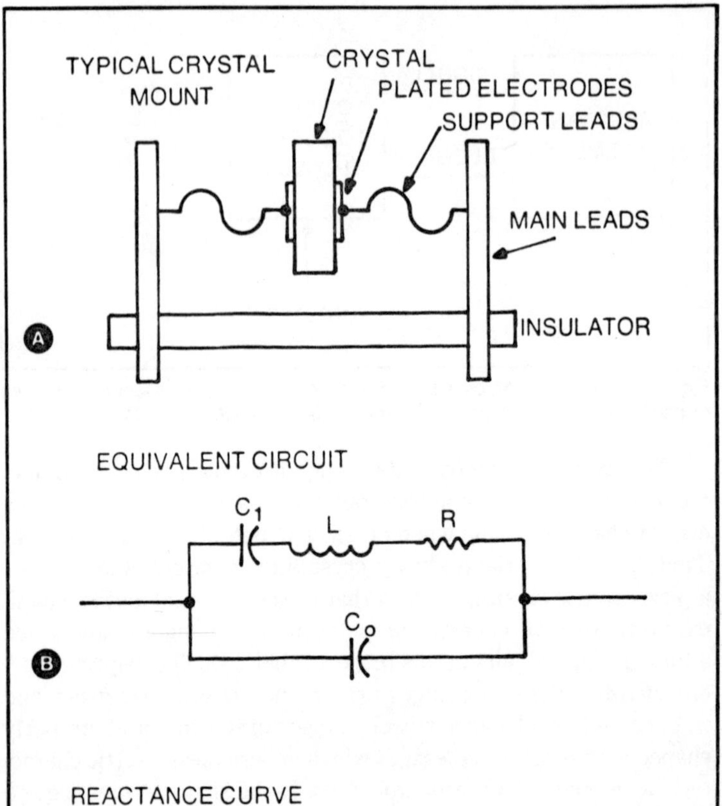


Fig. 2-8. Principles of piezoelectric crystal operation. The mount is typical of modern designs; in old units the crystal is clamped between two plates. All designs have the same equivalent circuit, and show series and parallel resonance. The frequencies vary with crystal cut, crystal size and mount design.

material as the dielectric. These elements are arranged as shown in Fig. 2-8(B). Because of the two paths between the terminals, there are two resonant frequencies involved, one due to the series elements L and C1 and the second due to these plus the parallel combination Co, the electrode capacitance.

The manner in which the resulting ac impedance varies with frequency is shown in Fig. 2-8(C). The point of zero reactance is the series resonance frequency of L and C1, and the point of maximum reactance is that of parallel resonance. The crystal may operate at either of these resonant frequencies, as determined by the external circuit, or it may operate a small distance away from the exact points of resonance, also as determined by the external circuitry. The frequency marked on the crystal container by the manufacturer usually is based on a particular external circuit having a known capacitance, 32 pF being the usual value. When ordering crystals, the resonant mode and equivalent external capacitance must be specified if the frequency is to be accurate.

The magnitude of the equivalent components shown in Fig. 2-8(B) varies with the frequency at which the crystal operates and with the method of alignment with respect to the piezoelectric axes, usually called the crystal cut. While there are many dozens of these cuts, it is likely that only three will be encountered in clock work. Each of these three has a set of characteristics which are best over a particular range of frequency, and as frequency changes, the cut encountered also changes.

The most likely types, and their characteristic at a specific frequency, are shown in Table 2-1. Here, it should be especially noted that the equivalent inductance of a crystal is extremely high, thou-

**Table 2-1. Typical Piezoelectric Crystal Characteristics.**

| Cut | Frequency<br>32 kHz<br>XY-Bar | 525 kHz<br>D | 2 MHz<br>AT |
|---|---|---|---|
| Vibration Mode | Length | Face Shear | Thickness Shear |
| R—Equivalent Resistance (ohms) | 40K | 1400. | 8.2 |
| L—Equivalent Inductance (H) | 4800. | 12.7 | 0.52 |
| $C_1$—Equivalent Series Capacity (nF) | 4.91 | 7.24 | 12.2 |
| $C_0$—Shunt Capacity (pF) | 2.85 | 3.44 | 4.27 |
| C—Ratio | 580. | 475. | 350. |
| Q—Figure of Merit | 25,000. | 30,000. | 80,000. |
| (Data from RCA Bulletin ICAN 6086) | | | |

sands of henries for the very low frequency crystal. In contrast, for normal capacitive-inductive circuits, the inductance will usually be millihenries at the low frequencies, and microhenries at the high frequencies. Of course, to secure resonance with these large inductors, the series capacitor must be extremely small. This is a way of saying that crystals are high impedance devices.

There is another factor of very considerable importance. This is the fact that the crystal resistance is relatively low, as a result, the ratio of reactance to resistance, the Q or quality factor, is very high. This is the factor which makes a crystal a stable resonator.

While the resistance is low it is by no means zero. An external circuit is needed to overcome the losses, and to supply power to external circuits. A family of simple, reliable circuits is shown in Fig. 2-9. Basically these circuits are just a multivibrator composed of two amplifiers, with the crystal replacing one coupling capacitor. The amplifiers for the three circuits shown on this figure are discrete component transistors, TTL NAND gates or MOS NAND gates. FETs or NOR gates could be used equally well. The transistor type is capable of operation over an extremely wide range, from the lowest possible crystal frequencies up to the highest possible, typically from 50 kHz to 25 MHz. The other two types are also reasonably wide range, depending on the exact gate types chosen.

Some additional crystal oscillator circuits are shown in Fig. 2-10. The first shows a NAND gate type oscillator, with simple amplification and inversion to supply two-phase output. The second shows use of a Schmitt trigger, which can be added to the above oscillator or used in place of the amplifiers, for better square-wave action. The final pair shows other types of oscillator of the discrete component class. In addition, there are a class of oscillators intended for great precision, plus many others intended for general use. The references should be consulted for these circuits.

One problem with crystals is that they shrink or expand as temperature changes. In the simpler cuts, this causes the resonant frequency to change with temperature. However, it is possible to arrange the cut with respect to the crystal axes so that the change in one axis cancels the change in another. In fact, over a range of temperature, it is possible to get essentially no change in frequency with a change in temperature. The range over which this "zero temperature coefficient" lasts depends on the cut. Curves for two typical cuts are shown in Fig. 2-11. Note that the temperature correction of the AT cut is essentially zero over a wider range than normal

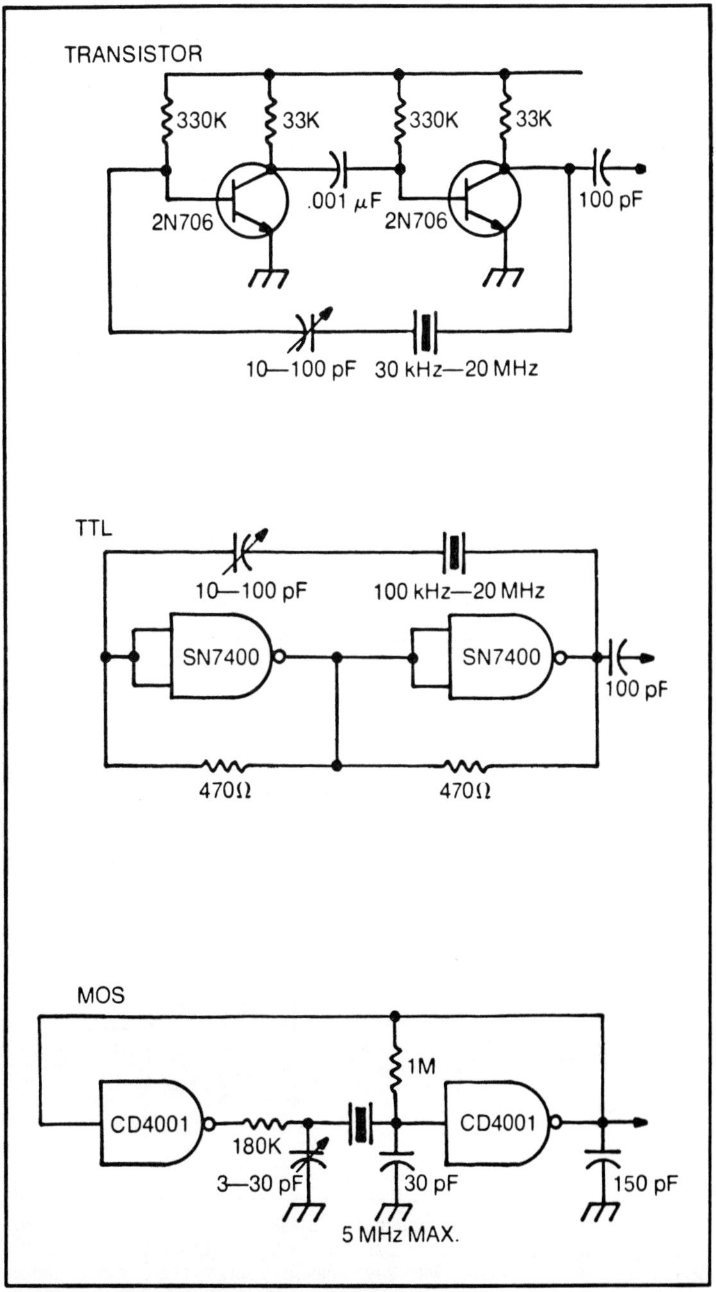


**Fig. 2-9. Crystal oscillator circuits based on the multivibrator. The circuits are simple, reliable, and suited to most uses.**

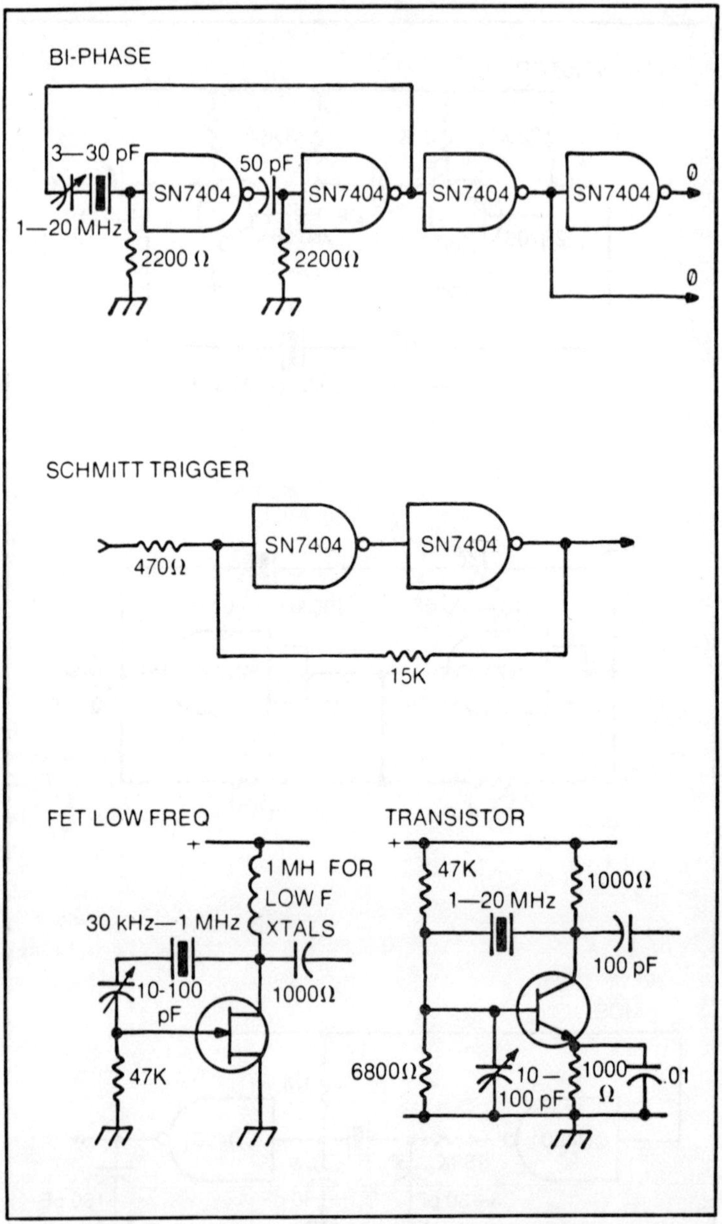


Fig. 2-10. Additional oscillator circuits. The top shows a design for "two-phase" output rarely used in timekeeping, but often required in other applications. The Schmitt trigger can be used with any oscillator design. The discrete component circuits are sometimes encountered. For maximum precision circuits, see the references to Chapter 9.

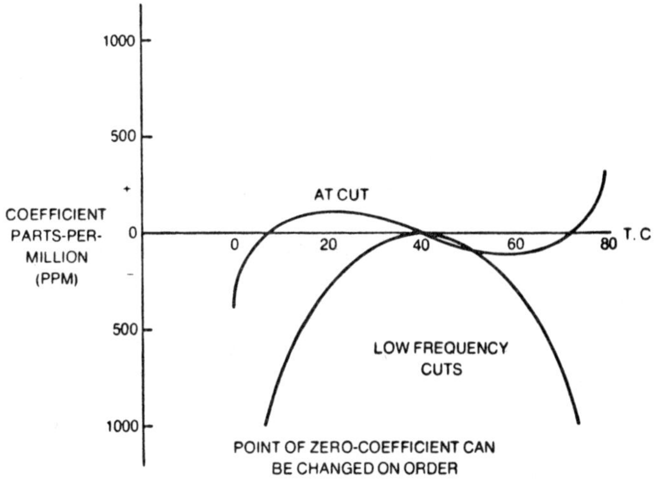


Fig. 2-11. Typical thermal characteristics of crystals of low coefficient cut. The best AT cuts will approach zero coefficient over a range of 30-50 C°.

room temperatures. This is one of the main reasons for preference for this cut. The exact temperature at which the temperature coefficient is zero can be moved, within limits. It is, however, difficult to get precisely a zero temperature coefficient at precisely the desired temperature, so practical crystals usually have a small coefficient. In really precision work, it is necessary to maintain the crystal at constant temperature, by means of an electrically heated oven.

Another problem with crystals is that small pieces flake off the surface, due to the fact that the crystal is vibrating. As a result the crystal frequency changes, or drifts, with time. Partial solution to this is to operate with the vibration amplitude very low, that is, with very low excitation voltages supplied to the crystal. Transistor circuits tend to do this automatically, one reason they are currently preferred in crystal oscillator design.

At one time the best clocks, those which kept the national time, were based on quartz crystals. However, as has been seen, it is not easy to maintain precise oscillation frequency and so effort on improved resonating elements continued. Currently the most accurate resonators use a beam of cesium atoms, which have vibrational properties at a precisely known frequency. With these types of resonators, accuracy of one part in ten billion or better can be obtained.

While oscillators of this type are completely beyond the capability of the home constructor, most homes today have the frequency derived from these sources available in the form of the color reference signal in the color television set. Use of this source is discussed later.

It is no great problem to make a crystal oscillator which has much greater instantaneous stability than the power line. However, care is needed in oscillator construction and in choice of components if the average accuracy is to be as good as the power line, even over a period as short as a day's time. Chapter 9 covers the construction of a middle grade of crystal standard, one which is not too difficult to construct and which should give better accuracy than the power line. For really accurate types of oscillators, consult the references.

## DIVIDER CHAINS

To keep track of time in a clock, pulses are needed at various time intervals. A typical relation of these pulses, and the division ratios needed to go from one time element to another are shown in Fig. 2-12. Here the timing chain is assumed to start with one second pulses. The first six dividers are present in all electronic clocks, to give the seconds, minutes and hours readings. The seventh is also usually present, to give some indication of A.M. and P.M., or for the twenty-four hour day format. The remaining dividers are associated with the calendar, and are not present in the smaller chips. In these chains, note that a relatively small number of division ratios are needed. For a simple clock, division by 2, 6 and 10 satisfies all requirements. Even in the calendar clock, only the ratios of 3 and possibly 7 are needed in addition.

These chain elements start with one pulse per second, but additional dividers are necessary to go from the power line or crystal oscillator frequency to this rate. Some typical chains of these preliminary dividers are shown in Fig. 2-13. For the power line input, the division ratios are not great; for 60 Hz a cascade of divide by 6 and divide by 10 accomplishes the job. For the 50 Hz line, a 2, 5, 5 divider chain would be ample.

With crystal oscillators the division ratios must be larger, since it is difficult to get crystal operation even as low as tens of kilohertz. One typical chain is based on binary division entirely, starting from the 1 Hz rate. About the lowest frequency which corresponds to satisfactory crystal operation is the 15th power of

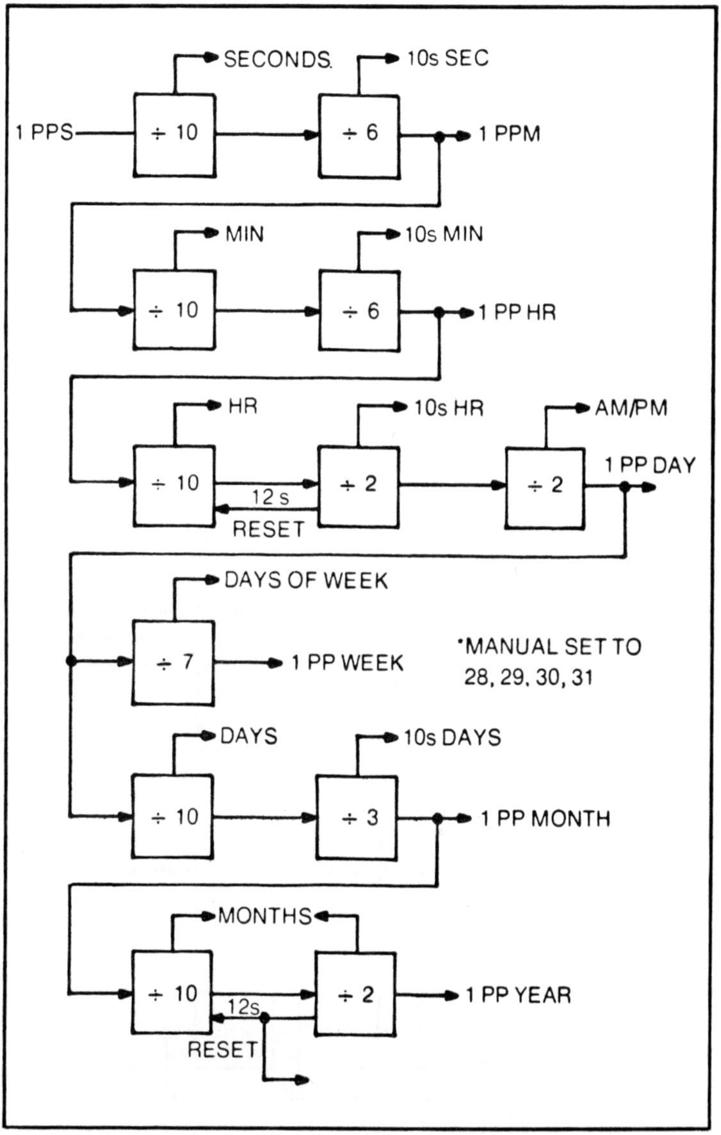


Fig. 2-12. The divider chain for a calendar clock, starting with a 1 pps rate. The first six or seven blocks are found in all digital clocks.

2, corresponding to a crystal frequency of 32.768 Hz. Electronic wrist watches often use this frequency. However, crystals become smaller and of better characteristics if they operate in the range of a few megahertz. The last two chains show dividers for this range, the first being based on binary division and the second being based on decimal plus binary division. Other combinations of ratios will be found.

For clock use, the exact crystal frequency and division ratio is less important than are factors relating to accuracy and simplic-

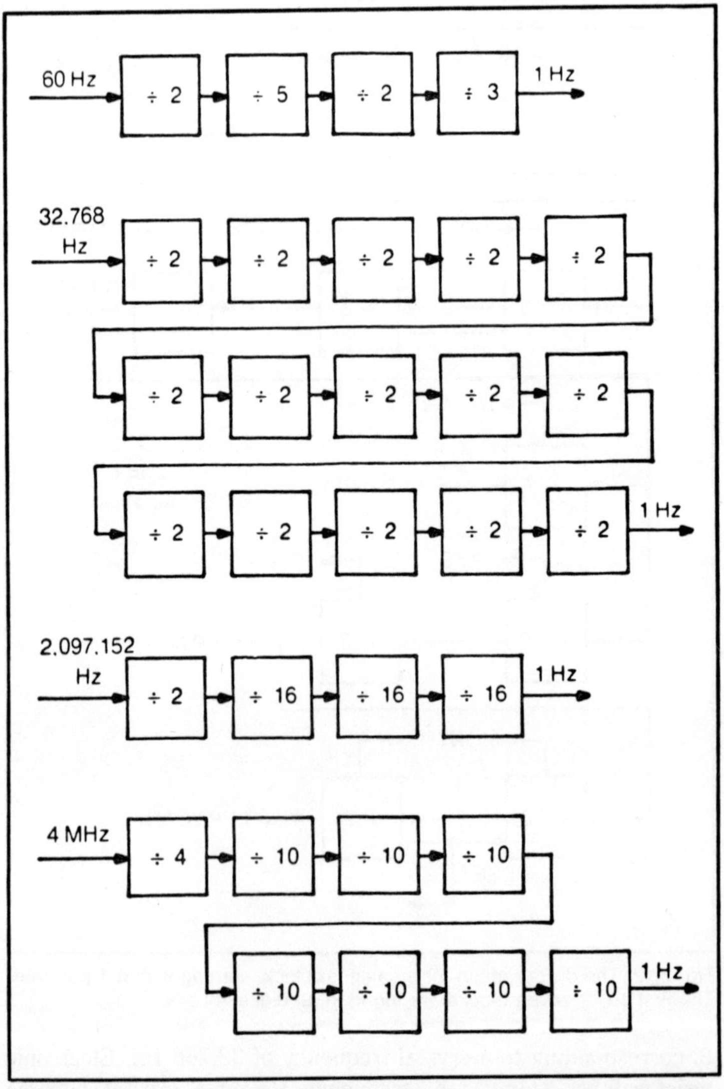


Fig. 2-13. Divider chains to develop the 1 pps rate from common sources. The 60 Hz chain is common in clock chips, the 32768 Hz in digital wristwatches based on a crystal oscillator.

ity of design. However, if the frequency of the oscillator is to be used for other purposes, for example in radio work, certain frequencies are preferred.

Despite the great range in divider ratios which may be encountered, the divider can usually be thought of as a cascade of a limited number of ratios. Division by 2, 3, 5, 7 are not uncommon. Occasionally a very large ratio will be encountered—88, for example. For these it is usually possible to factor the ratio into smaller steps. For example, 88 can be factored into division ratios of 2, 2, 2 and 11; however, since 11 is not divisible by any number, it is necessary to have available a technique of obtaining these prime number ratios as necessary.

The basic binary divider can be thought of as an assembly of gates, interconnected as shown in Fig. 2-14. In this simple divider,

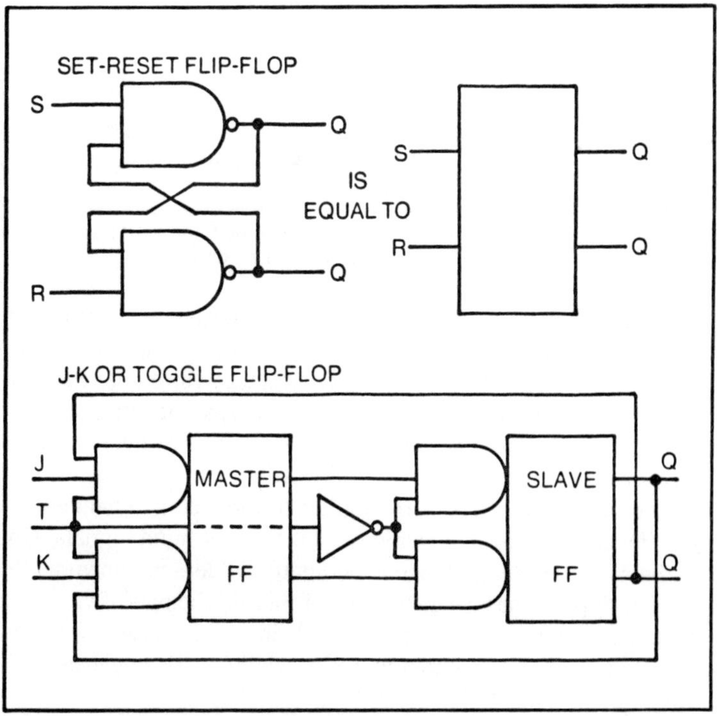


Fig. 2-14. Basic flip-flop connections. The S-R type is often used to "debounce" or remove false signals which occur at mechanical switch closing. The J-K Master-Slave type will change state or toggle if both the J and K inputs are high, and serves as a 2:1 divider element. There are other Master-Slave types which have different action. More complex types may also have preset capability.

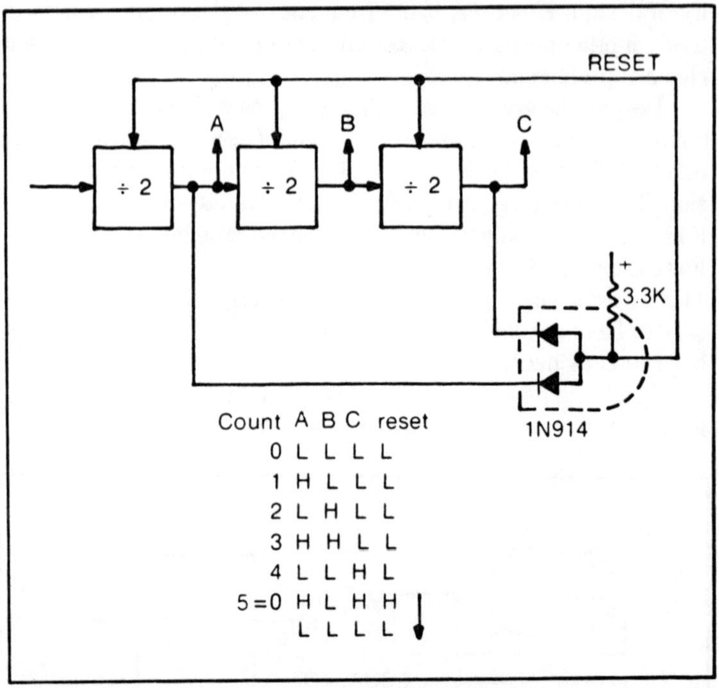


Fig. 2-15. Use of preset capability to give divide by 5 (or 4 & 1) action. The principle may be extended to give any value, i.e., divide by N.

when the input or clock pulse is high, the output rests at its last value. When the clock pulse goes low, one of the input gates also goes low, which in turn forces the output gates to change state. The next clock pulse reverses this process. Several of these counters can be cascaded to secure the successive binary stages, division ratios of 2, 4, 8, 16, etc. Also, additional gates can be connected in various parts of the circuit to force stage changes at a specific time, or for a specific combination of input signals. One of the most flexible of these added gate dividers is known as the J-K flip-flop, the basic construction of this also being shown in Fig. 2-14. These J-K flip-flops are available in a wide variety of forms, for example, with or without preset capability, with single and multiple inputs, etc.

Figure 2-15 shows the basic method of obtaining divider ratios other than 2. First, a counter chain is needed which is capable of operating at division ratios higher than the one finally needed. Suppose a ratio of 5 is desired. This ratio corresponds to a count of 4 + 1 and requires a 3 stage counter. To recognize when the

4 + 1 condition is reached an AND gate is connected to the output of the third counter, the 4, and to the output of the first counter, the 1. The output of the AND gate becomes high when both inputs are high, on the count of 5. This change to high is fed back to the preset inputs of the three counter stages, forcing them to reset to zero. The counting action then resumes, starting from the zero state, with reset again occurring on the fifth count. The truth table for the five counts, that is, the state of the three counters, is indicated in Fig. 2-15.

Some package counters have built-in gates, which can be used for reset to zero on signal. The TTL types 7490, 7492 and 7493 have a two input AND gate which can be used as shown in Fig. 2-16. The 7490 also contains a second AND gate which can be used to preset the counter to the 9 state. These counters can be used directly without external elements to obtain any count which can be described by the addition of any two of the numbers, 1, 2, 4, 8, subject to the maximum count limit of the particular type (10 for 7490, etc.).

It is also possible to use feedback from one stage to another to control the division ratio. Figure 2-17 shows connections for J-K counters for three common division ratios. One reference describes additional connections for ratios from 3 to 100 for this type of counter.

Some counters can be preset to a given value by external signals, and then count either up or down from this preset value. Figure 2-18 shows such a counter, using switches for the preset value. Note that the value in the switch is only transferred to the counter

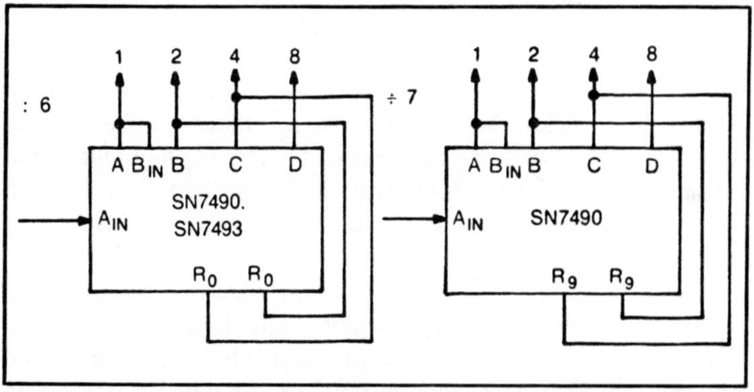


Fig. 2-16. Use of internal gates to give desired division ratios. The divide by 6 is obtained by reset on (4 & 2). Divide by seven is obtained by using gates which preset to 9, corresponding to (4 & 2) + 1.

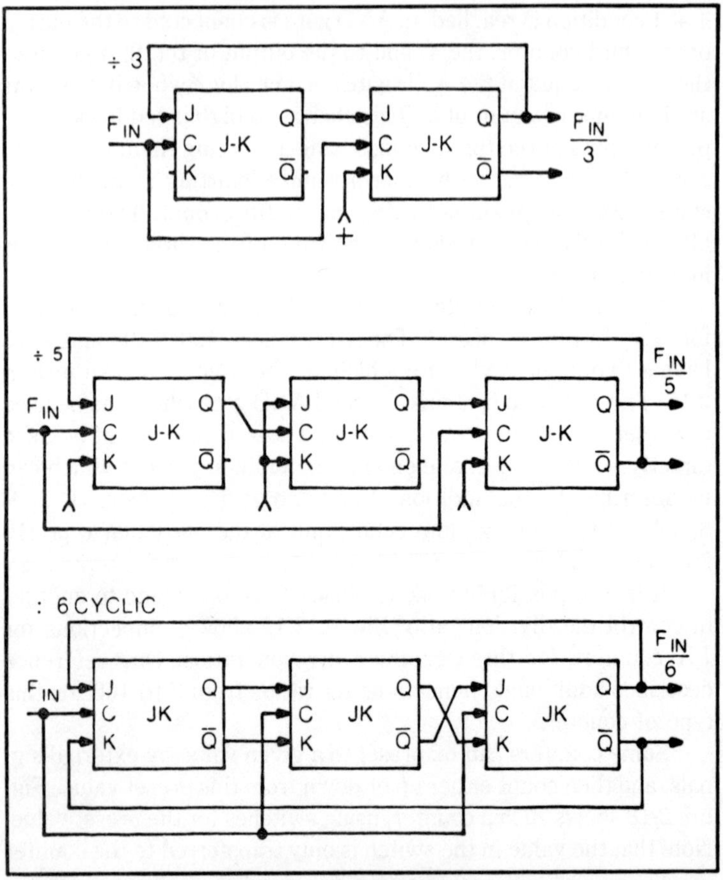

Fig. 2-17. Development of required division ratios by feedback. See the reference material for designs up to divide by 100.

when a separate line called LOAD is taken to the high condition. When this line is low the switches have no influence. The clocking signals may be connected to either of the two inputs up or down, causing counts to proceed say, from 6, 7, 8 . . . 3, 4, 5 . . . or from 5, 4, 3 . . . Reset can be obtained from the output, from AND gates, or from external signals as needed.

For experimental use, it is convenient to have a divider which can be connected to give any division value. Figure 2-19 shows a single stage of such a divider. It uses a divide by 16 counter and AND diodes in any of the four output lines, for division ratios from 2 to 15. If higher division ratios are needed, counters can be cascaded. This is done by taking the output of the highest active stage

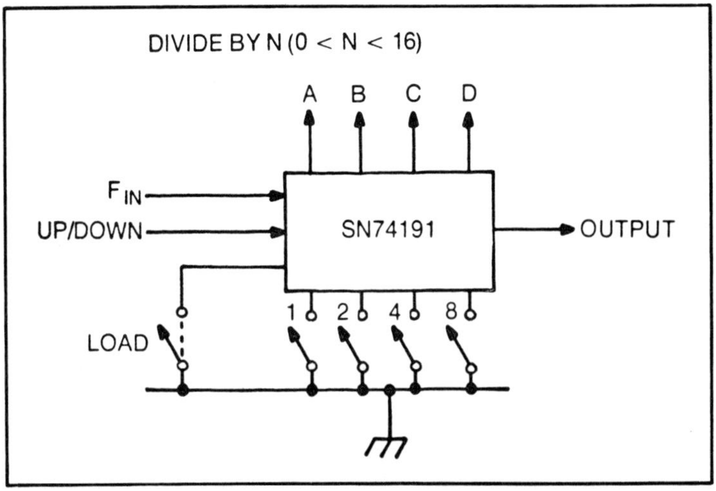


Fig. 2-18. A multiple-ratio divider based on the 74191 up-down counter. When LOAD is taken high (switch open), the counter is preset to the value given by the preset switches being open, or in the high state.

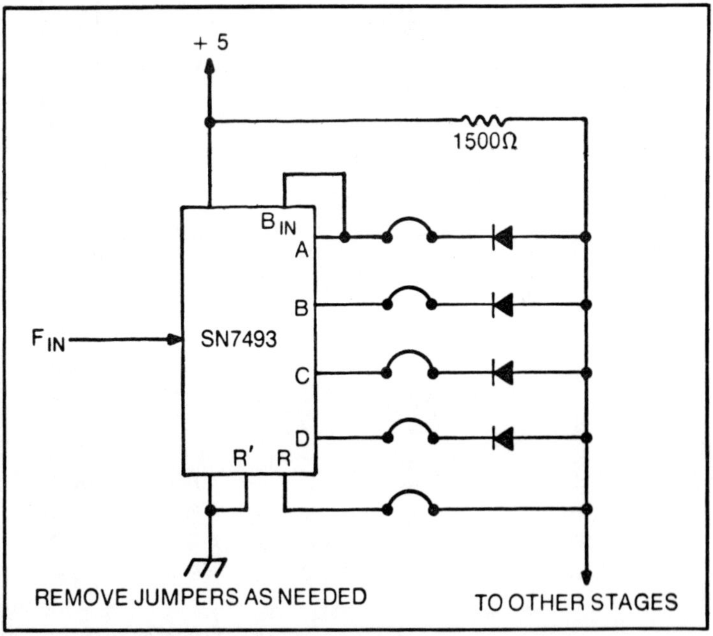


Fig. 2-19. A multiple-ratio divider using a diode AND gate. Several can be cascaded: connect the input of the following stage to the highest active output (example, if divide by 7, use C for output). The reset bus is in parallel for all stages.

to the input of the succeeding counter, and also connecting the reset buses in parallel. The parallel connection ensures that all counters are set to zero simultaneously. If the reset action is not needed, or if a stage is to always operate at full count, the reset should be grounded, as indicated by the dotted connection. The 7493 divider is a readily available low cost unit for this purpose.

If desired, provisions can also be made for using the 7493 as a self-gated divider. The basic connections are shown in Fig. 2-20. Here, dotted lines indicate typical external connections. The count ratios which are available by this technique are tabulated; if other ratios are needed, externally-gated dividers such as those of Fig. 2-19, or specialized dividers intended for preset gating, should be

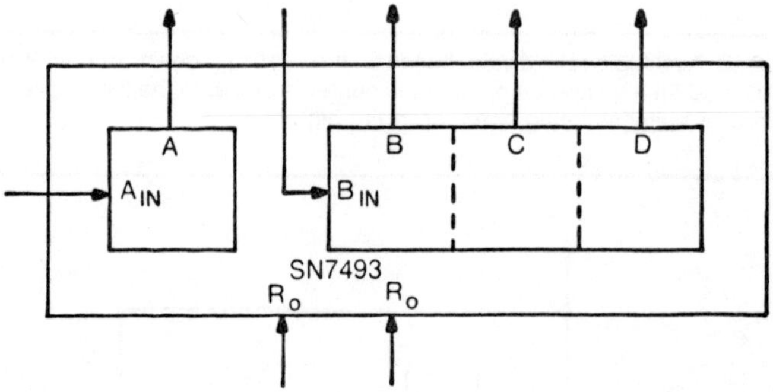


| COUNT MODULUS | INPUT | OUTPUT | CONNECT | RESET. $R_o \cdot R_o$ |
|---|---|---|---|---|
| 2 | $A_{IN}$ | A | | G-G |
| 3 | $B_{IN}$ | B | | B-C |
| 4 | $B_{IN}$ | C | | G-G |
| 5 | $B_{IN}$ | C | | B-D |
| 6 | $B_{IN}$ | C | | C-D |
| 8 | $A_{IN}$ | D | | G-G |
| 9 | $A_{IN}$ | D | A-$B_{IN}$ | A-D |
| 10 | $A_{IN}$ | D | A-$B_{IN}$ | B-D |
| 12 | $A_{IN}$ | D | A-$B_{IN}$ | C-D |
| 16 | $A_{IN}$ | D | A-$B_{IN}$ | G-G |

G = COMMON BUS

Fig. 2-20. Connections for the 7490 to obtain various division ratios. For other ratios, use the connections of Figs. 2-16 through 2-19.

used. Experimenters may find it convenient to wire-up a two or three stage breadboard using one or more of the preceding divider circuits.

## DISPLAY COUPLING AND THE READ ONLY MEMORY

Let us return to the accumulator, the divider chain of Fig. 2-2. As we now see, each of the stages of this chain is actually made of one or more binary dividers; 4 if the stage divides by 10, or 2 if it divides by 4 and so on. For each of these stages there will be a set of binary readings. For example, for the first or second stage of the chain, at a reading of 7 seconds, the four stages will read 0111, the most significant figure being read first. Other chains will have similar readings, some with a maximum value of 6, some with a maximum of 9 and others with a maximum of only 1, 2, or 3.

We could read the clock output in binary, but this has two serious disadvantages. The first is that binary is not easy to read, even with practice. For many people it is impossible since they have never learned the arithmetic involved. Then, for binary output a very large number of pin connections would be needed—for the seven stages which make up the basic twelve-hour clock we would need a total of 4 + 3 + 4 + 3 + 4 + 2 or a total of 20 pins. This large number of pins is both inconvenient and expensive.

To make the display easier, all clocks use a code which can be translated into a positional code which resembles common numbers. The details of this are discussed in the next chapter. For now, let us concentrate on the fact that a conversion is needed. The solution which is adopted in almost all chips is to use a read-only memory, as sketched in Fig. 2-21. Such a memory can be a collection of gates, having a set of inputs and a set of outputs. If the inputs are set at some value, say the number 7, the outputs are some other value, say a number series which we can call X. The outputs are never any other value for this output, but there may be more than one input which gives output X. The number of digits in the output can be greater or less than the number of digits in the input. Just for example, the diode AND gates of the force-to-reset counter of Fig. 2-19 form a simple read-only memory. This has a single output which is one value for one specific binary input value and a different value for all other input values.

Some chips provide only a binary coded output, intended to be used primarily in other than clock applications, or which can be used with an external read-only memory for clock applications.

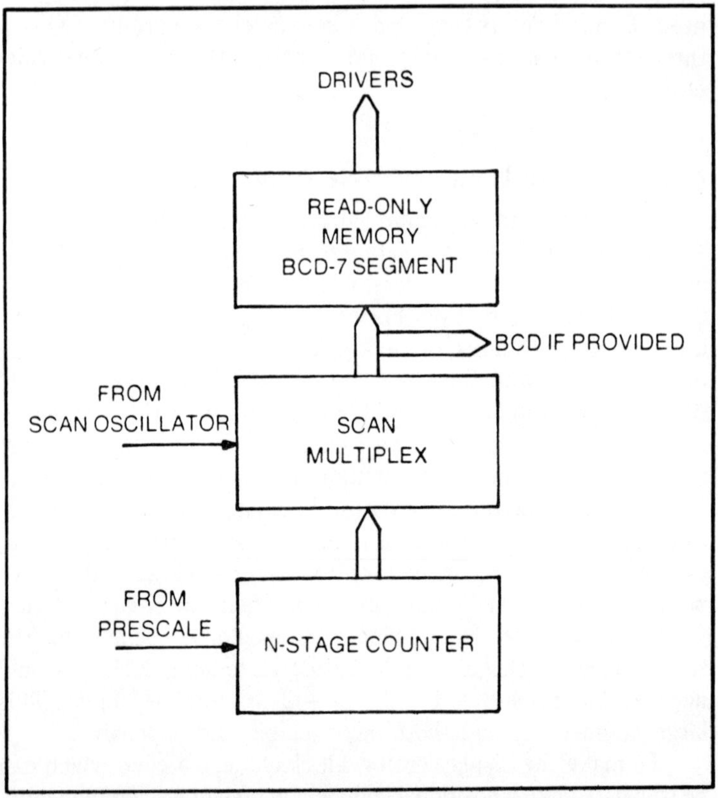


Fig. 2-21. Basic principle of conversion from the binary coding of counter chains to the coding required for visual presentation. Data on the drivers and display is given in the next chapter.

Others provide both binary coded and a display drive output. All chips known to be available on the market use a seven-segment code if it is not binary, this being the industry standard for display purposes.

The particular transformation involved is almost always that shown in Table 2-2. However, there are some variations; some chips use the inverse of this, that obtained by replacing ones in the output column with zeros and vice versa. A few also use a variation in the method of forming characters, especially for six and nine. A very large percentage of the chips also use multiplex operation, with a single set of output lines conveying the digital readings sequentially. This and some variations are discussed in the next chapter.

Table 2-2. Transcoding.

| Decimal | Weight: 8421 | 7-Segment abcdefg |
|---|---|---|
| 0 | 0000 | 1111110 |
| 1 | 0001 | 0110000 |
| 2 | 0010 | 1101101 |
| 3 | 0011 | 1111001 |
| 4 | 0100 | 0110011 |
| 5 | 0101 | 1011011 |
| 6 | 0110 | 0011111 |
| 7 | 0111 | 1110000 |
| 8 | 1000 | 1111111 |
| 9 | 1001 | 1110011 |

## DISPLAY INTENSITY CONTROL

There is one special factor for displays sometimes found on the chips, and sometimes made up as an external element. This is a display intensity control circuit, occasionally made automatic to adjust for variations in ambient light level. A typical manual adjustment circuit is shown in Fig. 2-22. Basically the circuit is very simple, being a simple adjustable voltage regulator. The output volt-

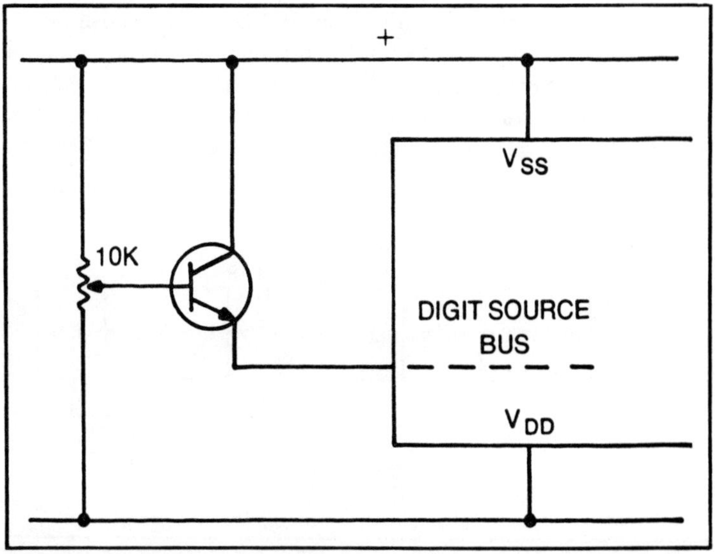


Fig. 2-22. A technique for providing variable display intensity. Found in some clock chips; for example, the MM516.

age supplies power to the display, and to display drivers, if used. This circuit can be added to the output of any chip which uses external drivers, or can be adapted to use such drivers. If automatic adjustment is desired, use a variable resistance type of photocell in series with a fixed resistance. The junction of these two elements would be connected to the transistor base. Values would need to be determined from photocell characteristic curves, or from experiments.

## TIME SETTING CIRCUITS

To keep proper time, it is necessary that the clock start in coincidence with true time, which requires some method of setting the various accumulators of the divider chain. One of the common methods of accomplishing this is to provide a set of three controls, normally marked HOLD, SLOW and FAST. These actually operate on the prescale circuit as shown in Fig. 2-23. The HOLD switch, usually a slide switch, interrupts the power line input so that the clock's readings do not change. The slow set and fast set inputs bypass one or more stages of the prescaling dividers. Usually the fast set causes the clock to run 3600 times normal speed and the slow set 60 times normal speed, but there are variations in this. Usually the connections are such that slow and fast set will both operate with the clock in either the HOLD or RUN position.

The setting operation is not too difficult. The best technique is to stop the clock, using the HOLD switch, then set to the correct hour using the fast set switch, followed by setting to the next minute ahead and to 00 seconds. The HOLD switch is then thrown

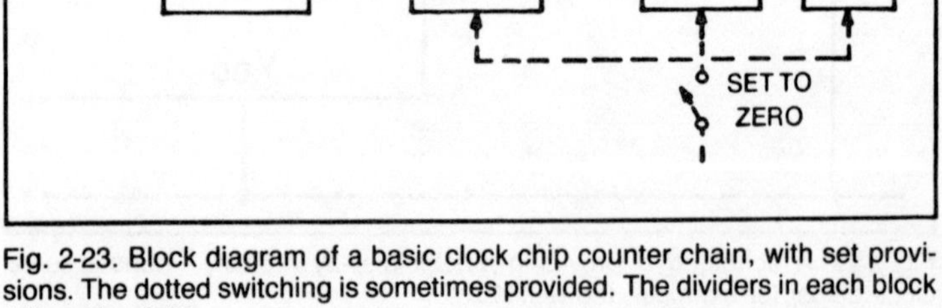


Fig. 2-23. Block diagram of a basic clock chip counter chain, with set provisions. The dotted switching is sometimes provided. The dividers in each block follow Fig. 2-12.

in synchronism with the minute time tick or reading of another clock. Actually, for line operated clocks, an error of four or five seconds in the setting is not very important, since clocks usually do not keep truly accurate time, a fact which should be remembered.

Some chips also provide for direct setting of the hour and the minute counters, some independently of other clock operations. Other chips provide for a clear operation, that is set to 00 hours, 00 minutes and possibly to 00 seconds. A variation of this is to set to 12:00. These set methods are indicated schematically in Fig. 2-23 by the dotted lines. A further variation is to use a pair of switches, one for hours and the other for minutes, with each closing of the switch advancing clock reading one count. With this type of circuit, slow and fast set controls are not provided. A version of this circuit is shown in Chapter 5.

## CONTROL VARIATIONS

Most clock chips use separate switches for each of the control functions, accepting the consequent increase in pin count. However, a few combine controls to give as much flexibility as possible while still keeping small chip packages. One chip, the CT7001, is unusual in that it uses a multiplex control circuit, a technique common in calculators but seldom found in other electronics. The basic circuit and the range of control possible is shown in Fig. 2-24. As can be seen, a given function is secured by feeding one of the digit scan signals of a multiplex output back to an input line. For most of the controls, closing the switch internally connects the counter affected to the 1 Hz bus, giving a form of slow set. As can be seen from the table associated with this figure, two switches can be operated at the same time, one selecting the internal operation and the second the counter on which it is performed. Most clocks which use this chip do not include all of the possible switches, but combine them in various ways to give a control which the designer believes will be simple and still give the desired function.

## ALARM CIRCUITS

Chips are available which provide alarm or other signals. As shown in Fig. 2-25, these signals are generated from an additional presettable register and a comparator. When the added register is set, and the alarm function energized, the comparator continually compares the reading of the alarm counter against the time

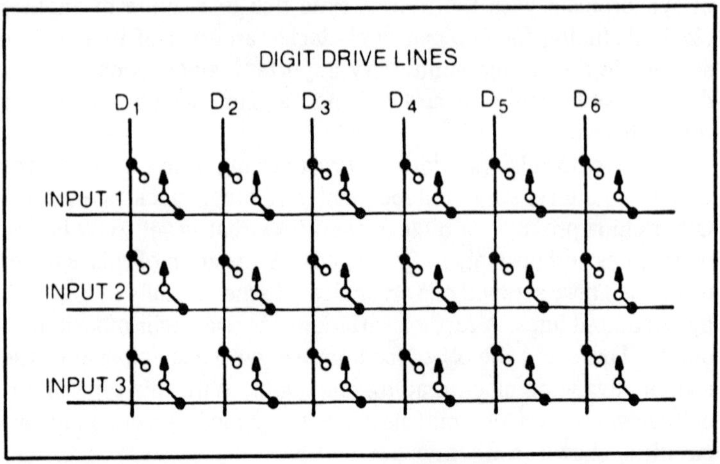


Fig. 2-24. Complex switching used in one multifunction clock chip. The digit drive signals are fed to multiplex inputs and decoded within the chip. See Table 2-3.

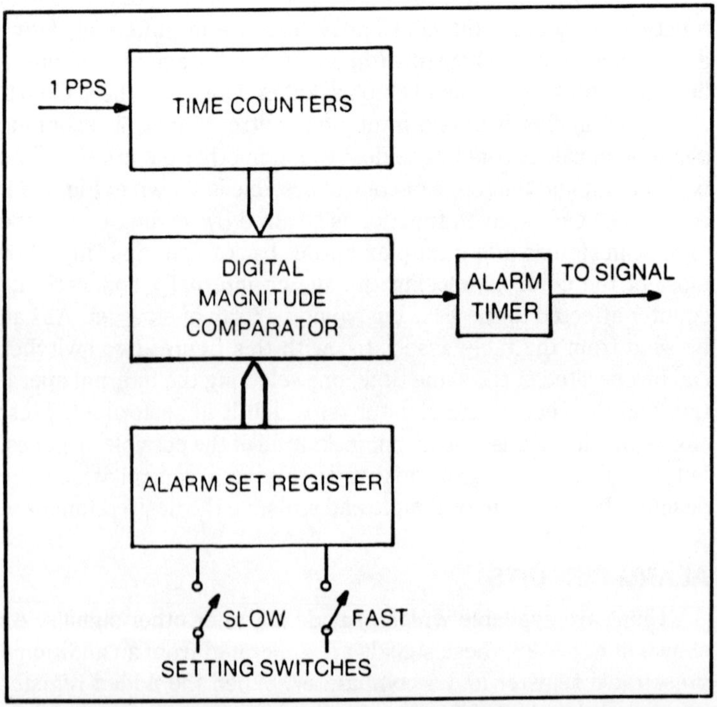


Fig. 2-25. Principle of alarm signal generation, provided by some chips. The circuit could be added externally for alarm or control use. See Chapters 7 and 10 for further data.

counters. When these have the same value, the comparator provides an output to an alarm timer, which initiates the alarm and causes it to run for a given period, or until shut off. Various additional controls may be provided, such as automatic repeat every 24 hours, automatic recycling every five minutes unless shut off a second time, and so on. Externally, the alarm signal may be used to energize a buzzer, or to turn on a radio or a light.

## AUXILIARY OUTPUTS

Some clock chips provide special outputs for such purposes as alarm, or to turn a radio off after an interval or for other control purposes. Where provided, it is most likely that the circuit has low current capability, possibly as low as a few microamperes. Therefore, interface elements must be used between a clock chip and the external circuit. Several of these interface elements are shown in Fig. 2-26. The first of these shows a typical output circuit for medium current applications, using a single transistor interface. The second circuit shows a variation of this, using a second stage for high current. For some applications, a relay may be preferable, as shown in the third circuit.

## SCAN OSCILLATORS

To save pins, many chips use multiplexing of the output signals, each digit being turned on for a short time, then off while the next digit is on (as considered in the next chapter). This process must occur at a rate which is high enough so that no flicker is visible to the human eye. Usually this means that the entire cycle must occur at least thirty times per second. For a six digit clock, this means that a change is occurring at least 180 times per second. Common practice is to include a basic scan oscillator on the chip, operating in conjunction with external frequency determining elements, usually a simple RC series circuit. A typical circuit is shown in Fig. 2-27. Values for this should be determined from the chip specification sheet, but 1000 pF and 490K are typical for the capacitor and resistor; respectively.

Since the oscillators used are a form of relaxation oscillators, the voltage at the pin connected to the common terminals of the RC circuit is a sawtooth. This can be used to synchronize external devices with the scan, providing not too much power is taken from the circuit. Each cycle of the sawtooth corresponds to the scan of one of the digits. One of the digit drive signals can be used to syn-

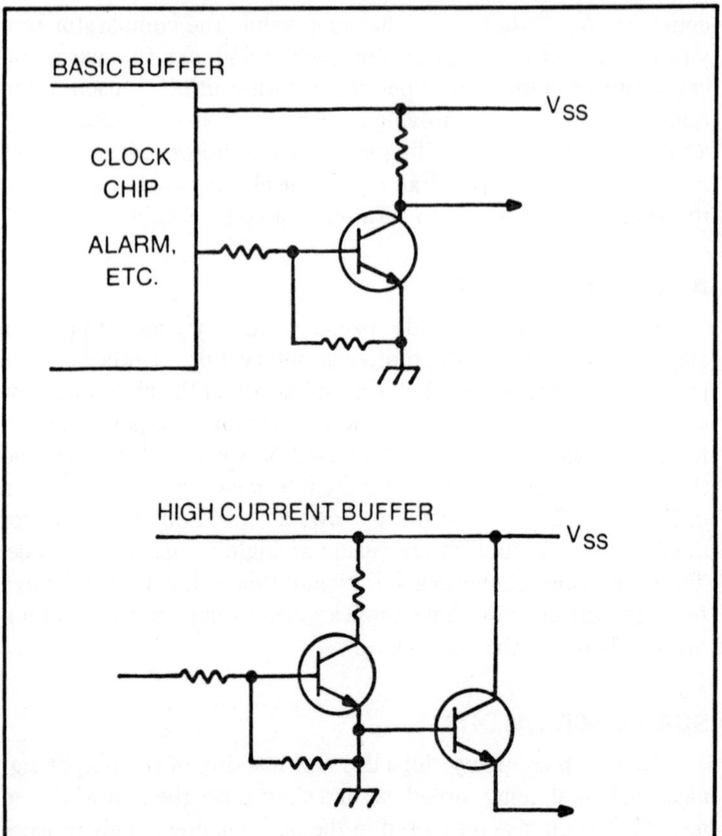


Fig. 2-26. Common output or "interface" circuits. Most clock chips have limited output capability, and require such buffers for alarm, radio, or other uses.

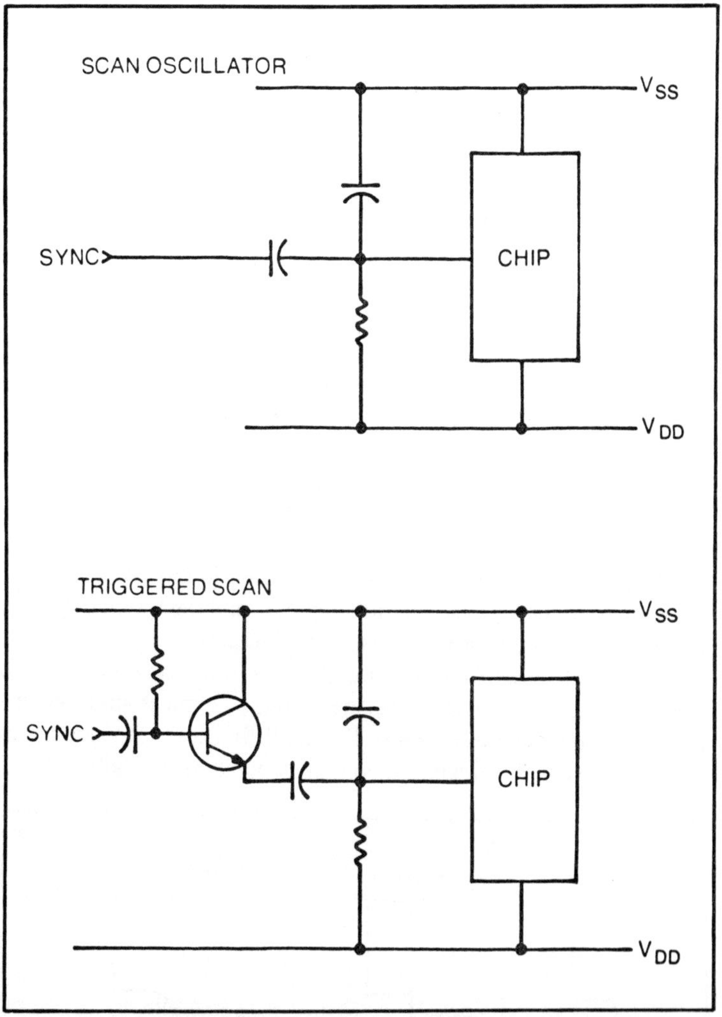


Fig. 2-27. Typical external connections for the scan oscillator, which is designed to free-run unless synchronization is required for special purposes. For values, see the chip data sheet.

chronize with the total scan rate. If used, do not forget that a buffer amplifier may be needed—a source follower or emitter follower would be suitable.

Relaxation oscillators have the property of locking to an external signal, so synchronization can be fed into this chip, to cause the scan signals to follow an external signal. The signal levels need to be watched—the sync input is usually a gate protected MOS, but damage is possible if the specification limits are exceeded. If no data is available, it is usually safe to feed in a signal which does not go higher than the power supply voltages, nor lower than the ground or common voltage level. A circuit which would be suitable for most applications is also shown in Fig. 2-27.

## OUTPUT CIRCUITS: INTERFACING

Typically, the output circuits of a chip are formed from open drain FETs. These have some current capability, typically on the order of milliamperes. Three circuits which may be encountered are shown in Fig. 2-28. The first two are simply open drain followers, the first a MOS-FET and the second a simple FET. The third circuit is sometimes called a push-pull stage and sometimes a totem pole driver. This circuit will usually be used where greater voltage swing or more current capability is needed, as would be the case for the digit drive. The specification sheets for the chip must be used to determine the current capability of these drivers. Typically, the open drain types will be used in circuits providing a few microamperes to a few milliamperes, and the totem pole type in circuits providing a few tens of milliamperes.

For many experimental uses the chip must be connected to other devices, often of the TTL logic family. Typical interface circuits for such use are shown in Fig. 2-29. In these, the clock chip is operated near its maximum voltage capability, typically with 17

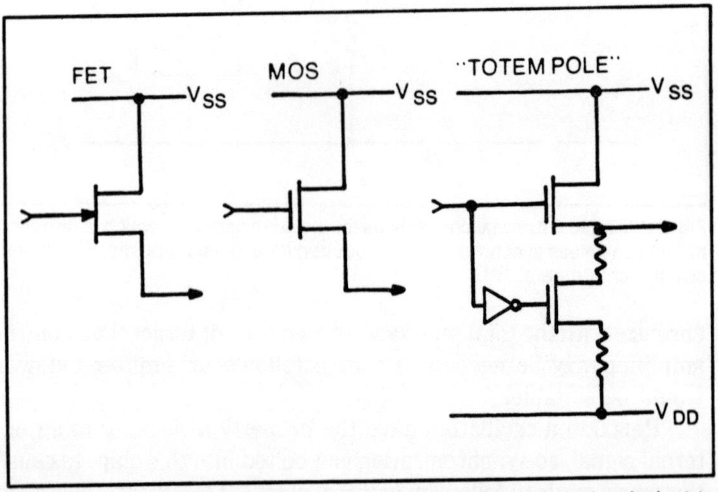


Fig. 2-28. Typical output circuits for the displays. Open drain (equivalent to the cathode follower) is most common.

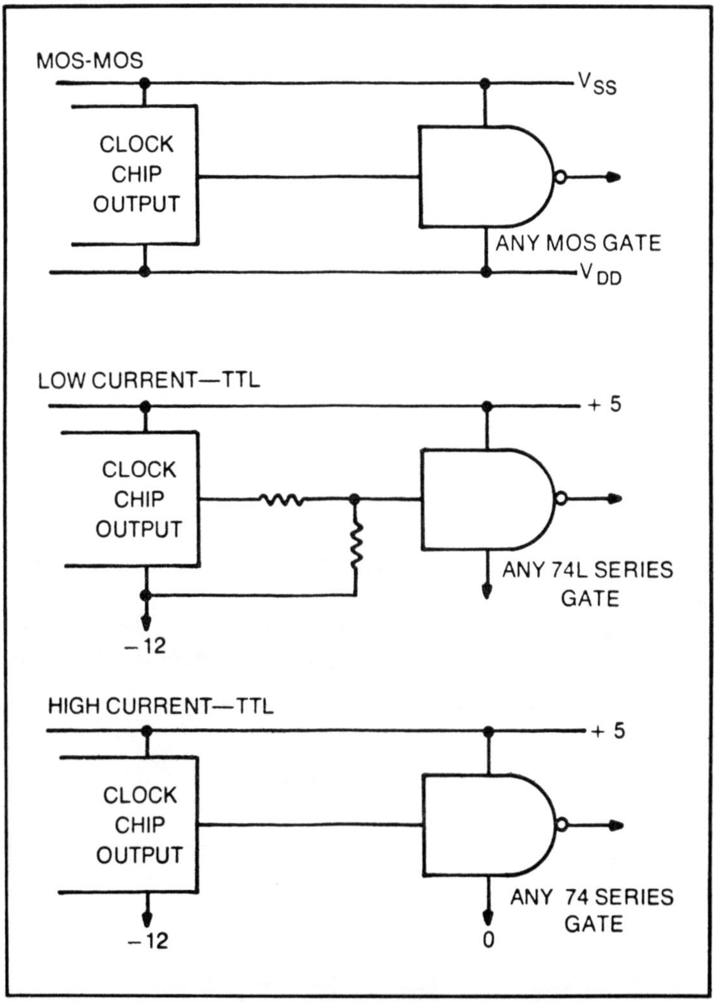


Fig. 2-29. Interface circuits for connection of display data to several logic families. See the data of Table 4-2 for an example of levels.

volts across it. However, this is accomplished by operating the chip from a +5 and a −12 volt supply, measured with reference to a ground point. The TTL circuitry is operated from the +5 volt circuit, operating to ground. The output voltage cannot exceed +5 volts, and so will not damage the TTL gate in this direction. In the negative direction, the input clamping diodes of the 7400 series of gates prevent the output voltage from going below zero; they have sufficient current capacity to work with typical clock chips without any protective elements. However, in the low power TTL series, series resistors and possibly a pull down resistor will be necessary to give adequate protection, as shown in the figure.

These interfaces are typical of the ones used with MOS logic families driving TTL circuits. See the references for a listing which gives a number of other interface circuits, for various combinations of logic family interconnections. Remember that in using these, attention must be given to both the voltage level and current capability of the chip. Typically the low current outputs, such as the segment output or the special control circuit will drive a single series 74L gate. The high current circuit such as usually provided for digit drive purposes will drive any 7400 series gate.

Occasionally it is desirable or necessary to use the display signals for other purposes, for example, for connection to a control element, or small microprocessor. In most applications this will require demultiplexing of the display signal. (A typical signal is shown later in Fig. 3-9). This can be demultiplexed into four separate seven-segment signals by using AND gates, or the equivalent decoder-demultiplexer.

The segment signals information must often be decoded. Decoders are available in the various logic families; for example,

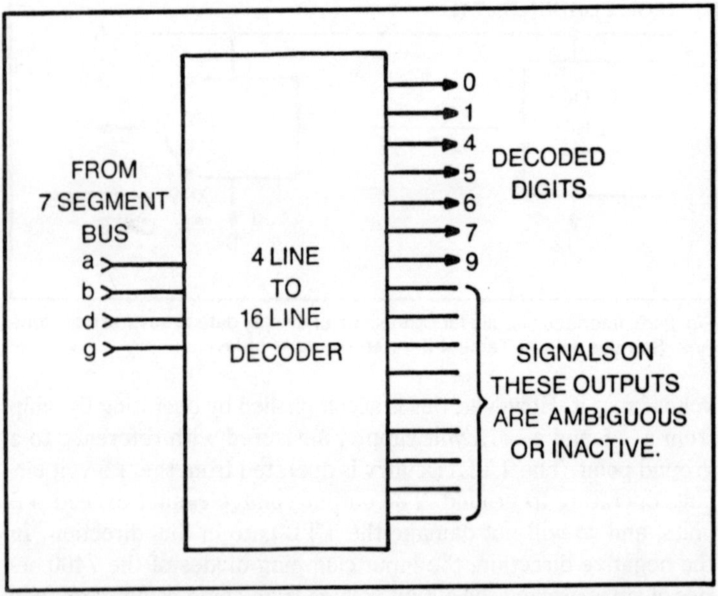


Fig. 2-30. Principle of demultiplexing the segment output signals. See Fig. 7-19 for complete demultiplexing.

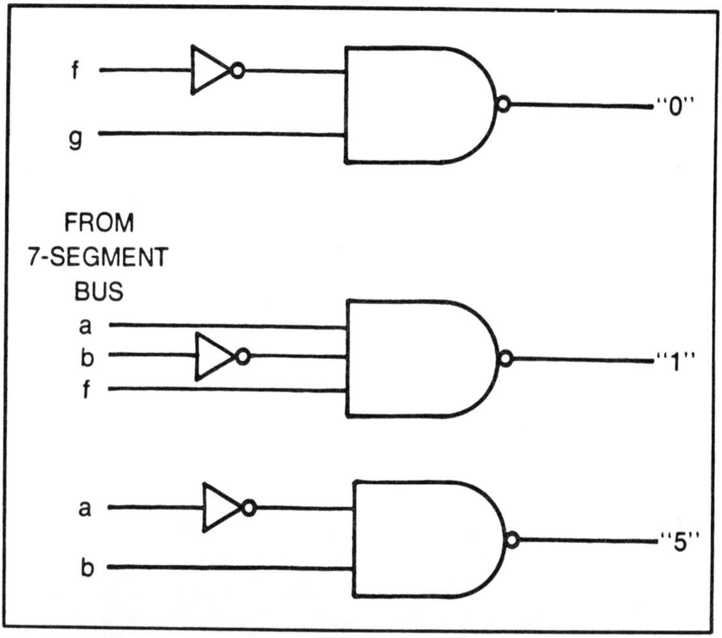


Fig. 2-31. Principle of recognition of a specific digit by use of AND gates. Inverters are used to reduce the number of required gate inputs. Depending on input polarity, the output may be true positive or true negative.

use of the 54154 four-line to sixteen-line demultiplexer of the TTL family gives the circuit of Fig. 2-30. Decoding can also be accomplished by the use of read-only memory chips, the inverse of the technique used within the clock chip to multiplex the signals.

In addition to demultiplexing, various character recognition circuitry may be used. Figure 2-31 shows typical circuits based on the use of inverters and AND gates for recognition of several common characters. The outputs of the character recognition circuits may be used in their multiplexed form, or may be separately demultiplexed; for example, to give a signal that the fifth hour has occurred.

Whenever these multiplexing and demultiplexing operations are to be performed, attention must be given to signal levels, current capability, and breakdown voltages.

## CHIP INPUT PROTECTION

All known clock chips use MOS transistors. The input circuit, or gate, of these can typically stand a maximum of 100 volts between the gate and the drain. Since static potentials can build up on the human body and on tools and instruments to levels of thousands of volts, some input protection for the chip circuits is necessary.

One of the complex circuits used for this purpose is shown in Fig. 2-32. The diodes clamp the input so that it cannot exceed the voltage of the power supply buses VSS and VDD, to within a fraction of a volt. The series resistor prevents too much current being supplied to the gate, the element most susceptible to damage, and also ensures that the voltage rise across the capacitor formed by internal capacitance is not so rapid that it exceeds the switching speed of the diode.

Not all of the chips use these four protective elements, but all are protected to some degree. However, it is still necessary to be cautious when handling these or other MOS chips. They should be kept wrapped in foil or stored in conductive foam. Tools and handling equipment, and the human body, should be discharged to the circuits in which the chip is to be mounted before removing the chip from its protective foil or foam. It is also a good idea to avoid touching the chip pins if possible. The problems with static electricity are far more severe in the desert areas of the country than elsewhere. In such areas it might be desirable to use a room humidifier to give high humidity when chips are being handled.

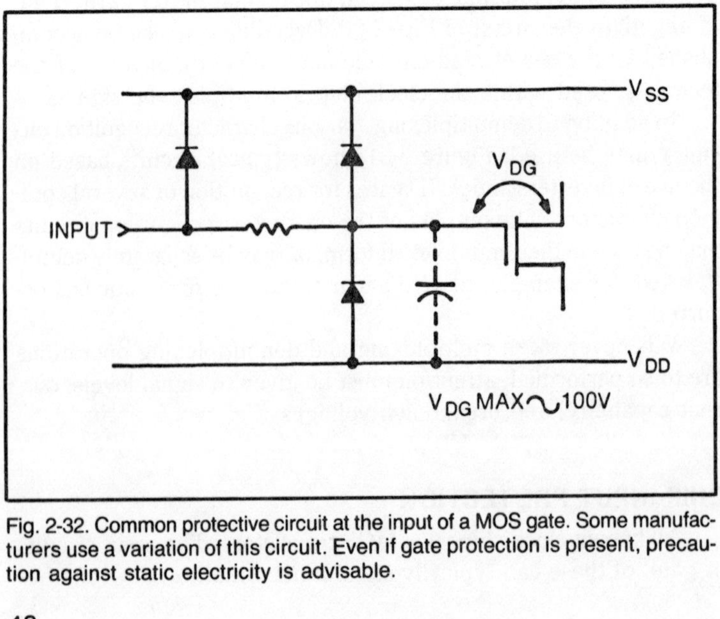


Fig. 2-32. Common protective circuit at the input of a MOS gate. Some manufacturers use a variation of this circuit. Even if gate protection is present, precaution against static electricity is advisable.

## TYPICAL CHIPS

The details of chip construction are buried inside the integrated circuits and are not available for study. Most specification sheets provide a block diagram to show the major functions and the interactions which can be externally controlled.

Figure 2-33 shows a typical chip diagram, this being one of a multiple chip family which provides time only. In this family all functions are sequential and follow directly from considerations outlined in this chapter. Figure 2-34 shows the block diagram of a different chip which contains clock, alarm control and calendar functions. This chip also has provisions for time continuance during power failure built in. Note that the complexity of the chip is appreciably greater, and that the interconnections involved have become much more complex.

The major features of a number of chips for clock and watch purposes are given in Table 2-3. This is believed to be accurate, but it is recommended that the specifications of the chip be checked before ordering or doing any design work.

In this regard, it should be noted that the IC manufacturers sell chips with some changes in internal structure, to suit individual users needs. These may be marked with a special number, or may have a standard chip designation with a special prefix or suffix added. These chips are sometimes found in surplus sales. The chip manufacturer will usually supply only the specification sheet of the basic family of these chips, the detailed changes being considered proprietary. Sometimes the changes can be worked out from the user's literature. If this is not available, it is likely that the chip has retained the basic power supply, clock functions and pin connections, but that there are variations in the control or special function circuits and pin connections. It may be possible to work these out using an oscilloscope, following the basic circuits described in this chapter.

## CHIP DESIGN CHANGES IN WITH TIME

As mentioned in the preface to this second edition, there are changes in any field, rapid at first, then more slowly, as the field matures. One consequence of this is that some designs disappear as do some manufacturers. Other designs converge toward a "most popular" design. This has happened with clock chips.

One consequence of this is that a number of chips listed in Table 2-4 are no longer available. However, it was decided that it

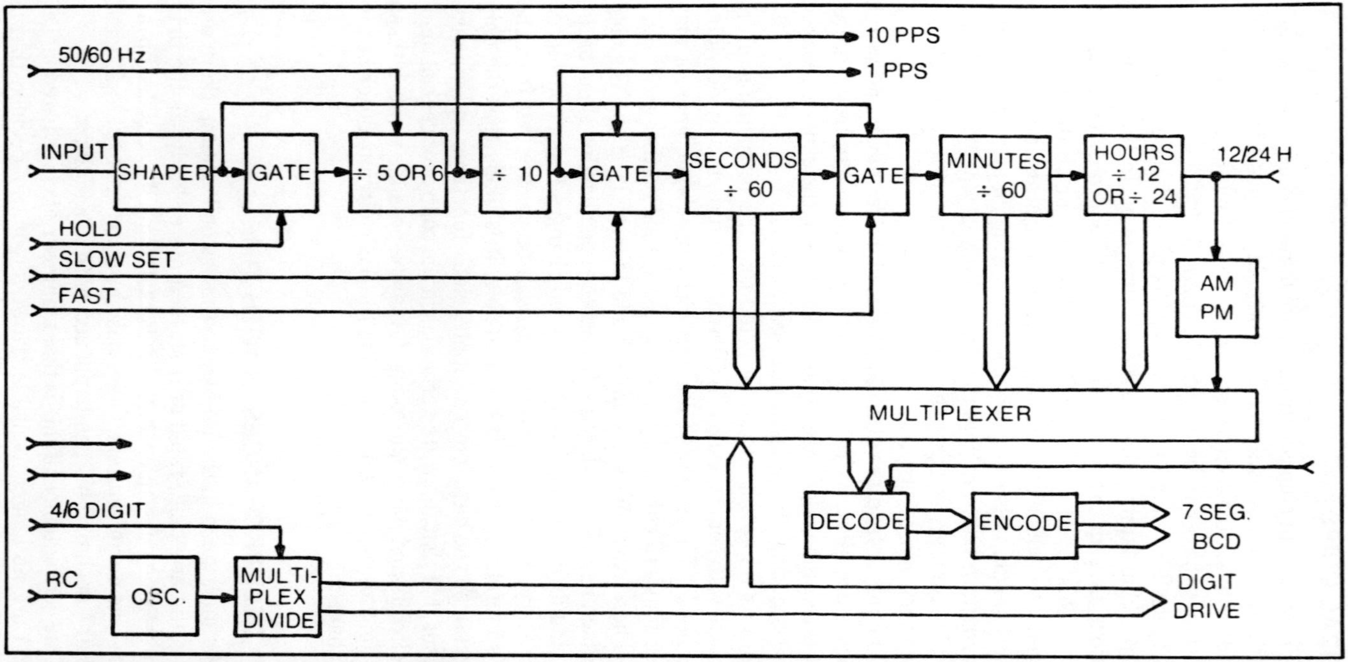


Fig. 2-33. Basic block diagram of a clock chip for time-only. Variations may eliminate one or more outputs, or such features as 50 Hz operation.

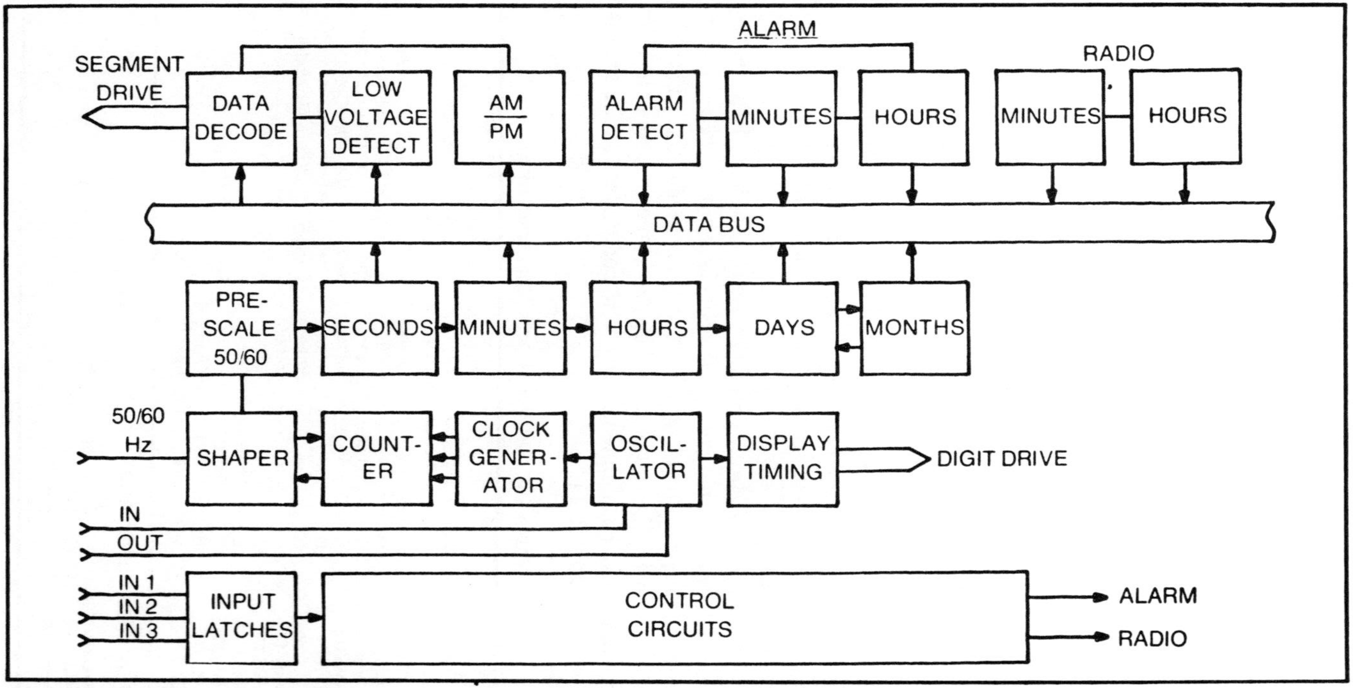


Fig. 2-34. Basic block diagram of a clock-calendar chip. This design uses the multiplex input scheme of Fig. 2-24.

**Table 2-3. CTT-7001 Set Functions, An Example of Multiplex Input Flexibility.**

| Input Pin | Scan Time | Input Name | Definition Connection | Definition, No Connection |
|---|---|---|---|---|
| IN1 | D1 | Set | Set Counter | Set Counter |
| IN1 | D2 | Set H/M | Set Hour or Month Digit | Set Minute or Day Digit |
| IN1 | D3 | Clock Radio Switch | Clock Radio Switch—On | Clock Radio Switch—Off |
| IN1 | D4 | Mode A | Mode A—Off | Mode A—On |
| IN1 | D5 | Mode B | Mode B—On | Mode B—Off |
| IN1 | D6 | 50/60 Hz | 50 Hz Input | 60 Hz Input |
| IN2 | D1 | Set Calendar | Set Calendar Counter | |
| IN2 | D2 | Set Clock | Set Clock Counter | |
| IN2 | D3 | Set Alarm | Set Alarm Counter | |
| IN2 | D5 | Set Clock Radio | Set Clock Radio Counter | |
| IN2 | D6 | Snooze Switch | Snooze Switch—On | Snooze Switch—Off |
| IN3 | D1 | Alarm Switch | Alarm Switch—On | Alarm Switch—Off |
| IN3 | D2 | 12/24 Hour | 24 Hour Operation | 12 Hour Operation |
| IN3 | D3 | C1 | C1 | C1 |
| IN3 | D4 | C2 | C2 | C2 |

would not be a good idea to delete these from the table. Their inclusion does give a picture of changes in time. Also, these chips may be encountered in remnant stock, and especially in clocks for repair (see Chapter 12).

New chips have been added. These tend to be in two classes. One is a four-digit chip with good drive capability and designs intended to give a useful clock with low component count and therefore low cost. The other class includes specialized chips, for example, the ones intended for use with computers. These are discussed in Chapter 11.

**Table 2-4. Sample of Chip Characteristics.**

| Type | Use | Digits | 12/24 hrs. | Features* | Remarks |
|---|---|---|---|---|---|
| CK3300 | C | 4 | Either | ARS | Standby Oscillator |
| TA8084 | W | 4 | 12 hr | | LCD, Xtal |
| MM5311-14 | C | Either | Either | BP | Some Have Strobe |
| MM5316 | C | 4 | Either | ARS | Variable Intensity |
| CT7001 | C | 6 | Either | CARS | Standby Oscillator |
| CT7002 | C | — | Either | CARSB | Standby Oscillator |
| MK5017 | C | 6 | Either | | |
| MM5375 | C | 4 | Either | A | |
| TMS3834 | C | Either | Either | A | |
| MM5309 | C | Either | Either | BP | Reset to zero |
| MK5030 | W | 4 | Either | C | |
| ICM7214 | W | 4 | Either | C, D | 4 year C |

*FEATURES ABBREVIATIONS

C—Calendar  S—Sleep  D—Date
A—Alarm  B—BCD
R—Radio  P—1PPS

Table 2-5. Additional Clock Chips.

| Type | Use | Digits | 12/24 | Features | Remarks |
|---|---|---|---|---|---|
| MM5387AA | C | 4 | Y | ASFP | LED Direct |
| MM5382 | C | 4 | N | ACDSFLP | LED Direct |
| MM5455 | C | 4 | Y | AFS | LED Direct |
| WD2412 | C | 6 | Y | ABCDP | See Chapter 11 |
| MM5377 | Auto | 4 | N | | Xtol<br>LED Direct |
| AY-5-1202A | C | 4 | Y | FP | Fluorescent<br>Direct |
| AY-5-1224A | C | 4 | Y | BF | LED Direct<br>LED Simple |
| CK-3300 | C | 4 | Y | ACFRS | Especially for<br>Clock Radios |

LEGEND

| | |
|---|---|
| A—Alarm | N—No |
| B—BCD | P—1 PPS |
| C—Calendar | R—Radio |
| D—Date | S—Sleep |
| F—Power Fail | Y—Yes |
| L—Light Level | |

Table 2-5 is a representative sample of newer clock chips. Note that, with one exception, these are four-digit units. The exception is essentially a super version of the above second class which has both good clock and good computer capabilities.

It is now uncommon to find clock chips in surplus. Manufacturers are watching their inventory too closely to allow this to happen. However, there is no real problem in obtaining chips. A good selection is available from the supply houses which cater to the hobby and experimenter market. Their ads can be found in the specialized magazines intended for these groups.

## CLOCK CHIP RATINGS

Specification sheets give most, but not all, of the quantities needed to prepare a design using the clock chips, or for experimental use. Further, at least one specification sheet value is likely to be understated, probably by a large amount. For the intended purpose, the omissions or conservativeness is no particular problem, but for experimental use, they are a nuisance.

The quantity which is almost always understated is the frequency capability of the chip. Typical upper limits given are 30-60 kHz, and some ignore the matter completely. Yet the chips are fabricated using a form of MOS technology, with individual elements capable of operating into the range of megahertz. The performance of each chip must be checked if the range is important.

The quantities usually not stated are the allowable series resistances in control leads, and the threshold voltages of the "ON" and "OFF" state. Usually, these circuits have an internal pull-up resistor, but this value is not given either. These values can be determined by experiment, without too much trouble, however.

The ratings which relate to maximum supply voltage, and to allowed voltage between pins appear to be accurate; at least, no failures have been encountered using the specification values, but chips have been lost by exceeding them.

Overall, the procedure of following these voltage ratings (or current, if specified) and measuring other ratings seems to work well. However, it should not be regarded as absolute insurance that no damage will ever occur—experimenters must accept an occasional lost chip.

# Chapter 3

# Displays

In general terms, a display is a device or assembly intended to present data to the human eye. A store window, the face of a clock, even advertisements in a newspaper are forms of display.

In electronics, and especially in clock electronics, the term display has a much narrower meaning. In this field, "display" is restricted to an electro-optical or electro-mechanical-optical device intended to show alphabetical, numerical or other symbolic data. For the amateur or home constructor, the display will almost always be an electro-optical device of a relatively limited number of types. Typical units available to the home constructor are shown in Fig. 3-1.

## THE DISPLAY PROBLEM

The design and construction of displays is no small problem, because of the many requirements that a satisfactory display must meet. First, there is the matter of recognition. The particular letters or numbers presented by the display must be recognizable at a glance. This is especially important in clock construction, since the usual method of referring to a clock is by a glance, to see if there is sufficient time remaining to undertake some task. This problem of recognition is partly solved by adopting displays which are more or less standard in shape—their style is a definite element in recognition.

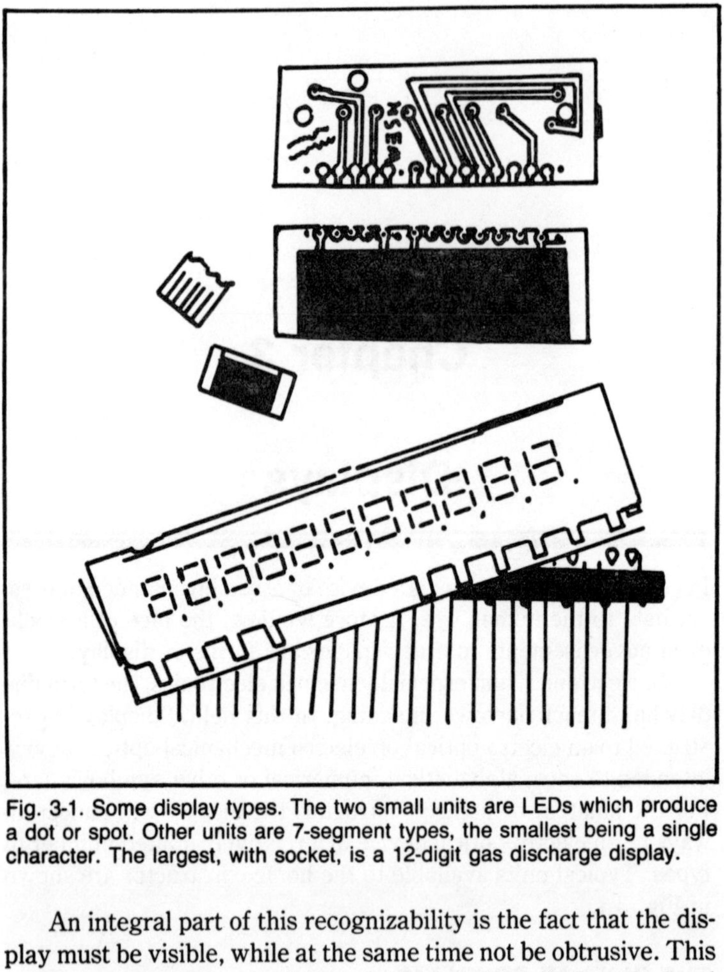

Fig. 3-1. Some display types. The two small units are LEDs which produce a dot or spot. Other units are 7-segment types, the smallest being a single character. The largest, with socket, is a 12-digit gas discharge display.

An integral part of this recognizability is the fact that the display must be visible, while at the same time not be obtrusive. This visibility is primarily a function of the size of the display, the level of illumination that it gives, and the contrast between it and its surrounding objects. The size is essentially independent of the exact mode of construction of the display, but the illumination level and the contrast do depend on the type of display used.

Finally, there is the important element of cost. For the type of displays used in the home, or for personal use, the cost must be extremely low—on the order of a few tens of cents per display digit. The fact that satisfactory displays are available within these cost limits is a great tribute to inventors, designers and the manufacturing process. This cost factor is considerably less important in specialized displays, the type used in public buildings, in airline terminals, or in such presentations as stock market prices, but even here cost is always a factor which enters into consideration.

At the present time, for electronic clocks and similar applications, the various factors which relate to the display problem have caused standardization essentially on only two general families. One of these families is based on the light-emitting diodes, the LED. The second type is based on a completely different principal, that of a liquid whose optical reflectance or transmittance changes under the influence of an electrical field; the liquid used has many properties of a crystalline substance, and is usually known by the term "liquid crystal," or LCD. Occasionally this is referred to by a more precise scientific term, "nematic liquid." The term means that the liquid molecules can be aligned in long strings, or filaments.

While displays of these two types may be encountered in all clock applications, it is more common to find the liquid crystal display in very small sizes as used in wrist watches, or in very large sizes as would be used in public buildings. The small size is used because the liquid crystal can be constructed to require negligible power, and the very large size because other types of displays become difficult to fabricate in the very large sizes. In the middle range, in the displays normally used for the home clock, LEDs are the most common.

It should be remembered that there are other types of displays, such as the hot wire, gas discharge, fluorescent, or cathode glow displays. These are often found in other electronic fields, for example, in the hand calculator or in specialized displays such as elevator floor indicators. At present, because of cost or performance factors, these other types are rarely or never encountered in clocks or watches. However, conditions may change as developments occur and, in any event, the experimenter may wish to try some of these additional displays, not only as an experimental item, but also as a way of changing the "decorator" characteristics of clocks.

In the following, the LED and liquid crystal types of displays are covered in some detail.

## DISPLAY SHAPE

For all intents and purposes, the electronic clock field has standardized on a single display shape. This is the arbitrary character form made possible by seven segments, arranged as shown in Fig. 3-2. Because of the arrangement, this is commonly referred to as the seven-segment display. The exact method of construction of the segments does vary from manufacturer to manufacturer. In

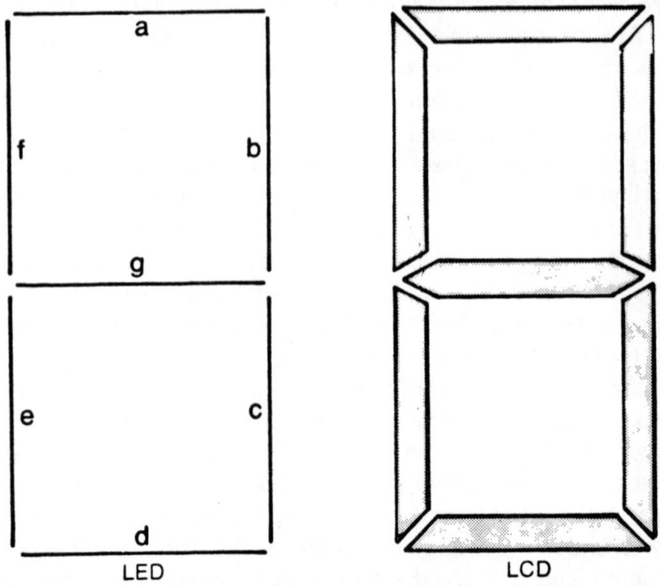


Fig. 3-2. Basic appearance of the 7-segment display. The light emitting diode type appears as bright, narrow segments, commonly in red. The liquid crystal may appear as broad segments showing by reflected light, usually greenish, or may appear to be illuminated from the rear. Segments are designated a through f, clockwise from the top, with g at the center.

some, the segments are made of a row of dots, each emitted by a small LED crystal. In others, the segments may be formed by a prism, illuminated by a single LED. All of the LED type displays tend to produce a relatively narrow line as compared to the length of the segment. In contrast, the liquid crystal displays are formed from a continuous bar, usually relatively wide with respect to the length. The reasons for this difference in appearance will be understood when the characteristics of liquid crystals are considered in detail.

For both of these types, in fact, for all types of seven-segment displays, the display nomenclature has been standardized. Each segment is designated by a lower case letter, also as shown in Fig. 3-2. Using the convention that a zero indicates that a segment is inactive, or OFF, and a one indicates that it is active or ON, the seven-digit code which corresponds to the numbers made possible by the seven segments is shown in Fig. 3-3. Note that there are two forms of the numerals six and nine, both being used.

Since seven binary digits are needed to specify the state of each segment of the display, it should be possible to produce 127 or 128 different characters. Only ten of these are used by the number dis-

| NUMERAL | A | B | C | D | E | F | G |
|---|---|---|---|---|---|---|---|
| 1 | 0 | 1 | 1 | 0 | 0 | 0 | 0 |
| 2 | 1 | 1 | 0 | 1 | 1 | 0 | 1 |
| 3 | 1 | 1 | 1 | 1 | 0 | 0 | 1 |
| 4 | 0 | 1 | 1 | 0 | 0 | 1 | 1 |
| 5 | 1 | 0 | 1 | 1 | 0 | 1 | 1 |
| 6 | 0 | 0 | 1 | 1 | 1 | 1 | 1 |
| 7 | 1 | 1 | 1 | 0 | 0 | 0 | 0 |
| 8 | 1 | 1 | 1 | 1 | 1 | 1 | 1 |
| 9 | 1 | | 1 | 0 | 0 | 1 | 1 |
| 0 | 1 | 1 | 1 | 1 | 1 | 0 | 0 |
| ALTERNATES | | | | | | | |
| 6 | 1 | 0 | 1 | 1 | 1 | 1 | 1 |
| 9 | 1 | 1 | 1 | 1 | 0 | 1 | 1 |

Fig. 3-3. Numerical characters of the 7-segment display, and the binary equivalent of the segment lines, a one indicating that the segment is active.

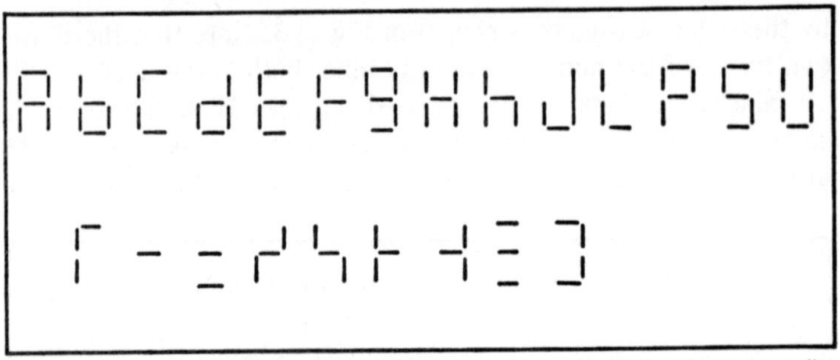

Fig. 3-4. Alphabetical characters which can be formed with the 7-segment display, plus some arbitrary symbols.

plays, so it should be possible to make a large number of additional characters. Practically, an appreciable number of additional characters are possible, but it is not nearly as great as might be expected at first glance. Partly, this lesser number is due to position redundancy within the display: for example, if only a single horizontal segment of a single digit is activated, it would be difficult to tell if this were the upper, center or lower segment, and similarly for the vertical segments. In addition, many of the possible arrangements of characters have no relation to any other standard form, and so are of limited value.

Figure 3-4 shows some of the characters which can be generated with the seven-segment display. The upper row shows upper case alphabetical characters which are possible assuming that the numerical characters must be retained without ambiguity. It also shows possible lower case letters. Using these upper and lower case characters in combination will allow many alphabetical presentations to be made, by using abbreviations or arbitrary assignments. The lower row of characters in this figure shows a few of the arbitrary symbols which are possible and identifies some of these with normal mathematical symbols.

Because of the relatively restricted range of alphabetical and arbitrary characters which are possible with the seven-segment display, a number of additional display formats have been developed and used. One of these retains the seven-segment display, but adds two vertical segments at the center of the seven segments, as shown in Fig. 3-5. This nine-segment system produces exactly the same numerical characters as the seven-segment, and additionally produces more alphabetical and arbitrary characters. For example, the nine-segment can produce, without ambiguity, the symbols A.M. and P.M. widely used in clock construction, and can also be used to produce such character pairs as MO, TU, WE, etc., as would be used to designate the days of the week.

In another variation, a set of lines extending from the center of the characters to the corners may be introduced. Usually the horizontal segments are divided into two parts also, giving a 16-segment display. The introduction of the diagonal lines makes possible better representation of alphabetical characters and of many arbitrary characters. A few such characters as usually formed as shown in Fig. 3-5. Still another variation which may be encoun-

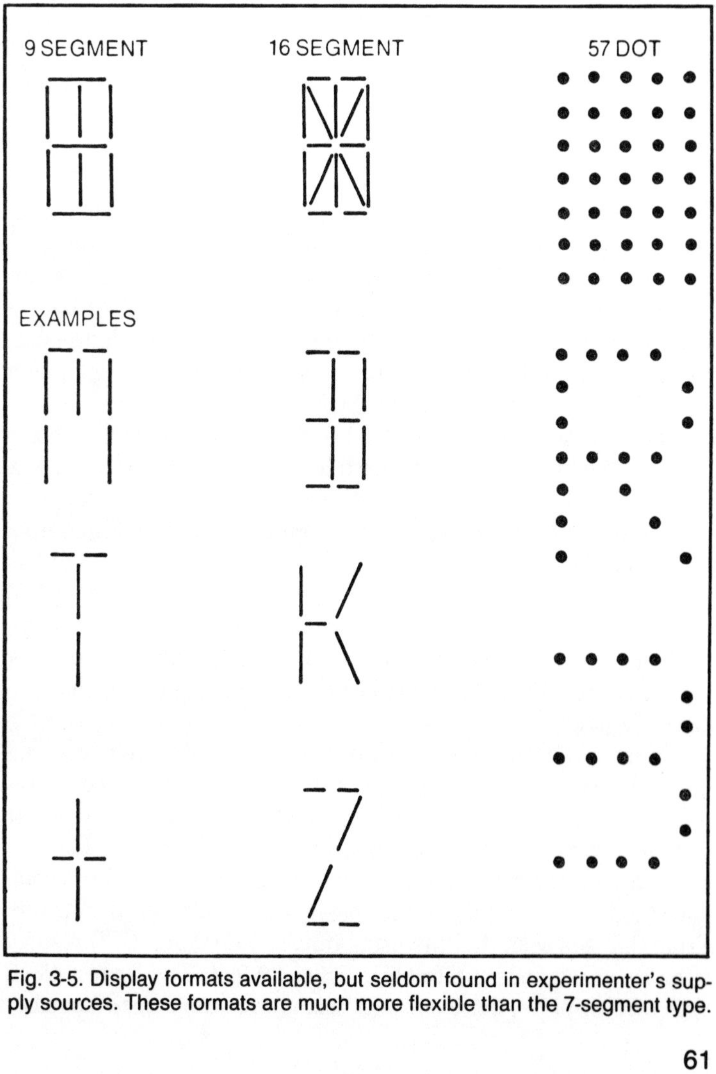


Fig. 3-5. Display formats available, but seldom found in experimenter's supply sources. These formats are much more flexible than the 7-segment type.

tered is the display formed entirely of dots, the usual arrangement having 35 LEDs arranged in a five by seven array. This display is extremely flexible in its ability to form characters, a few being shown in Fig. 3-5. A large number of Japanese characters, the Roman numerals, and many arbitrary or special characters can also be formed.

It is not too likely that these special displays will be encountered by the home experimenter. The usual place for application of the nine-segment display is in watches and clocks which keep track of the days of the week and the months. The 16-segment display may be encountered, but at present none are known to be used with watch-clock devices. The five by seven display has been used for clock purposes; in particular, for time presentation on the television screen.

## DISPLAY VISIBILITY

In addition to being readily recognizable, it is important that the display be readily visible. This visibility is primarily a function of two factors. The first is the size of the display, which enters into its visibility at a given distance. The second is independent and relates to the light characteristics of the display, i.e., the light emitted if it is of the LED type or the light falling on the display and its characteristics if it is of the LCD type.

The required size of the display can be determined from the fact that the human eye has a certain angle of acuity. This means that for equal visibility, distant objects must be larger than close ones. Related to this is the fact that there is a required minimum separation between two objects if they are to be distinguished as being separate; this minimum separation is also an angle, and so objects must be further apart as the distance increases if they are to be separated. These two factors combined mean that a display for use at great distances must be larger in all dimensions than the close display.

Most of the data sheets for displays give a recommended maximum distance. For example, the common 0.3 inch LED displays will usually show a maximum distance of 10 feet for use. At lesser distances the display is still perfectly usable, of course. The distance ratings of a number of display sizes and types have been compared and summarized by the distance-size relation of Fig. 3-6. From this it is seen that the usual range of LED and LCD display sizes cover distances up to about 25 or 30 feet. The largest clock type LCD displays, those one inch height, are usable up to distances

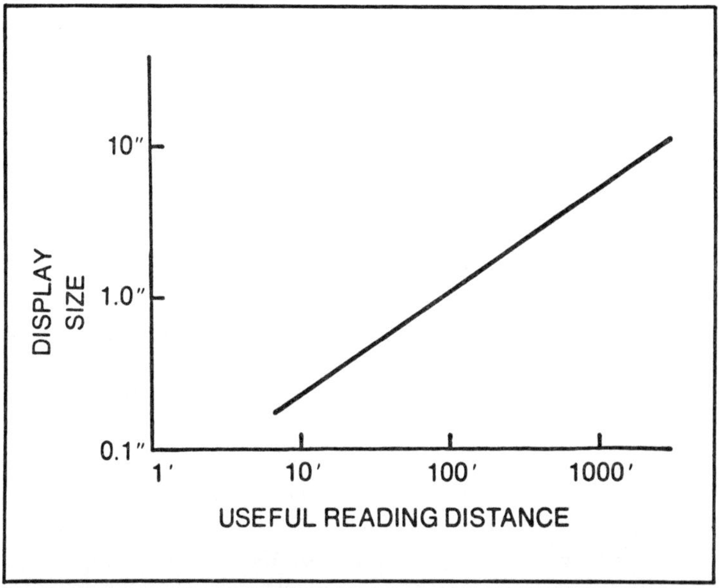


Fig. 3-6. Required display size versus reading distance. This assumes adequate brightness or contrast, and clear air.

of about 60 feet. The largest commercial displays, having 5 to 8 inch characters, are usable at distances of several hundred feet, distances such as would be encountered in sports stadiums, airports, and for advertising clocks. Still larger displays are possible but would have to be designed, a worthwhile project for the home experimenter.

The curve of visibility versus size shown in Fig. 3-6 is based on the assumption that the light characteristics of the display are ample. If not, usable distance of the display will be reduced. At present there are no standards for the required light characteristics and the selection must be based on experience or on tests of the particular device being considered. There are, however, a few guides to required light characteristics which can be developed from consideration of common light sources and the characteristics of the human eye.

Before starting a discussion of these characteristics, it seems necessary to spend a little time on the units of light measurement. This can be a very confusing field, partly because few people have taken time to straighten out the principles involved, and partly because of the great number of measuring units which are used in the field. In doing this it is necessary to remember that three different factors are involved, depending on the nature of the display. If the particular display has a very small source, essentially corresponding to a point source, then the luminous intensity is the important factor. However, if the display uses a broad area light source, approaching a size which can be completely resolved by the eye, a different factor, the brightness, is important. Finally, if the display functions by reflected light, the important factor is the contrast, including the contrast between the On and OFF state as well as the contrast between the specific character and its surrounding area.

Table 3-1 gives some calibration points for various levels of intensity and brightness. Data from the specification sheets of displays can be compared to these to obtain an idea of the illumination characteristics that will be obtained. However, for the home experimenter who is not especially interested in optical matters, the best procedure is to follow the manufacturer's recommendations closely, or to arrive at required illumination levels by experiment. Of course, in doing this, attention should be paid to the other characteristics of the display: for example, increasing the light output of a LED can only be done by increasing the current through the diode, which may cause a failure by overheating. These types of problems are the basic reason that the manufacturer's recommendations should be followed closely.

## DISPLAY MULTIPLEXING

Most clock chips use a form of display multiplexing to reduce the number of pins required in the chip package. The most com-

**Table 3-1. Selected Light Values.**

| Location | Intensity of Illumination |
|---|---|
| Open Area, Clear Day | 10.000 footcandles |
| Open Area. Full Moon | 0.03 footcandles |
| Reading Area, Minimum | 12 footcandles |
| Drawing Table | 25-100 footcandles |
| **Object** | **Brightness** |
| Surface of 40 W frosted bulb | 8000 footlamberts |
| Full Moon Seen from Earth | 1500 footlamberts |
| Page, reading fine print | 10 footlamberts |

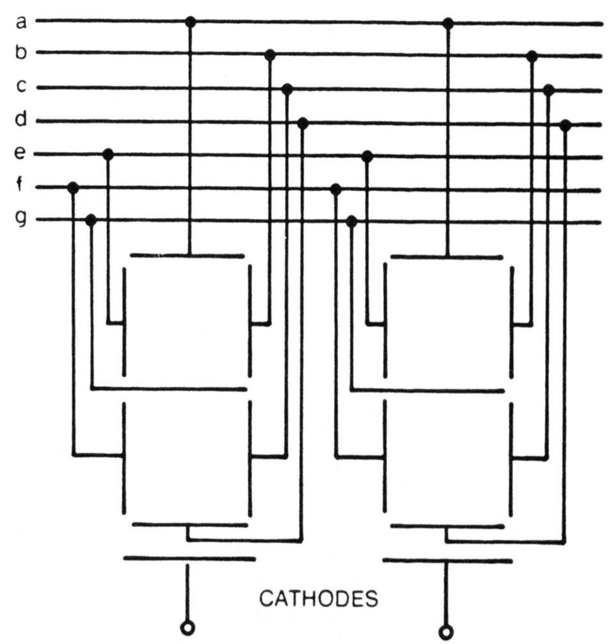


| NUMBER OF DIGITS | NUMBER OF LINES |
|---|---|
| 1 | 8 |
| 2 | 9 |
| 3 | 10 |
| 6 | 13 |
| 12 | 19 |

Fig. 3-7. Basic multiplex connections for common cathode digits. The saving in pin connections can be appreciable, 13 being required for 6 multiplex digits against 42 nonmultiplexed.

mon method of multiplexing is found in the standard seven-segment arrangement, wherein one element, anode or cathode, is common to all segments of a digit, and all "a" segments are connected, all "b" segments, etc. The chip data sheet should be checked, since some chips use the common anode arrangement and others the common cathode arrangement. In addition, there are several other methods of multiplexing.

The reduction in pin count by multiplexing is quite sizable, of course, as indicated in Fig. 3-7. For the usual six-digit clock, the pin count required for display of numerical characters alone is reduced from 48 to 13. This represents a great simplification in the packaging problem.

In the designs which are given later, some are based on use of premultiplexed displays, often called sticks. Others are based on the use of individual digits, with the digit multiplexing being done at the terminals. For these, a combination of printed circuits and wire jumping is used. This method was chosen to eliminate the need for double-side PC boards, or for very close spacing of leads. For those who prefer to eliminate the use of jumpers in the display multiplexing, Fig. 3-8 can be used to prepare the printed circuit boards. Boards of this type are available from several sources for most of the commonly used individual digit displays.

This method of multiplexing is not the only one used. One common technique is used only for the first digit of the display. At most, this needs to be only a 0, 1 or 2, since the hours count does not exceed 12 or 24, depending on the system used, and often with the 0 not being displayed. For the 12 hour type of display, a simple technique is to provide a signal which is on for the 10th, 11th and 12th hour. This signal is fed to segment D and C of the display. The signal is off for the interval between one and nine hours.

In some cases the chip will also provide a pair of additional signals to be connected to segments E and F, one being used to indicate the A.M. hours and the other the P.M. hours.

In another type of multiplexing format, four leads are provided for the first digit instead of seven. Diodes are used in conjunction

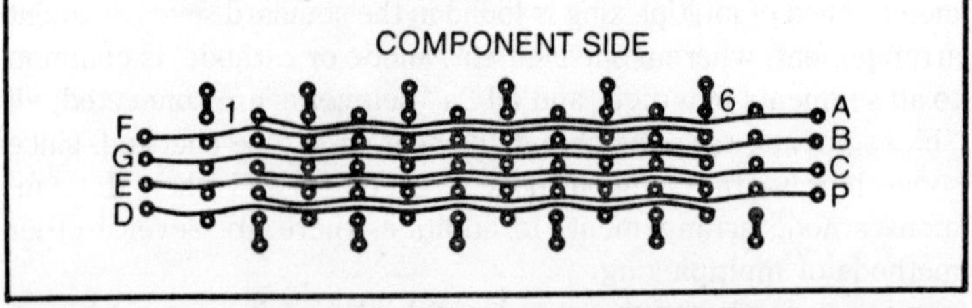


Fig. 3-8. Layout for a 6-digit multiplex board using FND70 digits, or equal. Photographic reproduction to 1:1 scale is recommended.

with these four leads to route signals correctly, to form either a one or a two. A common arrangement of this multiplex mode is shown in Fig. 3-9.

Still another multiplexing method connects the segments of two digits together. The anodes of these digits are driven from the ac line, alternate half-cycles driving one digit, then the other. Attention to the diode reverse voltage characteristics is necessary with this technique; if the inverse voltage applied exceeds the rating, breakdown will occur. This can be avoided by using rectifying diodes so that the idle half-cycle is clamped to zero, thus preventing the display from being energized.

## LINE FREQUENCY MULTIPLEXING

One of the techniques developed to reduce the number of leads from the chip to the display is to use the power line for duplexing. The principle is shown in Fig. 3-10. The display anode is fed with two pulsating dc voltages, generated by half-wave rectifiers. Since they are fed from a center-tapped transformer, they are out of phase, that is, only one is "high" at a time.

In the clock chip, the multiplexing circuits must "sort-out" the signals in synchronism with the anode voltage changes. In essence, this is accomplished by using the ac signal in the same way, to generate two gating pulses, which can be called TIME and $\overline{\text{TIME}}$.

Consult the chip manufacturer's data sheets for details of the connections.

## LIGHT EMITTING DIODE (LED) CHARACTERISTICS

The light emitting diode is a semiconductor junction whose material and assembly process is chosen so that light is emitted at the junction area when the diode is carrying current. Like all semiconductors, the allowable current which can be passed through the junction is limited by temperature considerations. Increased temperatures reduce the efficiency of light emission and if sufficiently high can completely destroy the junction.

A typical curve of current versus forward voltage for a medium-sized light emitting diode is shown in Fig. 3-11. Note that the threshold of conductance is even sharper than for most semiconductor materials. This fact is occasionally useful for other than light emitting purposes, the diode being useful in general electronic work.

One characteristic of light emitting diodes must be remembered. This is the fact that most types have very low reverse volt-

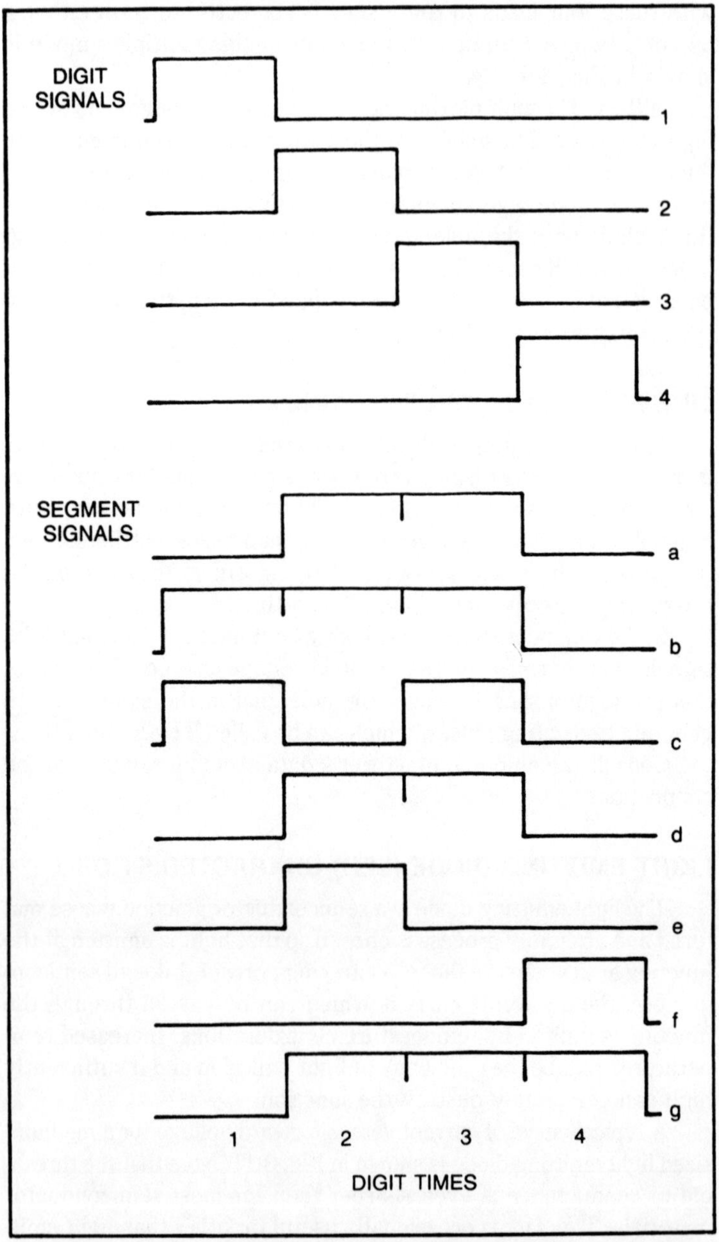


Fig. 3-9. Digit drive or cathode signals, plus segment drive or anode signals for a 4-digit display reading 1234, scanned from left to right. The active elements are determined by the AND of digit and segment signals. Many displays scan from right to left.

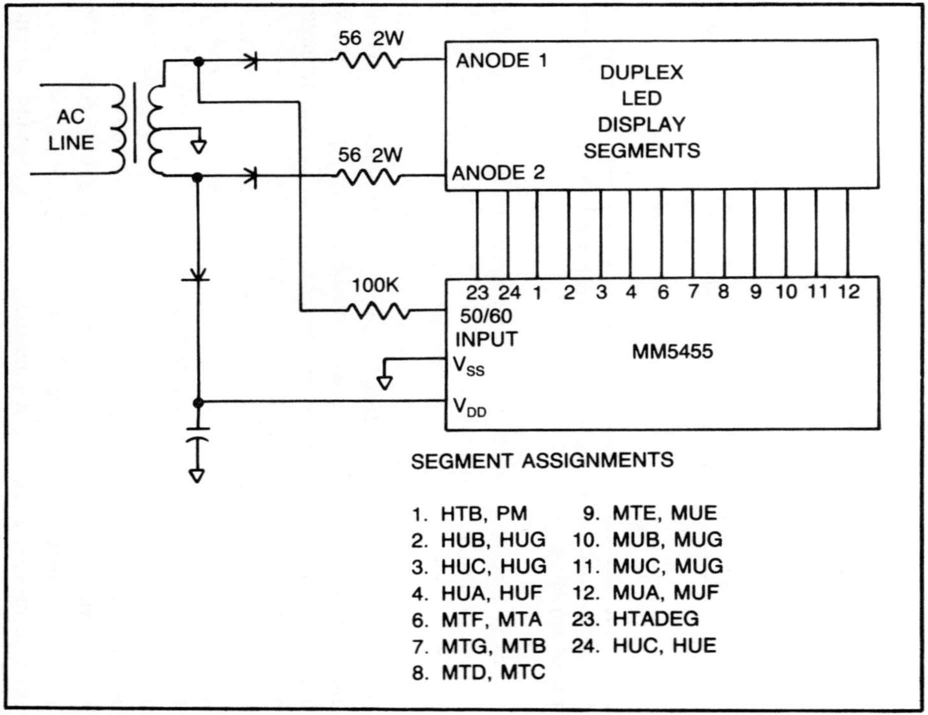


Fig. 3-10. Powerline duplexing in the MM5455 chip.

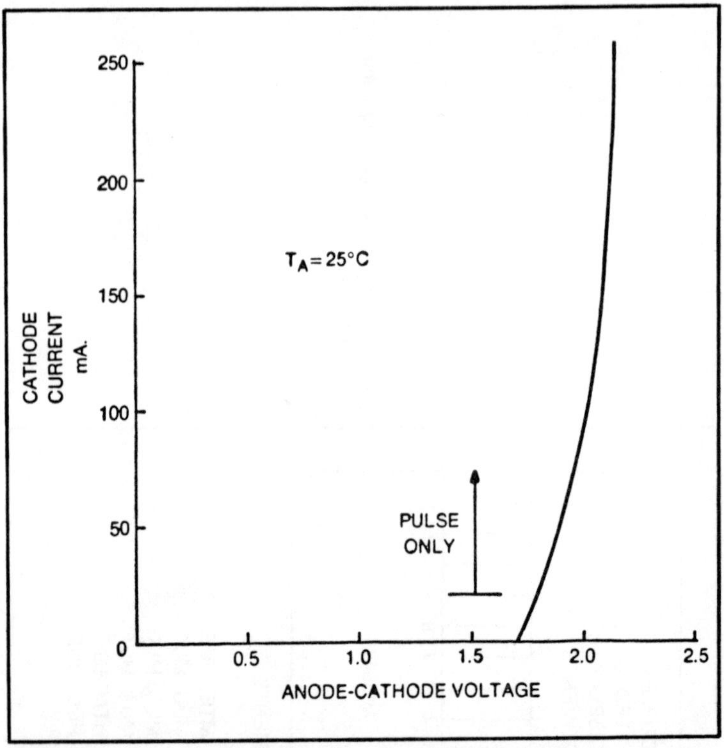


Fig. 3-11. Voltage-current relation of a typical LED. Within rating, the unit makes a good regulating element.

age capability—typical breakdown voltages are in the range of 3 to 5 volts. In experimental work this can be troublesome, and it is worthwhile to connect a conventional silicon diode across the LED in the reverse sense, as a way of preventing excess inverse voltage. If high back resistance should be important, a second inverse diode can be added in series with the LED, giving good front to back resistance plus inverse voltage protection. Typical arrangements are shown in Fig. 3-12.

Over a fairly large range, the light output of a LED is linear with current. However, two effects may introduce a nonlinearity. The first is that the efficiency of light emission is reduced as the temperature increases, so that high currents usually give somewhat less than linear light output as a result of heating. The second effect shows when pulse currents are used, as would be required in a multiplex display; for example, in a four digit display, the peak current through a segment will be four times the average current.

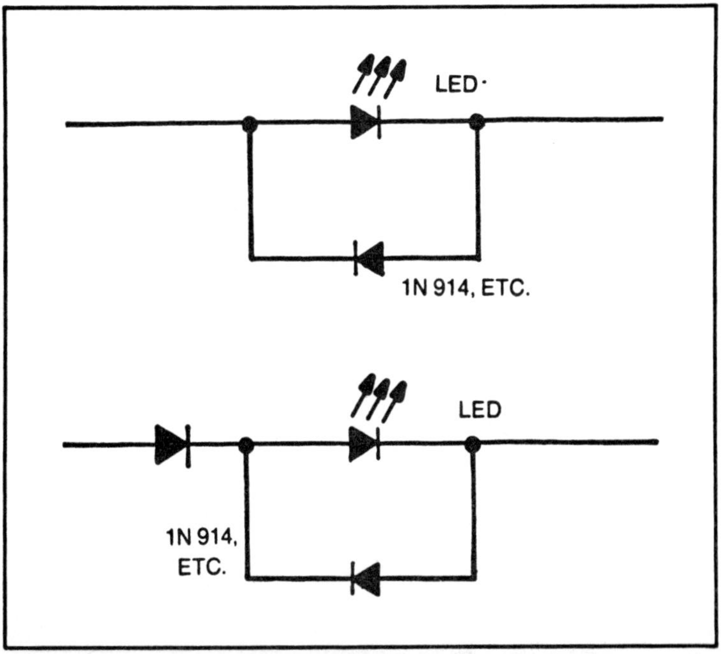


Fig. 3-12. Method of protecting an LED against high reverse voltage. The lower arrangement gives high reverse resistance.

Since the eye sees the total light output, the average current determines the apparent brightness. However, nonlinear leakage and capacitance make the light output nonlinear with current.

The result of these two effects can be seen in Fig. 3-13, which shows the luminous intensity of a typical medium sized LED versus current. The higher current ranges depart appreciably from linear operation. In addition, the effect of high peak currents increases as the average current increases. Generally the effect of this nonlinearity is unimportant, since it can be compensated for by a small increase in average current. Table 3-2 gives data on the voltages and currents of typical individual LED segments, or of individual LEDs. Note that the red LED is currently the most efficient; however, work on efficiency improvement for all types continues and individual specifications should be consulted.

Table 3-3 shows more complete characteristics for a number of common light emitting diode displays. The type numbers are those of the manufacturer and will usually be found in the advertisements and literature. Unfortunately, not all displays are marked with the type number, a manufacturer's code being used instead.

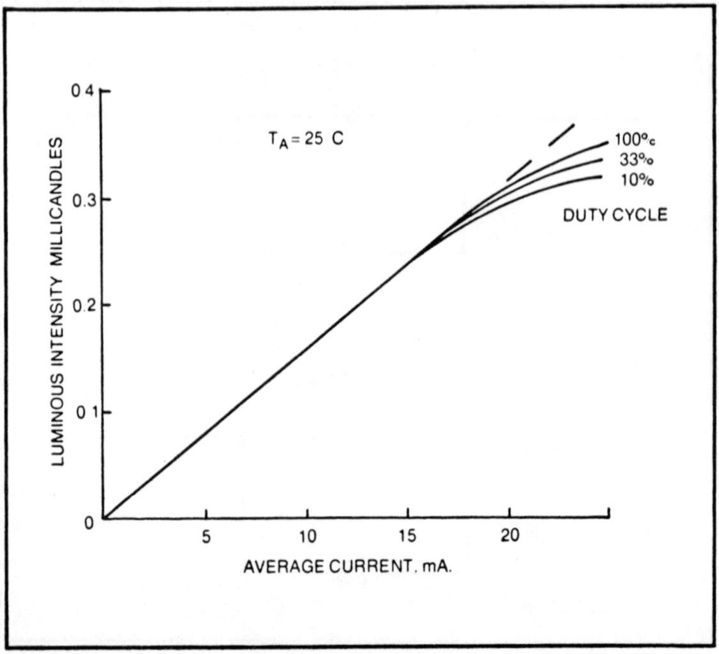


Fig. 3-13. Typical light output of an LED. The nonlinearity at high current is partly due to internal resistance, and partly due to electron-hole recombination.

If unspecified displays are purchased, as from surplus sources, it may be considerable trouble to obtain full identification. In this case the best procedure is to identify the anode and cathode by using a low current ohmmeter of the 1.5 or 3-volt maximum type. With the terminals identified, the rating can be determined by experiment, passing a small current through the diode, then increasing it until usable light emission is obtained. It may be desirable to sacrifice one of the diodes or diode segments to obtain the breakdown current. With this information obtained, it should be possible to obtain reliable operation by restricting the current to 50 percent of the breakdown value.

**Table 3-2. Some LED Characteristics.**

| Color | Red | Yellow | Green |
|---|---|---|---|
| Forward Voltage, V | 1.65 | 3.0 | 3.5 |
| Normal Forward Current, mA | 20 | 50 | 50 |
| Maximum Forward Current, mA | 50-100 | 50-100 | 50-100 |
| Maximum Reverse Voltage, V | 2-5 | 2-5 | 2-5 |

**Table 3-3. Selected LED Displays.**

| Type | Character Height in inches | Number Digits (*) | Light Output | Current MA | Voltage V | Reverse Voltage Rating |
|---|---|---|---|---|---|---|
| FND70 | 0.25 | 1 | 0.3 Red | 20 | 1.9 | 2 |
| FNA30 | 0.16 | 9 | 0.03 Red | 3 | 1.75 | 1 |
| TIL312 | 0.30 | 1 | 0.5 Red | 2 | 1.7 | 3 |
| TIL314 | 0.30 | 1 | 0.3 Green | 20 | 2.5 | 3 |
| DL57 | 0.32 | 1 (5×7) | 0.6 Red | 10 | 1.7 | |
| MAN3 | 0.125 | 1 | 0.4 Red | 10 | 1.7 | |
| 747 | 0.8 | 1 | | 8 | 5.0 | |
| 1720 | 1.02 | 1 | 1.0 Red | 20 | 1.6 | |
| DL416 | 0.16 | 4 (16 Seg) | 0.5 Red | 5 | | |

(*) 7-Segment unless indicated

## LIQUID CRYSTAL DISPLAYS

It should be remembered that the Polaroid sunglasses function because of a layer of a chemical crystal which lays in parallel lines in a thin film. Experimentally it has been found that there are a number of chemicals which will assume this array arrangement under the influence of an electric field. These chemicals are liquid, but because they can partake of a crystalline type structure they are usually called liquid crystals. The correct name is nematic liquid.

Since these liquid crystals can be made to change the amount of light polarization under signal, they can also be made to serve as a display. This is done by arranging the nematic liquid between two restraining sheets which are optically transparent, but also which are electrodes or electrically conductive. On each side of the plate is placed a polarizing film, as shown in Fig. 3-14. Now, when

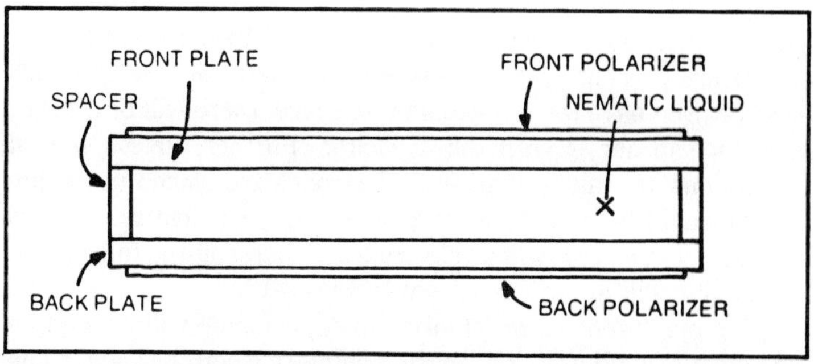


Fig. 3-14. Basic construction of a field effect liquid crystal, shown in cross-section. The dynamic scattering type does not use the polarizers. The display is formed by plating shaped electrodes on the front and back plates, these being thin enough to be partially transparent.

an electric potential is applied across the liquid, the amount of light transmission or light reflection changes. By arranging the electrode area in the desired shape, various symbols can be formed. While the most common is the seven-segment arrangement of Fig. 3-3, any arbitrary shape is equally possible. For example, the A.M. and P.M. associated with clocks, the date, or the ± associated with voltage measurement can be placed on a single display panel.

The concept of polarization has been used to explain the general construction of liquid crystal displays, but it will be found that they are grouped into two families. One is called the dynamic scatter family, since the effect occurs through the scattering of light in the display material. This type is front illuminated. The second is called the field effect type, since the change in light level is produced by changing the transmissivity. In this type the light must pass through the display once or twice.

Regardless of the specific construction and the material of the LCD display, it must involve a relation between the cell and a light source. Figure 3-15 shows a number of possible arrangements. The first two, the reflective and transmissive type, are used primarily with ambient light, that is, with front illumination. The other two use back or edge lighting from a separate light source. Sometimes the reflective and back-lighted modes are combined, for day and night operation.

## ELECTRICAL CHARACTERISTICS OF LCDs

Since a liquid crystal display is formed from two electrodes with a dielectric film between them, it will appear electrically as a capacitor. Also, since the dielectric is partially conductive, the liquid crystal display will appear as a leaky capacitor. A simplified equivalent circuit of such a cell is shown in Fig. 3-16. Actually further effects are present. Since the semitransparent electrodes which must be used have relatively high resistance, there will be a series resistance in series with the capacitor. Further, unless special precautions are taken, if several electrodes are placed in the liquid, as would be done to form a seven-segment display, there is leakage between the various electrodes. As a result the full equivalent circuit will appear as shown in Fig. 3-16.

The magnitude of resistance and capacitance will depend on the nematic liquid used, the spacing between electrodes, and some other factors, the most important being the size of the display. Typically in average sized displays of 0.5 inch characters, the capacity

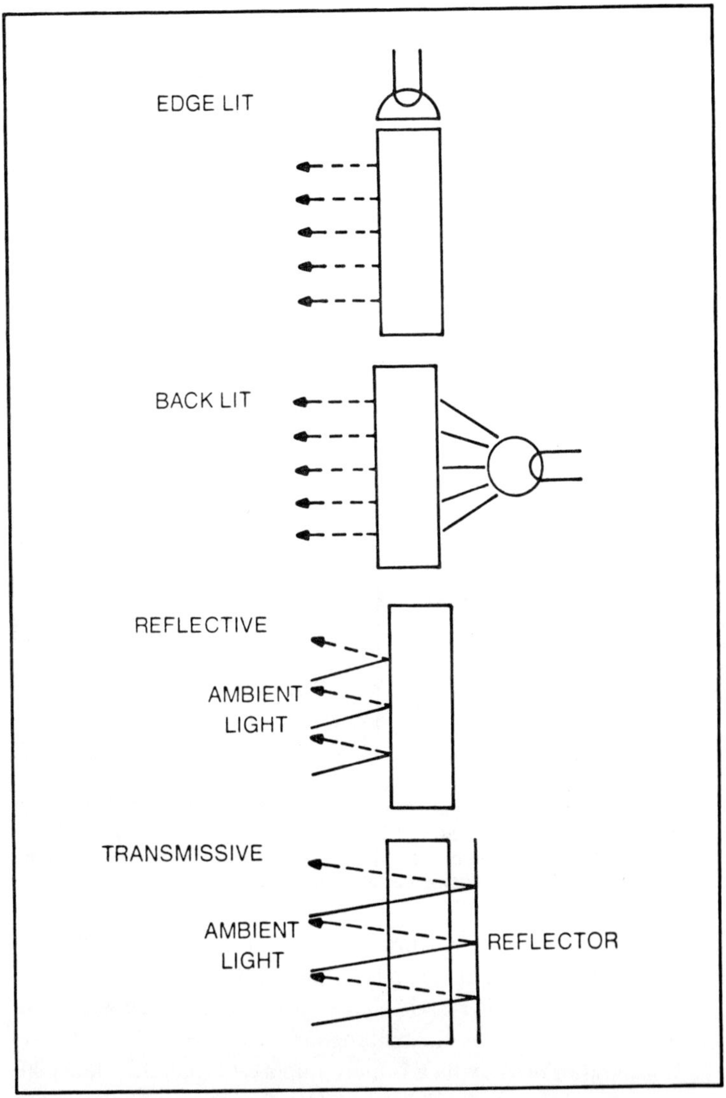


Fig. 3-15. Modes of operation with LCDs. The first two are usable at night, and in light of the self-illumination is sufficiently bright. The last two operate from the ambient light, and are usable in full sunlight, but not at night.

will be on the order of 50 pF. The leakage resistance will be on the order of a few megohms per segment, and the series resistances on the order of 1,000 ohms per segment. For a given construction the capacitance will be directly proportional to the electrode area

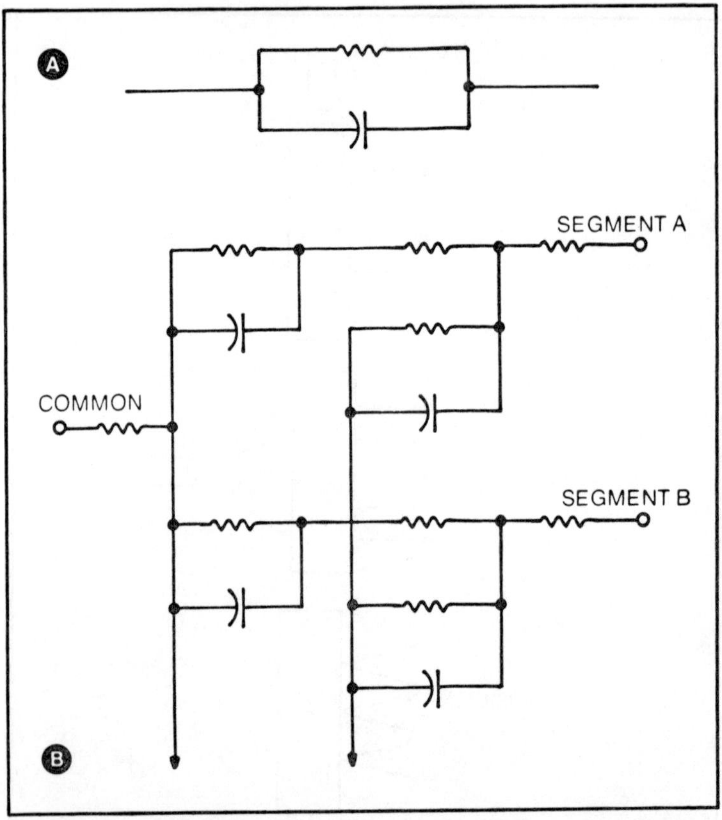


Fig. 3-16. Equivalent circuit of an LCD. Circuit (A) represents a cell with a single small electrode. The complex at (B) represents a normal display, and shows intersegment leakage, capacitance, and lead resistance.

and the resistance inversely proportional. Very large displays may have a capacitance of thousands of picofarads.

Because the liquid crystal behaves like a capacitor-resistor combination, it cannot have instantaneous response-time. Actually, there is another effect which is more important than this time constant due to capacitance. This is the fact that the molecules of the liquid crystal must actually be moved to change the display characteristic. The rate of this motion will be a function of the temperature. Figure 3-17 shows a typical response relation for a field effect type of LCD. Light transmission starts to increase as soon as excitation voltage is applied, proceeding along a curve similar to the curve of voltage across a capacitor. The rise time will be on the order of 0.2 second for the higher allowable temperature and may

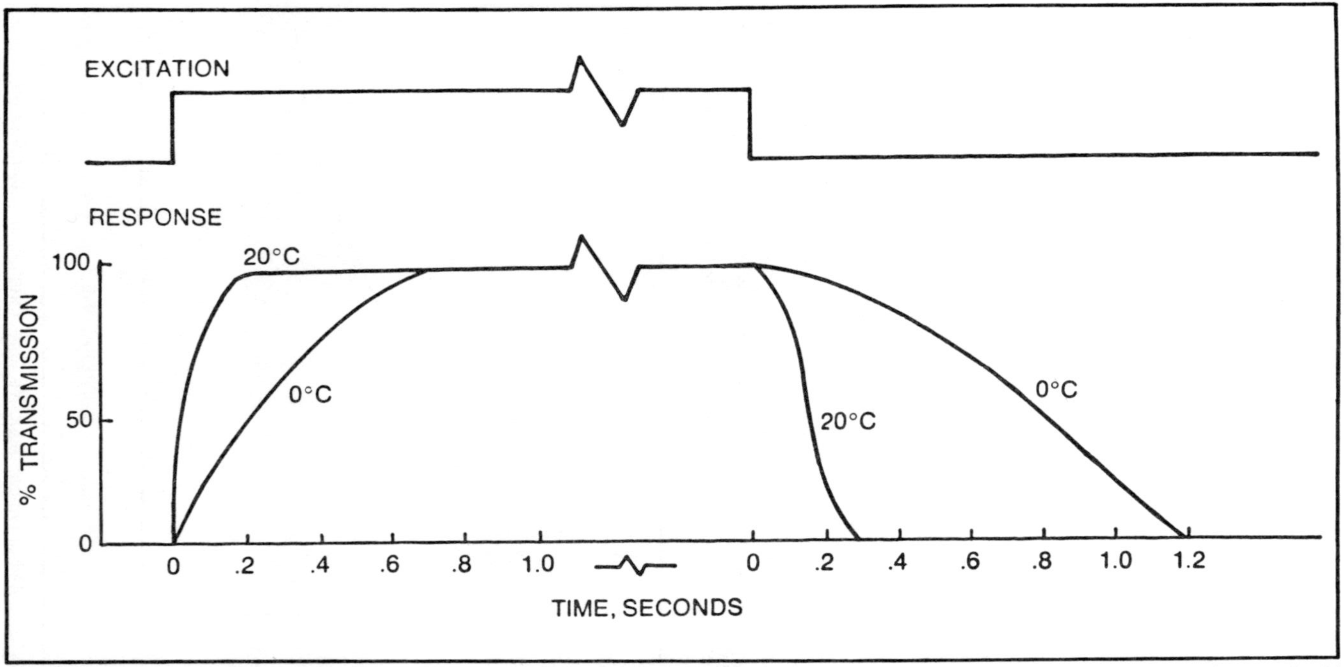


Fig. 3-17. Time of response of an LCD, with normal excitation levels. The operating temperature range can be varied by changing the liquid formulation.

be as long as one second for lower working temperatures. The fall time behaves similarly, except that it takes longer for the randomization process to occur. The effect of temperature on fall time is shown in Fig. 3-18. The rise time characteristic is not greatly dissimilar.

The voltage supplied to the cell also affects the response time, as shown in Fig. 3-19. Very low voltages have almost no influence on the display. As the voltage increases, the orientation of the liquid crystals occurs at a more rapid rate, so that the characters become visible quicker. There is a sort of a knee in the curve at about the 20 volt point for this type of display. For greater voltages the time continues to decrease, but the rate of change is lower. The time for a character to disappear is less affected by voltage change, but there is some small influence as also shown by the figure.

## OPTICAL CHARACTERISTICS

Since LCDs do not emit light, the optical elements measured for LEDs do not apply. Instead it is necessary to have a measure of the influence of the LCD on the light which falls on it. The usual measure is the contrast ratio—for reflective types this is the ratio

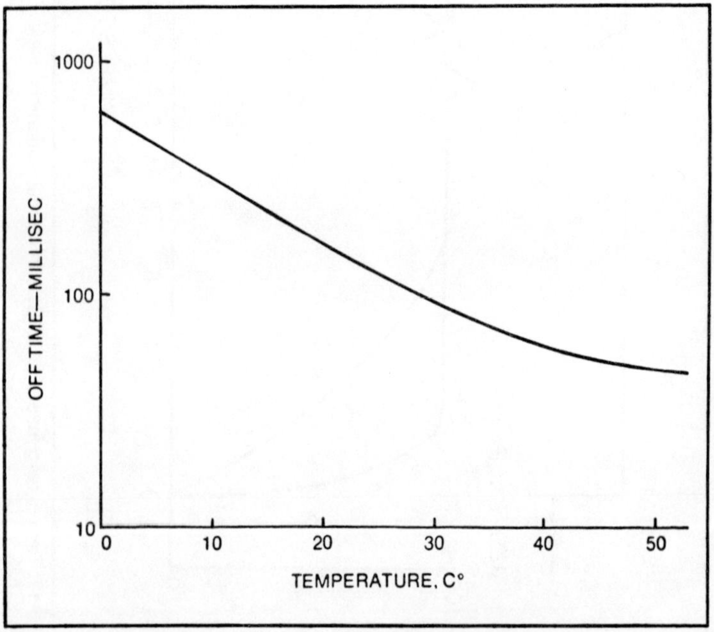


Fig. 3-18. Detail of typical response time variation with temperature.

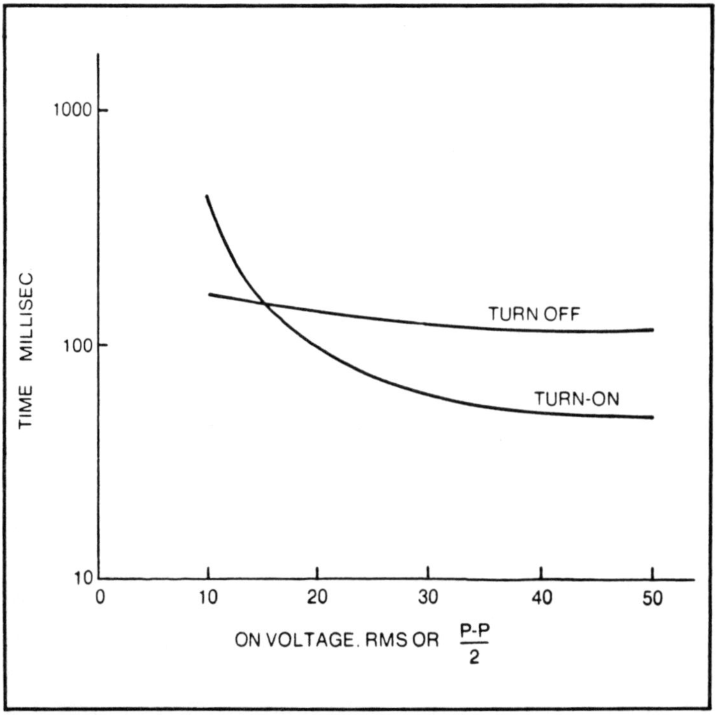


Fig. 3-19. Effect of excitation voltage on response time. With low voltage input, there is essentially zero response.

of the amount of light reflected when the cell is not energized, to the amount reflected when it is energized, assuming dark characters on a light background as is normal. Higher contrast means greater visibility, and in addition is generally more pleasing. Most people are familiar with the muddy, washed out appearance of low contrast photographic prints, and reject these if a high contrast print is available.

A major factor affecting contrast ratio is the applied voltage. As just seen for response time, too low a voltage will have no effect on the display, as shown in Fig. 3-20 for a particular display type. There is a threshold voltage. For greater voltages the contrast increases rapidly with time, although at a lower rate. Too high a voltage can puncture the cell, since it is really just a capacitor with close plate spacing. Normal practice is to operate above the contrast need curve, but well away from the point of voltage breakdown. The voltage recommendations of the manufacturer of the particular cell should be followed.

Contrast is affected by some other factors. An example is the frequency of the voltage applied to the cell, as shown in Fig. 3-21. There is a broad optimum—the fall off is fastest on the high side. However, it is not practical to go to low frequencies for two reasons. First, flicker will appear whenever the scan frequency falls below about 30 Hz. Secondly, it is also undesirable to go too close to zero frequency, since the lifetime of the cell becomes very short if there is a dc voltage across it. The decrease in life is due to deplating of the electrodes. Best practice is to operate near the maximum of the contrast ratio curve, a frequency of a few hundred hertz, using circuits which keep the average voltage to zero, with an error of no more than a few tens of millivolts.

## OTHER DISPLAY TYPES

While there are many other display types, these are not at present commonly encountered in clock designs. It should be expected that the number of display types encountered will increase in the

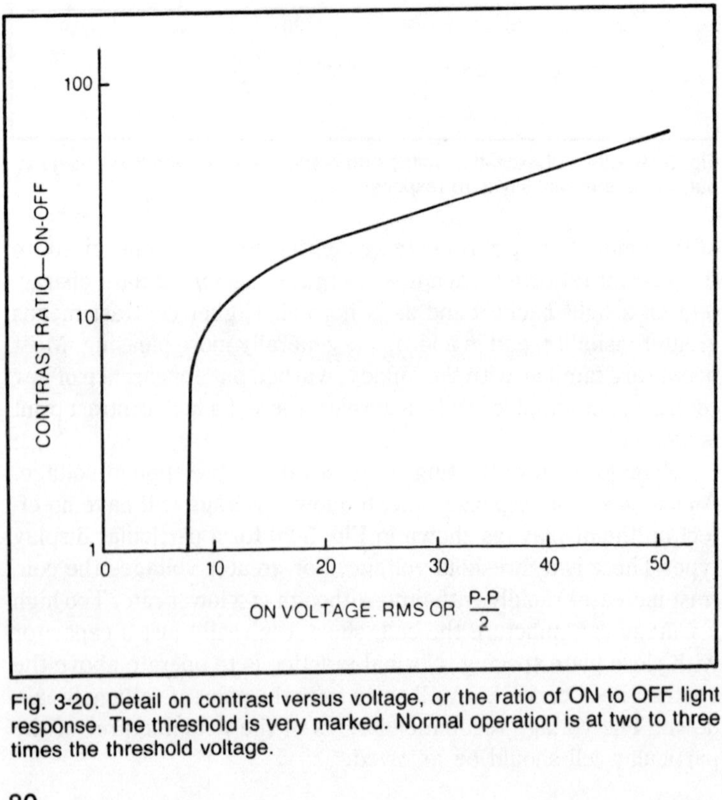


Fig. 3-20. Detail on contrast versus voltage, or the ratio of ON to OFF light response. The threshold is very marked. Normal operation is at two to three times the threshold voltage.

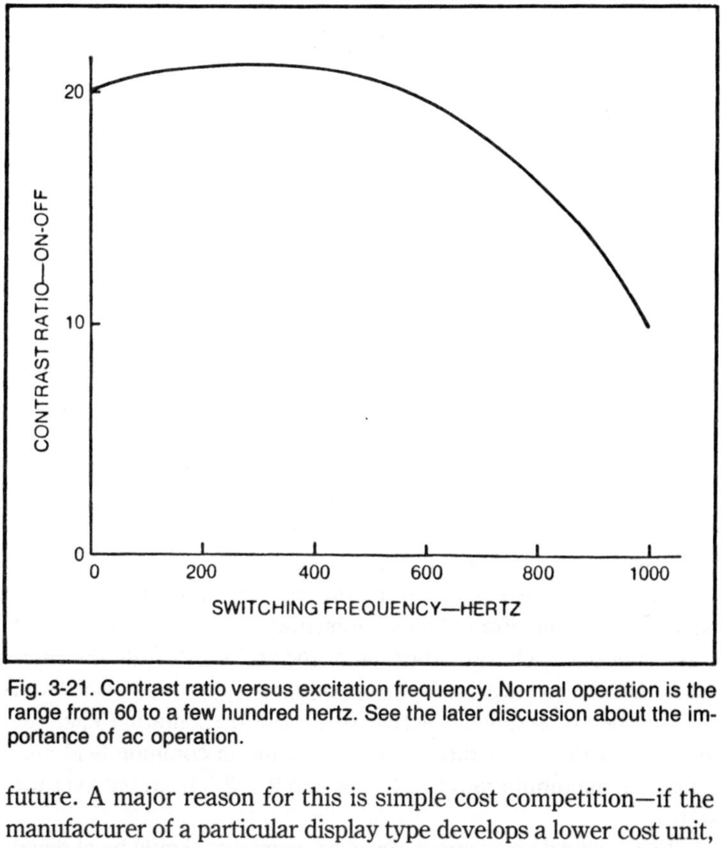


Fig. 3-21. Contrast ratio versus excitation frequency. Normal operation is the range from 60 to a few hundred hertz. See the later discussion about the importance of ac operation.

future. A major reason for this is simple cost competition—if the manufacturer of a particular display type develops a lower cost unit, it will be adopted by other manufacturers in short order. There is also another reason for expecting greater variation in display types—the decorator problem. Glowing red LEDs are highly visible but limited in decorator capabilities. Variations in color are especially demanded, and there is indication that variations in shape from the current seven-segment style are appearing. It is anticipated that sixteen-segment displays will become more common, and will be used in the clock types which combine time and calendar information. The major characteristics of some other display types are shown in Table 3-4. For details, and for application information, the manufacturer's literature should be consulted.

## DISPLAY DRIVERS

Because LEDs require more current than many chips can supply, and because LCDs require an ac supply if service life is to be

**Table 3-4. Selected Additional Display Types.**

| Type | Operation Basis | Typical Voltage | Typical Current | Typical Character | |
|---|---|---|---|---|---|
| | | | | Height | Color |
| SP-101 | Gas Discharge | 180 | 70 μA | 1.0 | Orange |
| FG-159A | Fluorescent | 35 | 4.5 mA | 0.37 | Green |
| DR2000 | Incandescent | 4.5 | 24 mA | 0.60 | Near White |
| 5853 | Multiple Cathode | 170 | 15 mA | 0.60 | Red |

satisfactory, there may be an intermediate element between the chip and the display. This element is usually called a display driver.

We can see some of the driver problems of LEDs from Fig. 3-22. This shows the seven-segment signals for the successive numbers, 0, 1, 2 . . . 8, 9. Since clock presentations use these numbers in sequence, this figure is also the amplitude-time plot of the seven lines which carry the display information. In the plot the convention of high equals a bright display is assumed. Note that the time scale is quite long—the numbered intervals are either one second, one minute, or one hour in duration.

There is quite a range in the number of segments which are on at a given time. For a "ones" presentation only segments b and c are energized. On the other hand, for an "eight," all are energized. On the average, about four segments are on. Variation in the number of segments activated as the presentation changes means that there is a rather large variation in common lead current. The maximum range is 4:1, or about ±2:1 from the average state.

For a good display appearance, all segments should be of equal brightness. Further, for good appearance, there should be no change in brightness of an on segment when the segment remains on while changing to another character; for example, when going from 1 to 2, segment b remains on, and should continue at the same brightness level. To avoid cross talk in the brightness or brightness blink, any element in the common lead must have essentially constant drop. This consistency would not be true if there were resistance in the leads, but it is reasonably satisfied by transistors operating in saturation.

An associated problem is produced by the multiplexing of digits to save pins. The digits must be energized in succession, to steer segment information to the correct digit. With a six-digit display, each digit is on for 1/6 of the total time. Since the eye averages light, the multiplexed digits must have the same average current as a nonmultiplexed digit would have to appear equally bright. To obtain this it must have a peak current which is this value multiplied by the number of digits in the display. The peak current buildup is quite large. At 10 mA per active segment, the average current per digit ranges from 20 to 80 mA. For a six-digit display, the average current ranges over the same value, but the peak current is six times larger, covering the range from 120 to 480 mA.

A related factor is that all unused segments in a given digit, and all unused digits, must be off, at least to the eye. If this is not

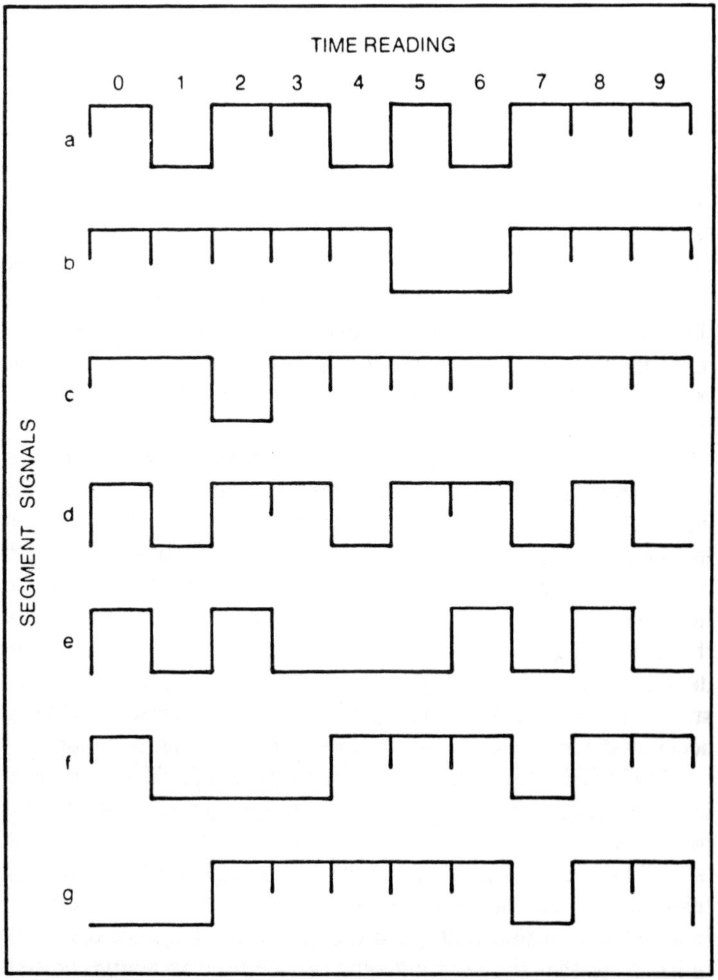


Fig. 3-22. Signals on the seven segments as a function of time for the 10s of hours, minutes or seconds digits. It is helpful to memorize the pattern, as an aid in troubleshooting.

satisfied, faint ghost digits will appear, reducing the readability of the display. Resistance in the common lead is especially troublesome here and leakage from one segment to another or from digit to digit must be kept low, whether in the display itself or in the driving devices.

For LEDs, two families of drivers are widely used. One, quite common in clocks, uses discrete components (transistors and resistors) in a series connection. Figure 3-23 shows the connection for one segment for a common cathode display, and Fig. 3-23 shows the connection for a common anode display. In both the connections for other segments and other digits are indicated schematically. As noted, the requirements for the transistors are good saturation characteristics, low leakage, and adequate current capability. On the other hand, speed of operation is in the low audio range, voltages are relatively low, and there are no inductive elements to produce over voltage. The transistor requirements are not especially severe, and a large variety of NPN and PNP audio transistors are usable. Typical ones are 2N4403 PNP and 2N3904 NPN types. Equivalent types can be determined from specification sheets, considering puncture voltage, saturation voltage and current capability as the primary factors. High gain, of course, is convenient, although most chips have more output than is necessary for average modern transistors.

The fact that high saturation current transistors are used means that current limiting resistors are needed, to set the current to the value required to produce the desired illumination. The calculation for these resistors is shown in Table 3-5. The desired current for the transistor and its voltage drop are determined from specification sheets. The other factors are determined by the designer. Note that the drop across the LEDs is assumed to be constant, independent of the number of segments which may be on. This same assumption is also made for the drop across each transistor, this being accounted for by the constant 1.5 volt term in the equation.

Packaged integrated circuit drivers are available and are used in clocks, although they appear to be more common in calculator design. The circuit of two typical driver elements is shown in Fig. 3-24. These use Darlington pair drivers, a method of reducing the tolerance demand on the transistors. The segment type is supplied in a package of four units, and the digit type in a package of six units. Note that the segment type can be used to source or sink current, whereas the digit type is only for a current sink. Normally, the segment drive circuits will require current limiting resistors just

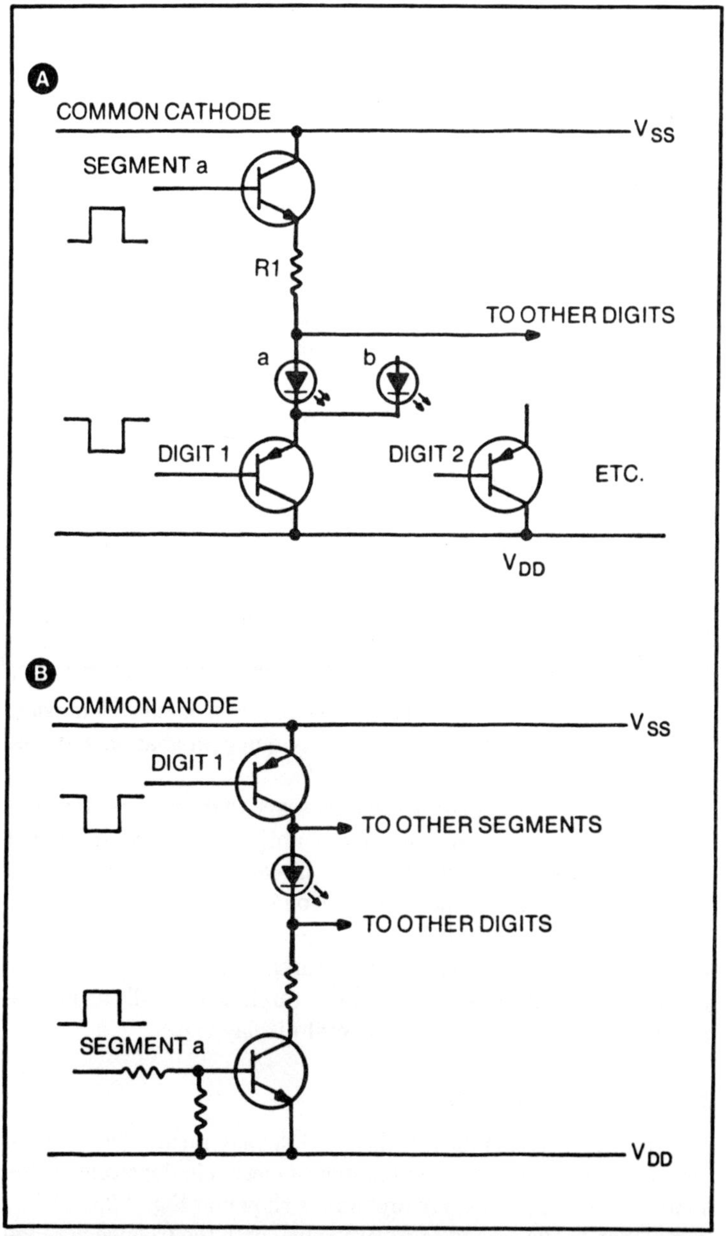


Fig. 3-23. Basic driver circuits for common cathode (A) and common anode (B) displays. Transistor choice is not critical; a wide range of silicon audio transistors are usable. The segment type is preferably chosen for continuous current capability, and the digit type for pulse operation.

**Table 3-5. Resistor Calculation for LED Drivers.**

For Fig. 3-22

$$R1 = \frac{V_{SS} - V_{DD} - Vf - 1.5}{N \times If}$$

R1 = Driver Load Resistance
Vss = Source Supply Voltage
Vdd = Drain Voltage
Vf = LED Forward Drop
If = Desired Average Current
N = Number of Digits Multiplexed

Example: FND70

Vf = 1.9 at If = 10 mA

Asume 6 Digits

Vss = + 5, Vdd = − 12

$$R1 = \frac{+5-(-12)-1.9-1.5}{6 \times 10 \times 10^{-3}}$$

= 220 ohms, nearest value

as the discrete component type. Of course these, or equivalent packages of other manufacturers, can be used in combination with discrete components, if desired.

It is sometimes desirable or necessary to drive other types of displays. For those which involve simply an increase in current level, the simple drivers of Fig. 3-23 can be changed to use a Darlington pair of transistors, the second one being chosen for the current level required. If high voltages are involved, as for glow discharge displays, see the manufacturer's literature. The specification sheets of the clock chip manufacturers will usually show connections and interfacings for several display types.

## LCD DRIVE CIRCUITS

As has been mentioned, the LCD display must have ac drive to avoid severe reduction in life due to electrode deplating. Schematically, the LCD display problem is shown in Fig. 3-25. The ac signal may be sine-wave or square-wave, with the frequency range between 30 and 300 Hertz typically and with voltage excursions between 15 and 100 volts. The individual segments of the display must be connected to this source by a switching mechanism as

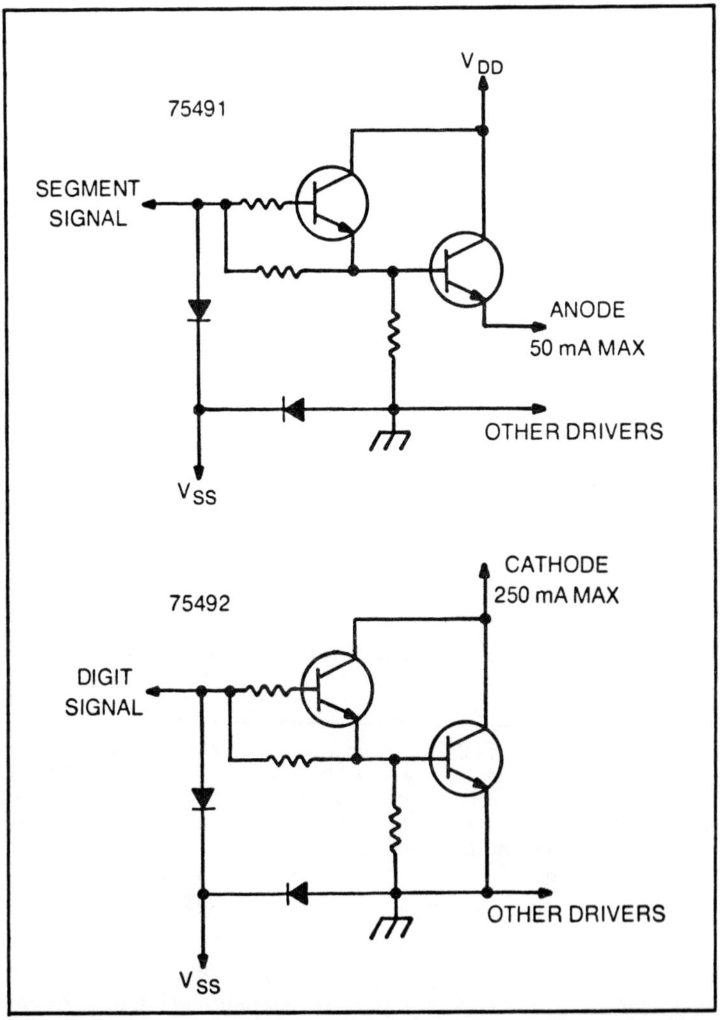


Fig. 3-24. Basic circuits of a pair of drivers available as integrated circuit. The circuits can be duplicated with discrete components if high drive capability is needed.

shown. Practically, the problem may be a little more difficult. It is likely that the clock chip to be used with the display was designed for dc output. Thus the display interface must provide both for switching, and for the generation of the ac signal from the dc signal supply.

A method of accomplishing this is to use a set of exclusive-OR gates as switches, connected to the display as shown in Fig. 3-26.

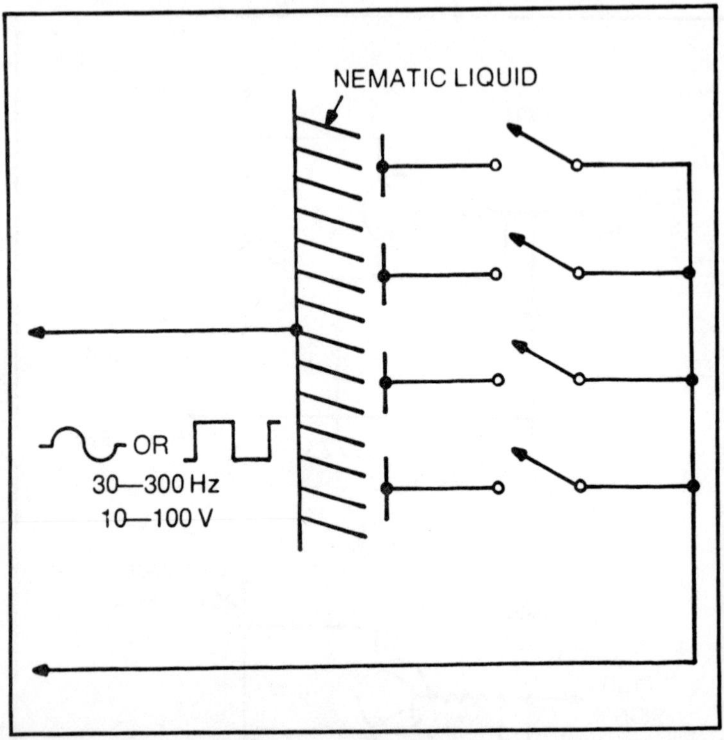


Fig. 3-25. Principle of the LCD driver. An ac voltage is placed across the segment or symbol to be activated by a switching system. The supply may have any wave shape which gives symmetrical positive and negative excursions. Any residual dc component causes the plated electrodes to be removed by electrolysis, causing loss of useful life.

The display common and one of the exclusive OR gate inputs are connected to the ac circuit. The other exclusive OR gate input is connected to the segment drive signal. The driving action is shown in the figure for the two conditions of segment OFF and segment ON. For the segment ON state, the display signal and the segment signal are in opposite states at the same time, causing the voltage across the segment to vary between VDD in the positive direction and an equal value in the negative direction. For the OFF state, the display signal remains the same, but the segment signal is inverted in phase. Now the display and segment inputs to the exclusive OR are either both on or both off so that in principle there is no voltage across the segment; in practice, small transients will be observed, since the various signals will not have exactly identical rise time. This display method requires that the dc output of the segment signals be converted to a square-wave whose phase can be inverted.

For displays which have a special bus for intensity variation, such as the MM5316, a relatively simple circuit can be used to accomplish the same result. This circuit is shown in Fig. 3-27. The common bus is driven by a signal, swinging approximately between VSS and VDD, being generated by a switching transistor fed from an ac line. Simultaneously this signal is fed to the common electrode of the LCD. Because of the phase inversion, the voltage across the segment swings between VSS positive and an equivalent negative voltage.

The liquid crystal display does not lend itself to multiplexing

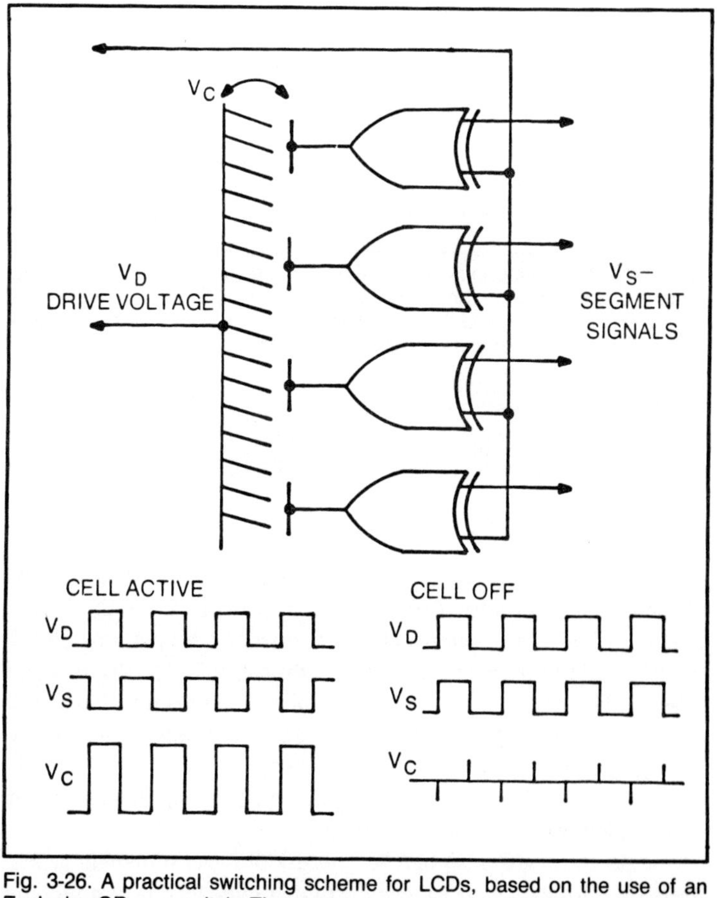


Fig. 3-26. A practical switching scheme for LCDs, based on the use of an Exclusive-OR as a switch. The segment signals are out-of-phase for the ON condition.

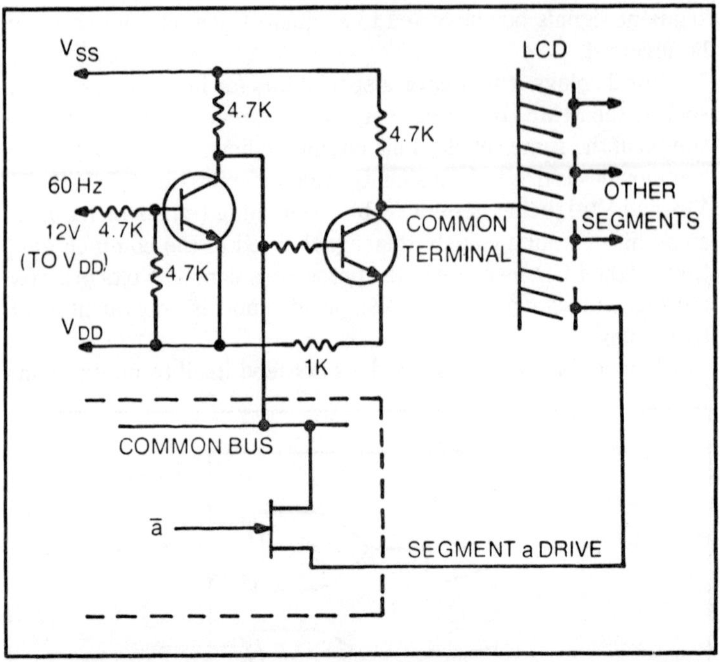


Fig. 3-27. The LCD interface used in the MM5316 clock chip uses the 60 Hz power line signal to drive the LCD Multiplex system. The value of the emitter resistor in the second stage is chosen to give good balance between the positive and negative excursions across the LCD cell.

because of the requirement for zero dc voltage, but it can be done with care. The clock chips which have individual display digits outputs are, of course, easy to use with LCDs.

## DIRECT LCD DRIVE

Some chips are designed to drive liquid crystal displays directly, either as an alternate to other displays, or as the only design display. These chips will usually have special features intended to extend display life. There may be modification of the drive signal to let the liquid state relax to its normal or null presentation condition after one multiplex state ends and before the next one starts. In displays with this feature, numbers appear to change relatively rapidly and clearly, whereas in simpler types the numbers appear to flow slowly from one to another.

Some chips are designed to use any of several display types, often by adding a few external components. For example, the MM5316 specification sheet suggests the interface circuit of Fig.

3-28 for LCDs. The purpose is to give an ac drive with no net dc component. As has been noted above, the dc component is the major reason for short display life.

## BRIGHTNESS CONTROL

Emitting displays which are bright enough to read easily during the daytime are usually too bright for nighttime use in a sleeping area. Some form of brightness control can be a practical necessity. A simple brightness control for LED displays of the common cathode type is shown in Fig. 3-29. This is a simple emitter follower current control. This gives satisfactory operation since the brightness of an LED is a function of the current through it. The brightness of fluorescent displays is also a function of current. By using a high voltage pass transistor, the anode current can be adjusted for the best illumination. It is also possible to use the grids for intensity control, but the circuits will necessarily be more complex.

The display-chip combinations which use ac for multiplexing

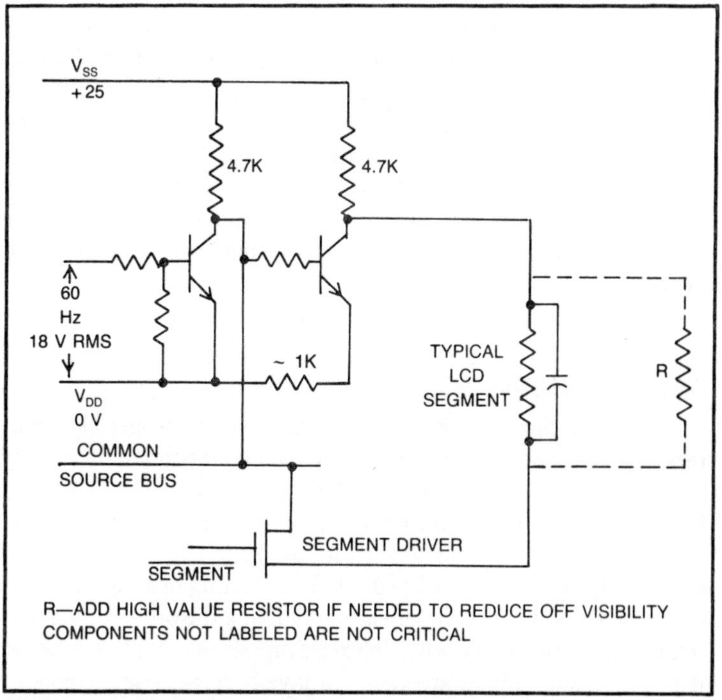


Fig. 3-28. Typical LCD interface as usual with the MM5316 chip.

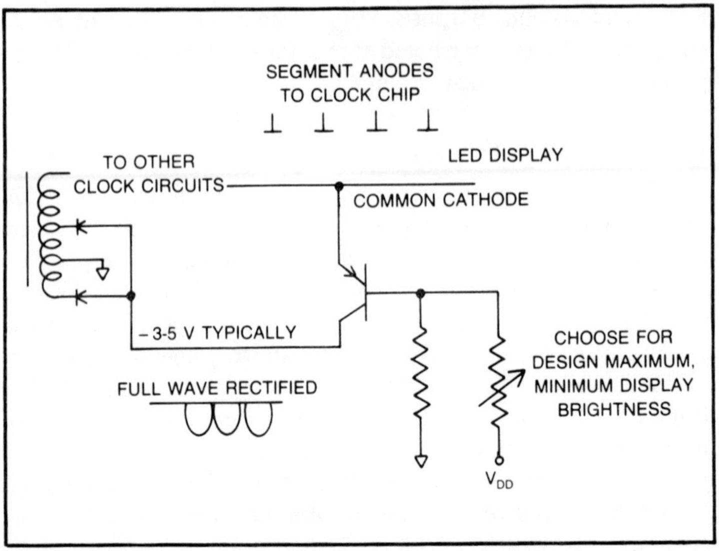


Fig. 3-29. Varying current through an LED varies brightness. This design uses a transistor as a variable resistor in series with the LED.

can also have brightness control, but again with somewhat more complex circuitry needed. Figure 3-30 shows an automatic type of control for the GK3300 chip. The current passed by the light sensitive resistor, an OAP-12, is amplified by two transistor stages and becomes the display current. Since the light-dark range of the resistor is very large, it is shunted by a fixed resistor, which sets the minimum brightness. The circuit can be simplified by replacing the light sensor and the first transistor by a simple potentiometer for manual control, or by replacing the sensor and transistor by a resistor switched in and out for day-night levels.

## SPECIAL CHARACTERS

It is quite common for clock chips to make use of special characters. Typical ones are the colons between hours and minutes, and the A.M. and P.M. indications. Colons may be added to a display board by using two small LEDs, connected across the power supply with a series resistor to set the intensity. Colons can also be made with some type of LED digits by inverting one digit to place its decimal point at the upper right, matching the other digit's decimal point at the lower left. The resulting display will look better if the individual digits are tilted to give a vertical semicolon, as shown in Fig. 3-31. As shown here, be sure to compensate for the

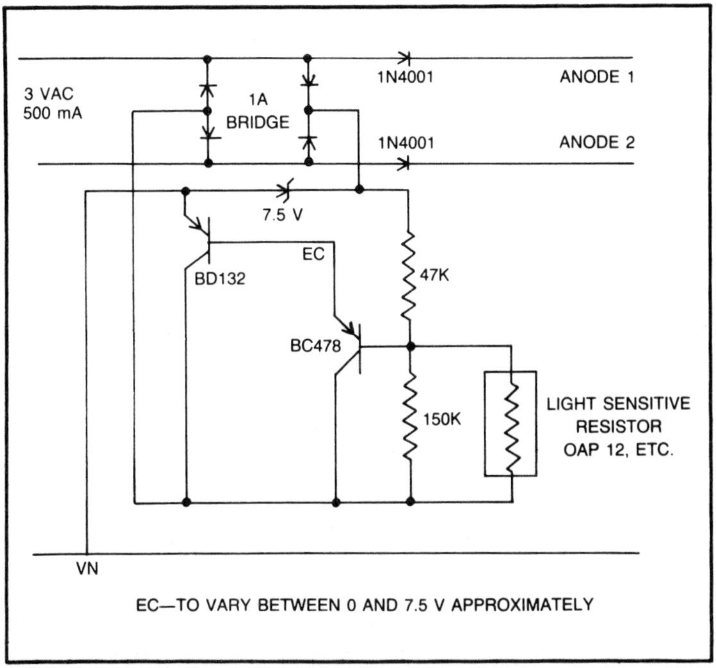


Fig. 3-30. An extension of the adjustable resistor design of Fig. 3-29 to give automatic brightness control. The current through the light sensitive resistor (a Cd-S cell is a common type) decreases when the ambient light decreases, thereby decreasing the LED current and light output.

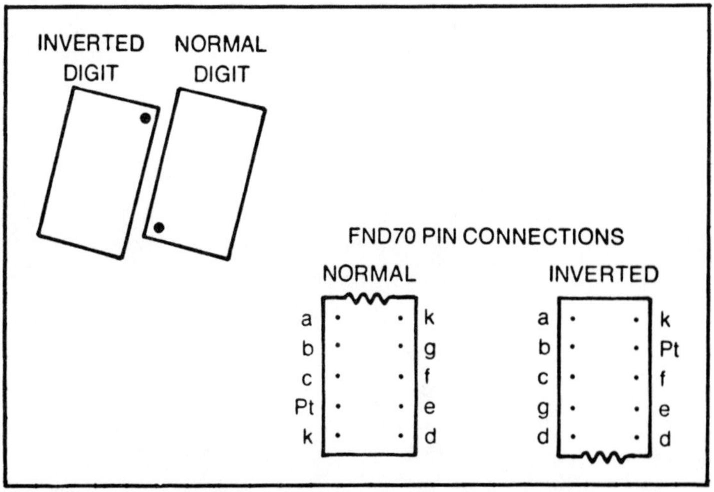


Fig. 3-31. Colon formation by inverting one digit. The appearance is best if the colon dots are vertical.

change in segment position due to the digit inversion.

At present, special "clock face" displays are more common in the LCD family. The reason is that the display format is made with a single silk screen electrode for the front panel. A special display made in quantity costs no more than a standard one, and offers the clock manufacturer more scope for design individuality. These special clock face displays are beginning to appear in the sources available to experimenters. If desired, they can be used with any clock chip by paying attention to the requirement for zero net voltage across the display. If they are to be used with many chips, demultiplexing of the chip out signal will be required. This can be combined with the drive elements needed by LCD displays.

Some special characters can be formed by turning the seven-segment display on its side. Use of these, or of such elements as 16-segment displays, will require some form of read only memory plus a driver. Here is a good field for experimentation.

# Chapter 4

# An Experimenter's Clock and Some Variations

The basic idea of the experimenter's clock is to take a simple clock chip and place it on an experimental board, making all functions and terminals available. With this availability, the experimenter's clock can be used as a component in breadboard tests of application ideas, as a source for time information, or as a source for various signal forms.

In line with its experimental use, the clock should have ample space for additions or extensions. It should not be miniaturized in the usual sense. A relatively open arrangement of components should be used, and the components should be of adequate size, to allow additional lead connections by soldering or by test jumpers. Also, there should be space for additional plug-ins, or for connection to additional boards or equipment.

The choice of the clock chip for the experimenter's clock is of some importance. If this is the first use of a clock chip circuit, it is desirable that a relatively simple chip be used. On the other hand, a fair degree of flexibility is desirable, in line with intended experimental applications of the clock. Weighing these factors, it seemed that one of the National Series MM5311 clock chips would satisfy the need. These chips are flexible, have several possible modes of operation, and additionally are almost always readily available from experimenters' sources.

## DESIGN DECISIONS

The MM5311 series of chips have several functions which are combined differently in various chips of the series. The basic functions are shown in Table 4-1. This table provides the data needed to select the chip closest to satisfying the experimental needs of the clock. For example, if a one pulse-per-second output is considered to be an important element of the experimental board, the MM5309, MM5312, or MM5313 chips should be employed. Alternately, it may be more important to have a hold function, in which case one of the other chips might be employed. Note the variation in the other functions for the selection of these chips.

For the prototype clock the MM5314 chip was chosen. This does not have the one pulse-per-second output or binary coding, but does have output strobe provisions and provisions for holding the count. BCD output or one pulse-per-second output can be generated from the seven-segment output, if this is desired. The MM5314 chip used in the prototype is listed in recent catalogs, and is readily available from distributors who specialize in hobby and experimenter's orders. It is not the only chip which can be used, however. You might prefer to go to one of the very complex chips in order to have the widest range of capability available.

In this connection, note that it is not necessary to have a microprocessor available to use one of the specialized chips. The signals from the clock chip can be read by a bank of LEDs and a bank of switches can generate signals to the clock chip. Some buffering and debounce circuitry might be needed, and the circuitry can be easily worked out from specification sheets and application notes. The cost of the complex chips might be two or three times as great as for a simple chip, but still a reasonable expenditure.

In addition to the basic clock chip, the display device and the type of display driver must be chosen. For experimental use, the

**Table 4-1. Functions of MM5311 Series Chips.**

| Type | No. Pins | Hold | Strobe | 1PPS | BCD | Digits |
|---|---|---|---|---|---|---|
| MM5311 | 28 | x | x | | x | 4/6 |
| MM5312 | 24 | | | x | x | 4 |
| MM5313 | 28 | x | | x | x | 4/6 |
| MM5314 | 24 | x | x | | | 4/6 |
| MM5309 | 28 | a | | x | x | 4/6 |

a—has reset to $oo^h\ oo^m\ oo^s$

very large displays did not seem necessary since the clock would usually be on the experiment work bench. Equally, the very small displays did not seem desirable, since often the major emphasis would be on some factor other than the clock reading. For this reason, a display size of around one-quarter inch was considered acceptable. The display chosen was the Fairchild FND70, a single digit display with a one-fourth inch character. These are also widely available from experimenters' sources.

For the display driver, it seemed reasonable to use the discrete component transistor driver system shown in the data folder for the clock chip. An advantage of this transistor drive is that all points are available for experimentation. For example, this arrangement allows for converting from the single transistor to a Darlington pair, in case high current operation is desired. A plug-in display was selected for its greater flexibility. To save pin connections, the connections are made at the output of the driver transistors, with the display digits and their multiplexing being on a separate small board.

Still another factor is the arrangement of the various control circuits. It was decided to provide switches for each of the controllable functions, and to provide for external connections for the most commonly used ones. The other connections would be available at solder points if needed. Also, it was decided to make the scan multiplex circuit variable, and to provide for synchronization of scan by an external signal or alternately to provide a synchronizing signal to an external circuit.

Considering the experimental use of the board, it was decided to arrange the parts on a relatively large piece of printed circuit board, with ample space for components. Space would be left for a set of board edge connectors at the rear of the master board, for displays and other experimental point pickoffs. Chips and transistors would be at the center of the board and controls at the front. One board edge connector would be wired for display plug-in. This could also serve as an output for some experimental use, and other connections could be added as needed.

A block diagram summarizing the above decisions is shown in Fig. 4-1. While complete, this diagram is misleading in one respect—each block is actually very simple. The major reason for showing this breakdown is to emphasize that terminals are made available for use, thereby giving a flexible experiment board. This simplicity is illustrated by the detailed circuit diagrams which follow.

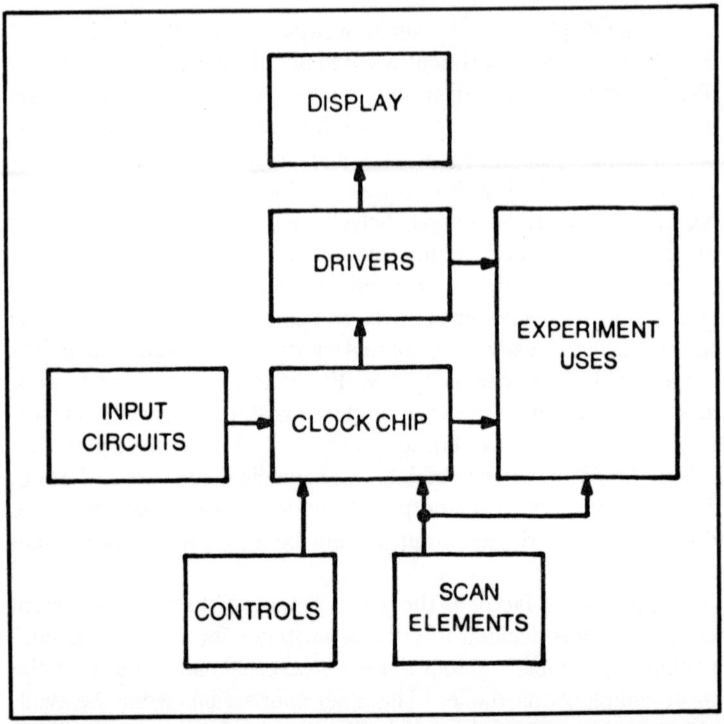


Fig. 4-1. The basic concept of the experiment board. While the assembly can function as a clock, its main purpose is connection to experiment assemblies.

Of course, for some purposes, it is desirable to build up a special clock, based on work with this experimental board. These clocks can be built using the same chip as the experimental board, but with unnecessary controls and elements omitted. Examples of these specialized boards are shown later.

## INPUT CIRCUIT

For reliable operation, the input circuit requires that the peak-to-peak excursion be about 80% of the total power supply voltage. However, the peak-to-peak excursion should not cause the input circuit voltage to exceed either the positive or negative power supply voltages, the voltage on $V_{SS}$ or $V_{DD}$. For this board, it was decided that an external amplifier would be used if small values of the input signals were desired. Thus, direct feed to the chip would be used. Over-amplitude protection could be provided by Zener diodes.

The chip is rated to operate to 60 kHz, and as noted in Chapter 2, operation to frequencies as great as 10 times this can be expected with at least some chips. However, experience has shown that direct connection to many sources, such as the power line, do not work well, because there is too much chance for stray signal pickup. High frequency signals on these circuits typically cause the clock to run extremely fast, but sometimes even cause the clock to stop. A solution to this is to use a low-pass filter, but this limits the input circuit frequency response. Since fast response is often desirable for experimental purposes, if only to speed up checkout, it was decided to have two inputs, switch selected. One would have a low-pass filter and the other would be a direct feed. The complete circuit, showing this and the provisions for diode protection, is shown in Fig. 4-2.

## DRIVE CIRCUITS

The chip drive circuits are rated to give nominal 5 milliamperes output for segment drive and to sink a nominal 10 milliamperes for digit drive. These values are on the low side for display use, and are very restrictive for experimental use. Therefore, drives are required. For this it was decided to use the discrete component drive suggested by the chip data sheet. The circuit resulting is shown in Fig. 4-3(A). The value of the series resistor is calculated using the relations of Table 3-5. The segment current was kept low, since the clock was intended for lab use only.

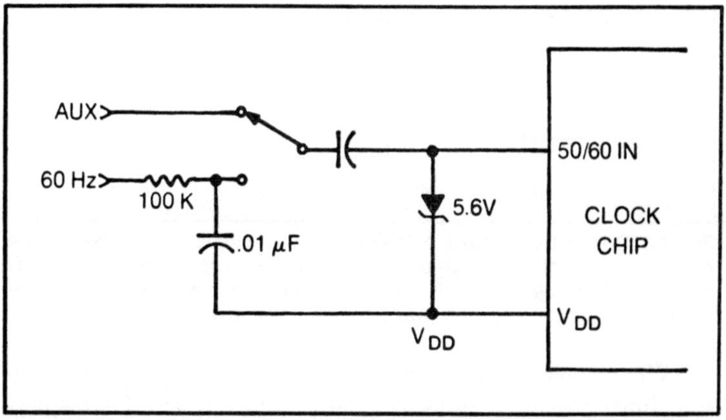


Fig. 4-2. Input circuit. A normal clock filter is provided for line operation or with noisy circuits. A separate input allows use of the full frequency capability of the chip.

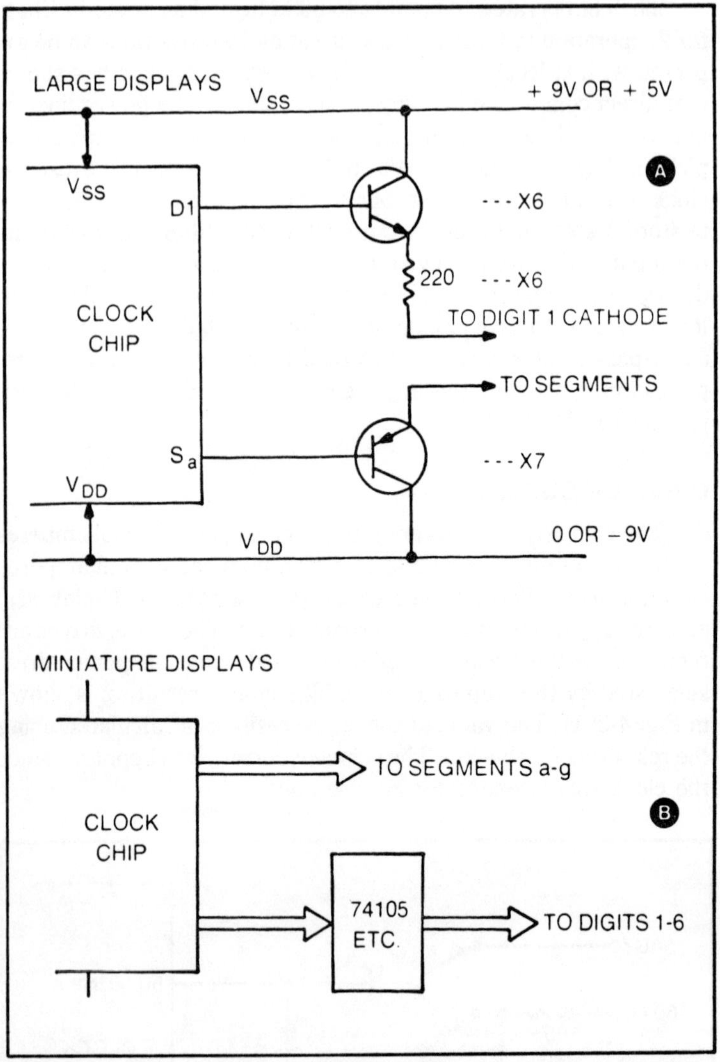


**Fig. 4-3. Drivers for the clock chip used. The top circuit (A) was used for the experimental clock. Note that the voltage can be made compatible with TTL packages. The lower driver (B) can be used with a miniature board, intended for the designs of Chapter 9 (watch signal polarities).**

When considering versions of the clock for special use, it is sometimes possible to use small, low-current displays. For displays around 0.1 inches in size, 4 to 5 mA of segment current is ample and direct segment drive from the chip is possible. For these displays a suitable drive circuit is shown in Fig. 4-3(B).

## SCAN CIRCUITS

The scan circuit is rated to operate from dc to 60 kHz, and can be expected to operate up to approximately twice this frequency. Also, as described in Chapter 2, this circuit can be synchronized or externally driven. It has seemed desirable to provide for a wide range of operation. The circuit chosen for this is shown in Fig. 4-4. Two switches provide three ranges of approximately 10:1 excursion each. The operating frequency is approximately logarithmic with resistance at the lower frequencies, but will approach a limiting frequency with the smaller capacitance value, the frequency depending on the particular chip. If an experiment requires a very low frequency scan, capacitors may be added externally. The approximate frequency of operation can be determined by the relation shown in Fig. 4-4, although the accuracy of this is quite poor near the upper frequency of operation. Note that no provisions are made for dc input to this multiplex circuit, or for removing the relaxation capacitor. If either of these conditions is required for experimental use, temporary removal or jumpering is possible, or other switches can be added.

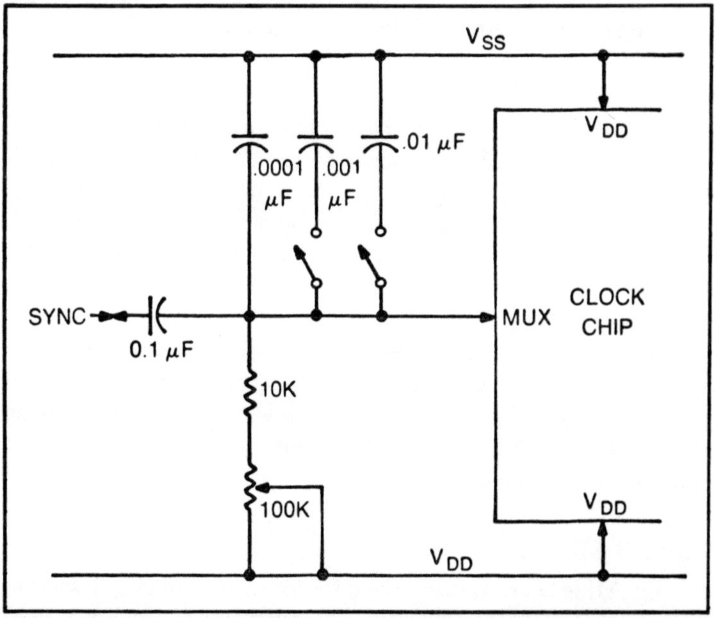


Fig. 4-4. The scan circuit utilized. The scan frequency is adjustable from about 10 per second to nearly 100 kHz. The synchronizing lead can be used for input or output.

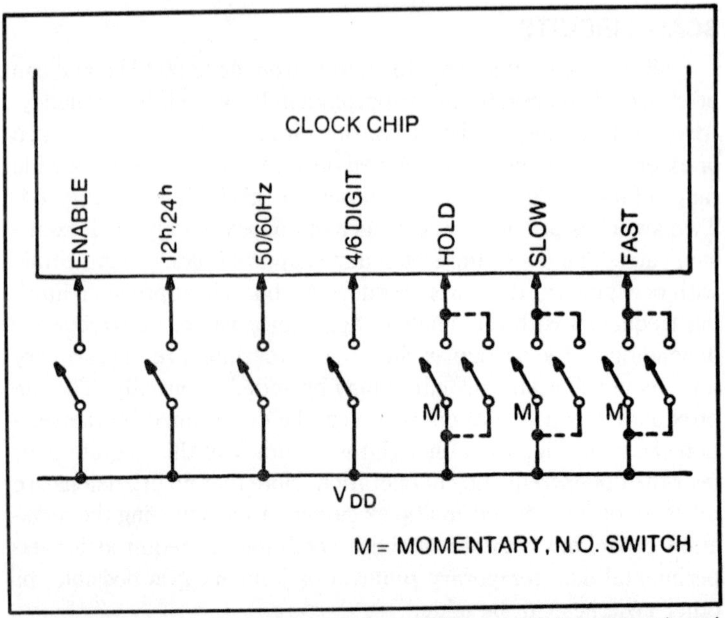


Fig. 4-5. The control circuit. Three of the switches are NO (momentary), and it may be desirable to parallel these with regular switches. The control leads may also be used as chip inputs for some purposes.

## CHIP CONTROLS

The other controls for the clock chip are normally switch functions. For great flexibility a separate switch is provided for each circuit. Some circuits, for example the slow and fast set, preferably use momentary contact switches, and are shown in this way in this figure. For experimental use it may be desirable to install two switches in parallel, one momentary, or to use a three-position switch with spring return.

Figure 4-5 shows the chosen switching. No special provisions are made for external switching, or for protection in case dc inputs are fed into the switch circuits. However, extra solder pads are provided at each of the switch points, simplifying addition of external connections.

## POWER SUPPLY

Since this board was intended for experimental use, it was anticipated that the laboratory power supply would be used. Any voltage between about 9 and 18 volts is satisfactory, with the range 9 to 12 volts best.

## COMPLETE SCHEMATIC AND PC BOARD LAYOUT

The complete schematic resulting from these decisions is shown in Fig. 4-6. It is simply an assembly of the individual schematics already presented.

The PC board layout used in the prototype is shown in Fig. 4-7. The space at the rear of the board allows for up to six board-edge connectors. One row could be replaced with posts for clip lead connections, or tipjacks. In front of this area the drivers, for the digits on one side and the segments on the other. Connections to the edge connectors are by jumpers.

The chip area is in the center of the board. Note that each chip lead has at least one extra connection point. Clip leads can be connected to these easily if a short, stiff wire is soldered into the board. This is a fast, inexpensive connection for experimental use.

The front area of the board is used for mounting controls. These are arranged in a row, based on simplicity of board layout rather than logic of use. However, because of the pin-out arrangement of the chip, the controls do group conveniently for normal clock use.

The component placement is evident from this figure. In mounting components, a socket for the integrated circuit clock chip is especially recommended. Sockets for the driver transistors could be added.

## THE DISPLAY BOARD

The exact detail of the separate display board must be determined by the displays and edge connector used. A minimum of 13 pins is required and more are desirable. The prototype used 18 connector pins, to match an edge connector which came from surplus. However, the prototype board was laid out to allow use of connectors of up to 25 pins without changing the board size. Intermediate sizes are handled by changing the notching at the edge of the board.

The prototype was intended for use with FND70 digital displays, readily available and of a reasonable size. It provides multiplexing for these. However, it was decided to use jumpers for multiplexing rather than using printed circuit for interconnections. This allows for changes, and also for use of other displays having the same pin spacing. With small changes in layout, other digits can be used. While this board takes some time to complete, it is actually easy to build up, although multiplexing of the FND70 is difficult if hand production methods are used. However, it is no

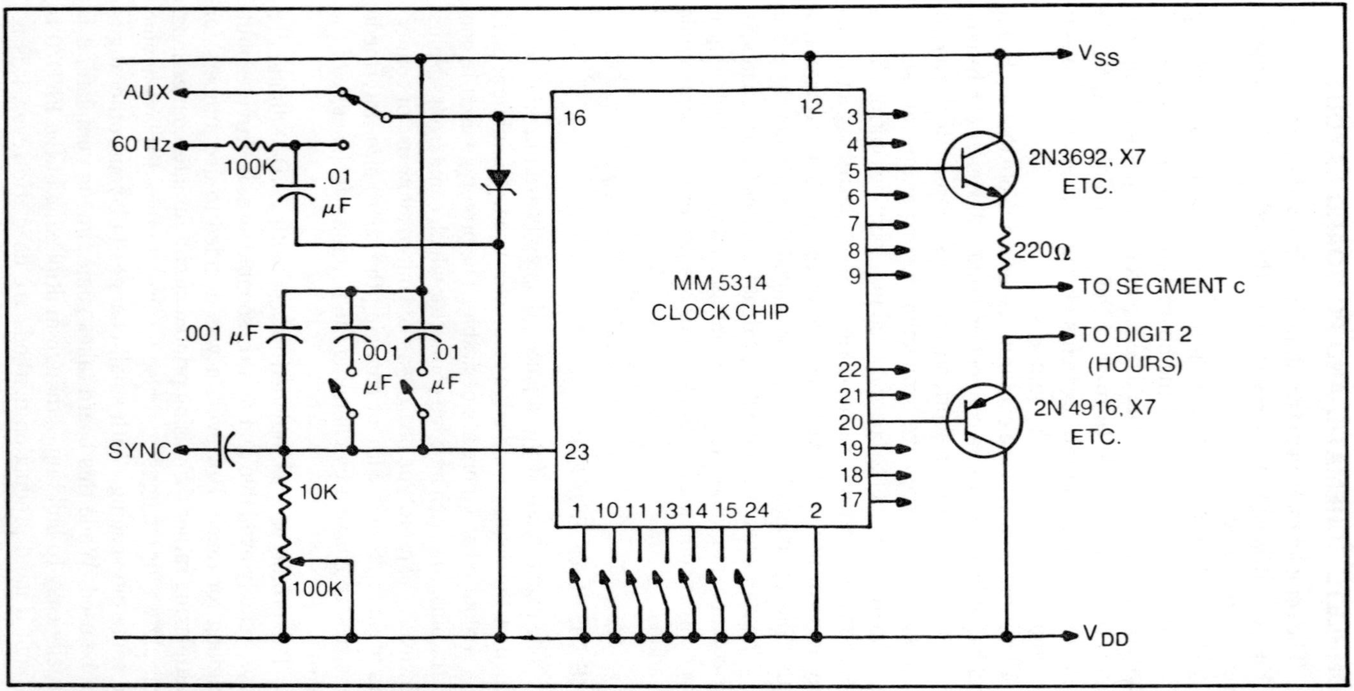


Fig. 4-6. The complete circuit diagram. All functions provided by the chip are available.

problem if photographic layout techniques are employed.

In the prototype, the FND70 units were soldered in. They could be mounted using Molex connections, or special sockets could be made by sawing up the low profile type of IC socket. The prototype board layout is shown in Fig. 4-8. When finishing this board, it is good practice to wire first the jumpers, connecting the "a" leads of all digits to a bus and so on. Remember to reverse the bus designations if the board is to be reversed in the connector so that the display faces away from the switches rather than towards them, as in the prototype.

Several boards can be made up with different displays or for various experiments. One such special board is shown in Chapter 7. A board with the edge connector and solder pads is a convenient way of using a cable for connection to another experimental board.

## ASSEMBLY AND CHECKOUT

Because of the relatively large area and the wide component spacing, board construction and final assembly is easier than for most PC projects. The only areas requiring care are around the transistors on the main board and around the display digits on the auxiliary board. Complete all other wiring and mounting and then install these elements last.

Because of simplicity, it is unlikely that there will be any problems if the normal inspection of soldered bridges and cold joints is made.

Assuming that this is your first work with a clock chip, it is desirable to spend some time getting acquainted with chip signals, voltage levels and so forth. Recommendations—keep a notebook of these results and make a practice of keeping the notebook up-to-date as you progress. There are two reasons for this. It is a useful reference for future experiment and design. Also, it provides a permanent record if you should come up with something new—it forms the basis for a patent application or an article and may even be necessary to defend a patent. It is also important in developing expertise.

## CHIP SIGNALS

In the following, the various chip signals as taken from the prototype board are reviewed. The first is the chip output digit drive signal, shown in Fig. 4-9. This is taken directly at the output of the chip, where there is essentially no loading, since high current

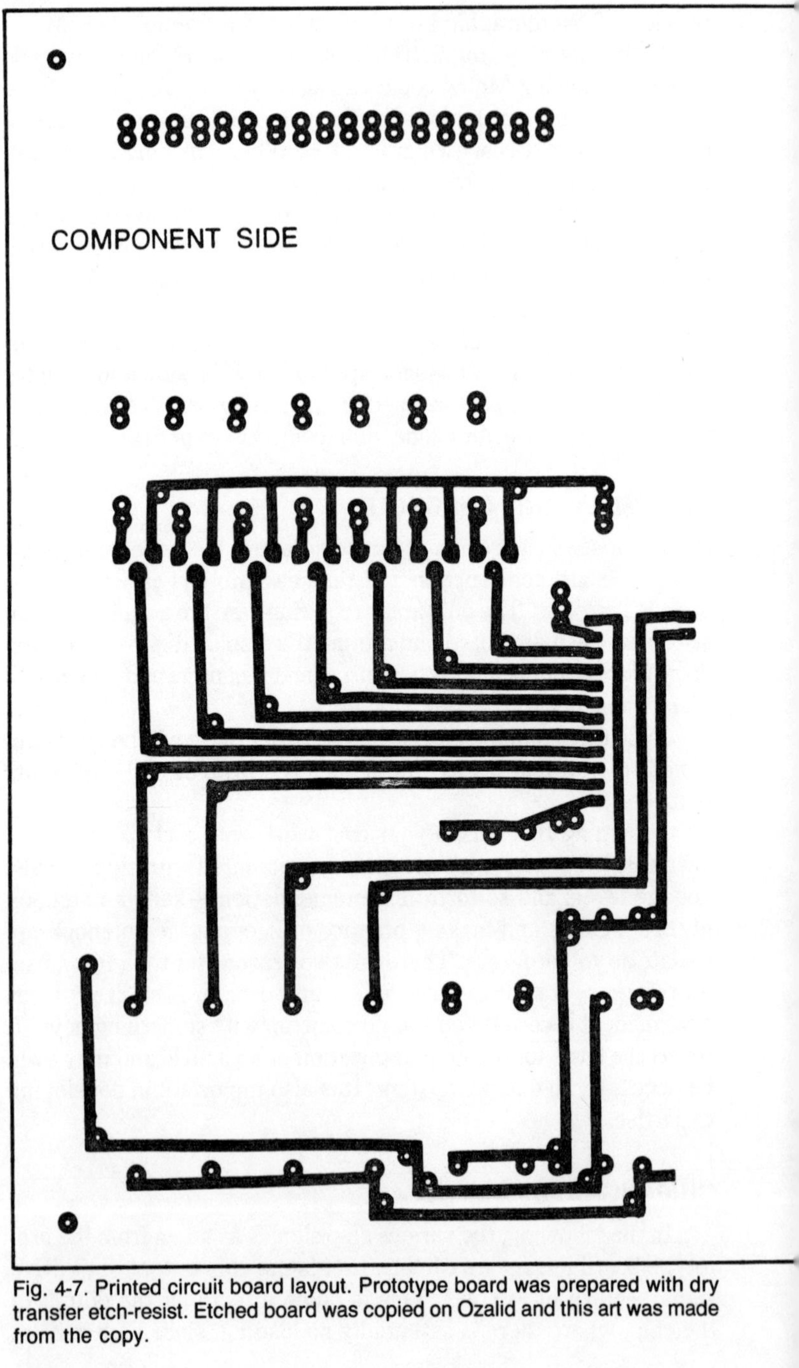


Fig. 4-7. Printed circuit board layout. Prototype board was prepared with dry transfer etch-resist. Etched board was copied on Ozalid and this art was made from the copy.

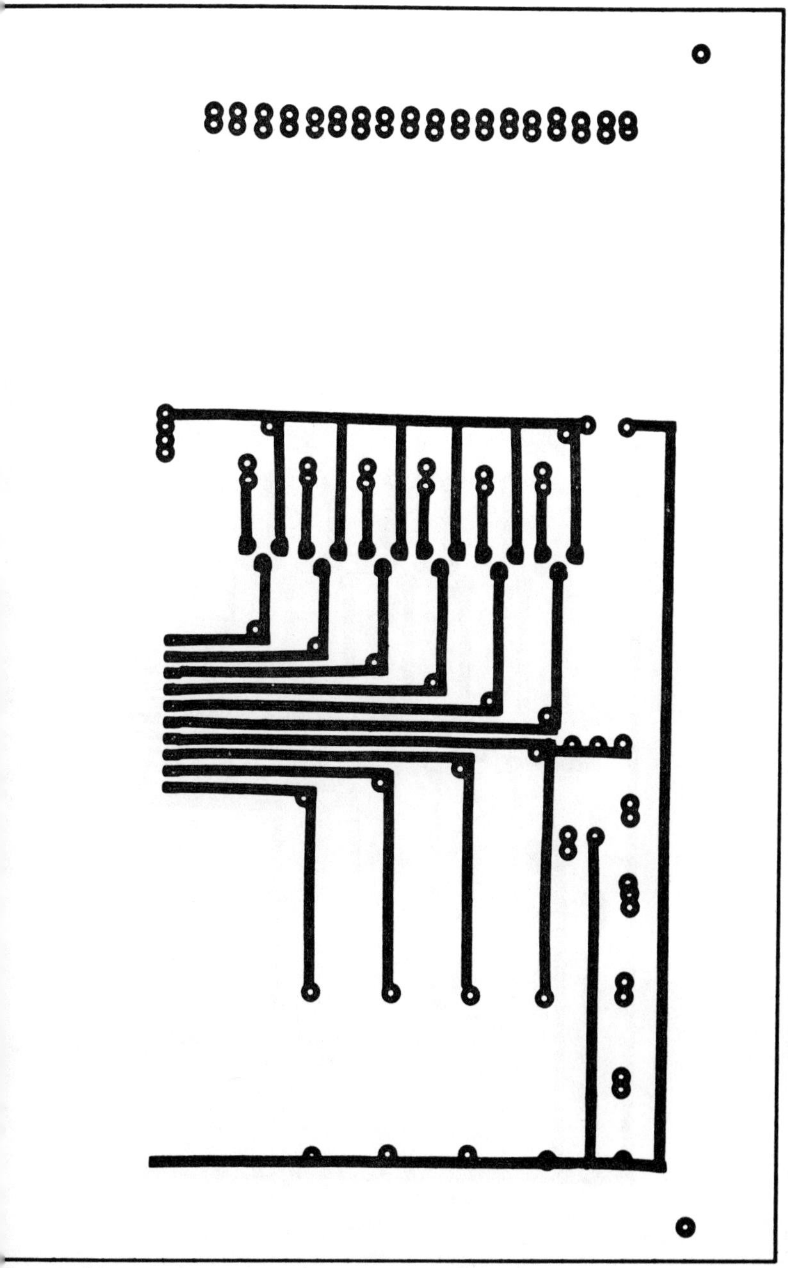

FOIL SIDE

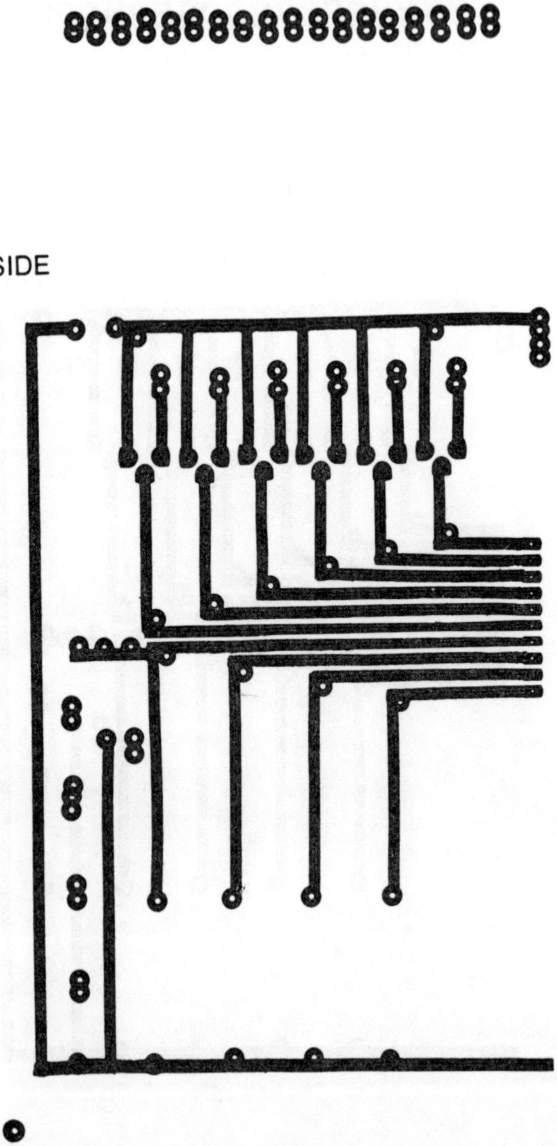

Fig. 4-7. Continued.

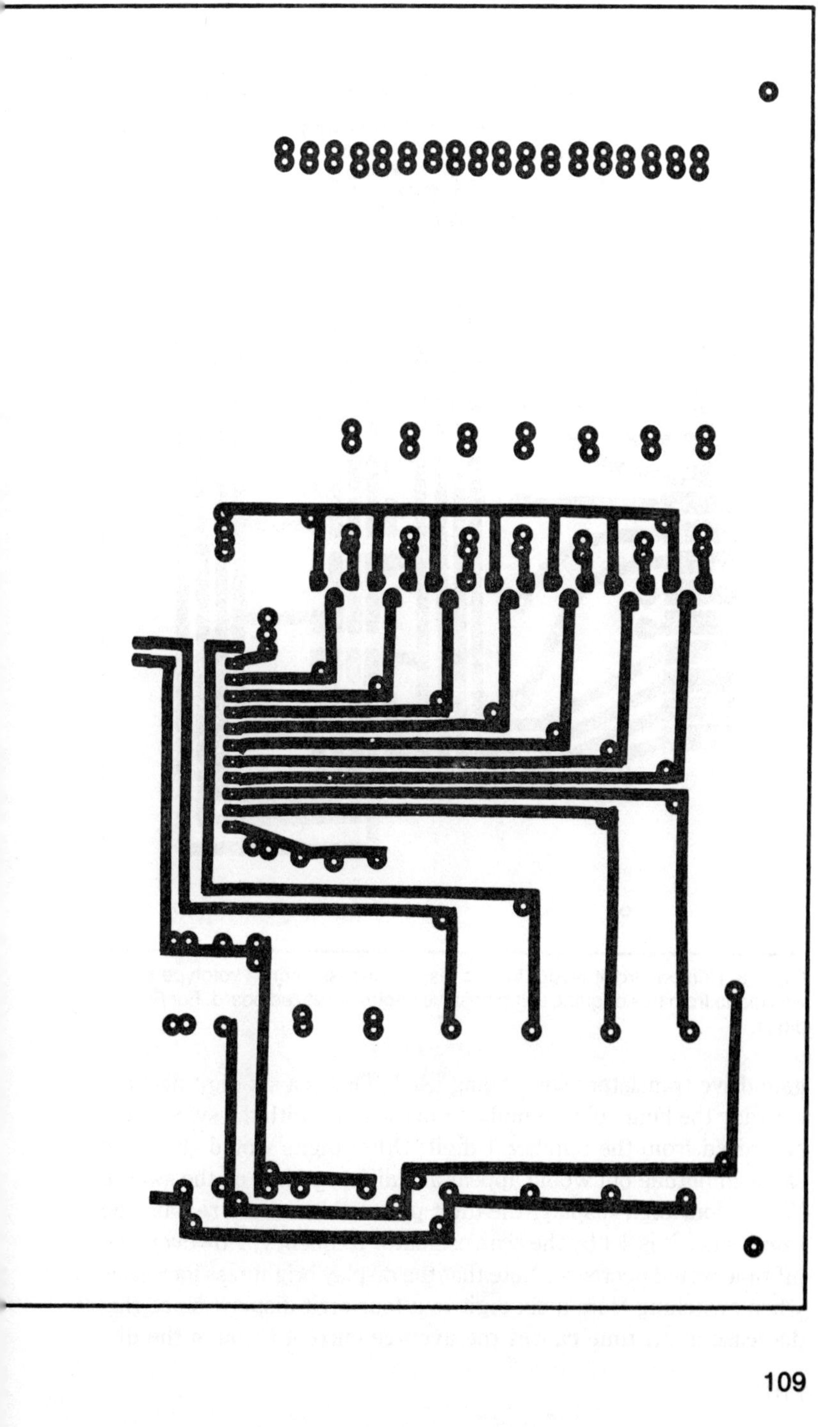

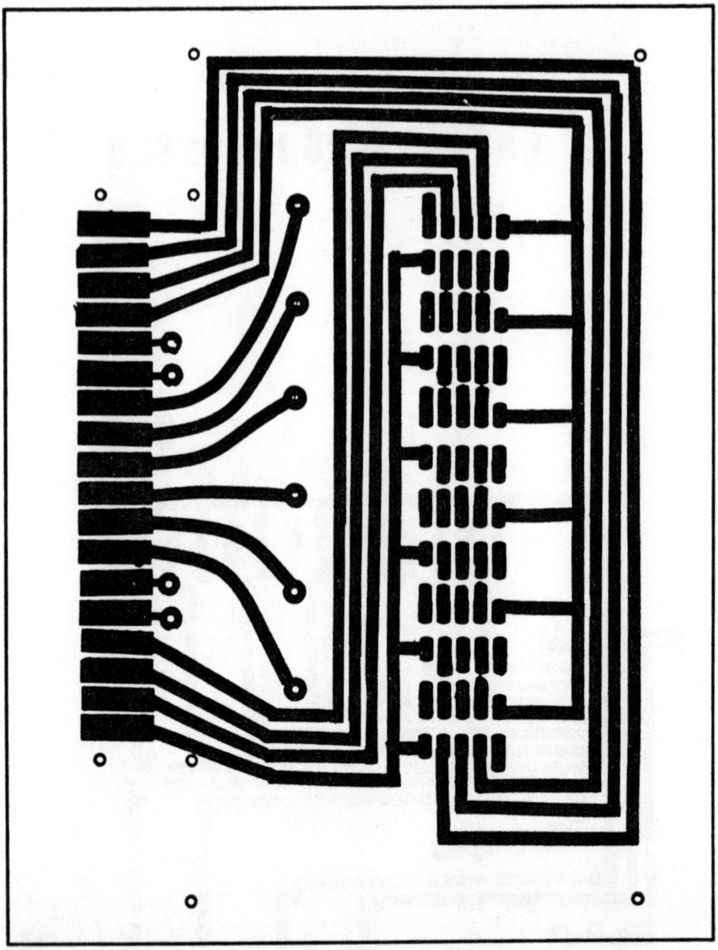

Fig. 4-8. Printed circuit layout for the display subassembly. Prototype board was made from this original, using positive photo-sensitized board. For FND70 digits.

gain drive transistors were being used. This is a six-digit display, showing the hours digit, number 2 in the scan, with the sweep synchronized from the number 1 digit. Other digits would show the same amplitude but would appear at a different time on the sweep. With a four-digit display, the digit pulse width would remain the same, since it is set by the scan oscillator frequency. However, the off time would decrease. Note that the display brightness increases when switching from a six-digit to a four-digit display, since this decrease in off time causes the average current through the dis-

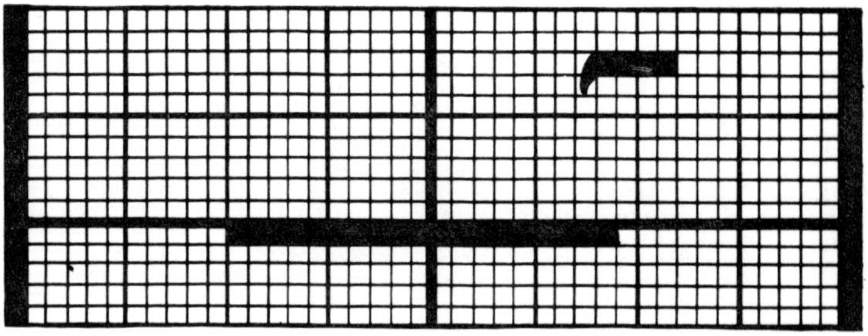

Fig. 4-9. The digit drive signal for digit 6 at chip output. There is a small amount of cross talk with other digit signals. See Table 4-2 for levels.

play segments to increase by the ratio 6:4. This is the preferred signal for interfacing or output, but a buffer amplifier should be used if appreciable current is needed.

The signal at the digit cathode is shown in Fig. 4-10, for a four-digit display. The signal amplitude is less, being just the voltage drop across the driver transistor. This changes somewhat with the number of active segments, since the saturation voltage of a transistor is not precisely a constant. Note that there is some cross talk from other driver elements, and some switching transients, which vary from one pulse to another.

The chip output segment drive signal is shown in Fig. 4-11. Again, this is directly at the chip output with essentially no loading. There is very little evidence of cross talk, but the switching transients are evident. Corresponding signal levels at the transistor output are shown in Fig. 4-12. Note that there is essentially no cross talk from other segments, but there are still some switching transients. This is the preferred signal to use for interfacing or output, and can supply appreciable current. In critical external circuits, an equivalent of debounce is probably needed; in many cases a small

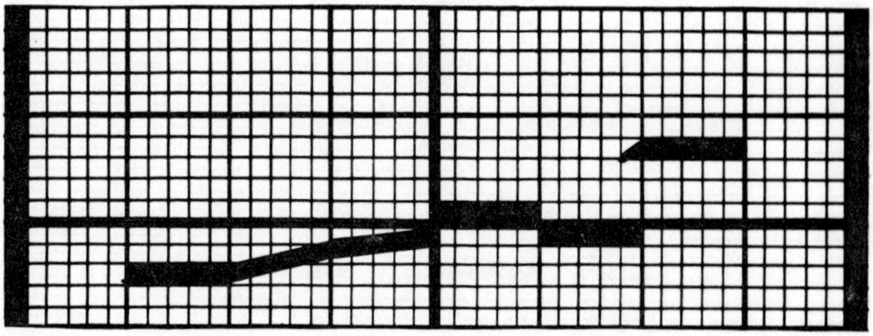

Fig. 4-10. The digit drive signal at the LED cathode. The cross talk is more evident. See Table 4-2.

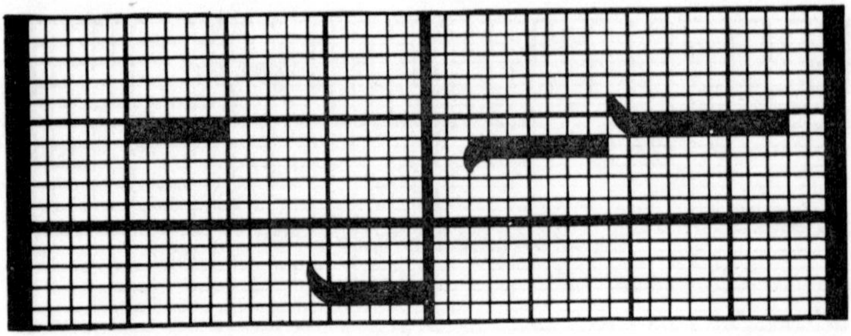

Fig. 4-11. The segment drive signal for segment g, at the chip output, with a four-digit display. See Table 4-2.

amount of integration by a series resistor, shunt capacitor combination will remove the switching transient without seriously affecting the desired signal.

The segment anode signal is shown in Fig. 4-13. Again, this has low amplitude . . . that due to the drop across the LED plus the drop across the digit driver. There is some cross talk, and also an appreciable switching transient. Some of the switching transient is probably due to switching delays in the LED diodes.

Figure 4-14 shows the free running scan oscillator signal. The signal is a sawtooth with some nonlinearity, with a fall time of approximately 10% of the total period. As in the case for all simple relaxation oscillators, the frequency can be pulled over a considerable range by an external synchronizing signal, or locking operation can be secured if the scan frequency is near an integral multiple of the synchronizing frequency. This last effect causes wave forms such as shown in Fig. 4-15 to appear. However, the visual display is unchanged.

Figure 4-16 shows an important parameter—the amount of drive current which can be developed between the two driver out-

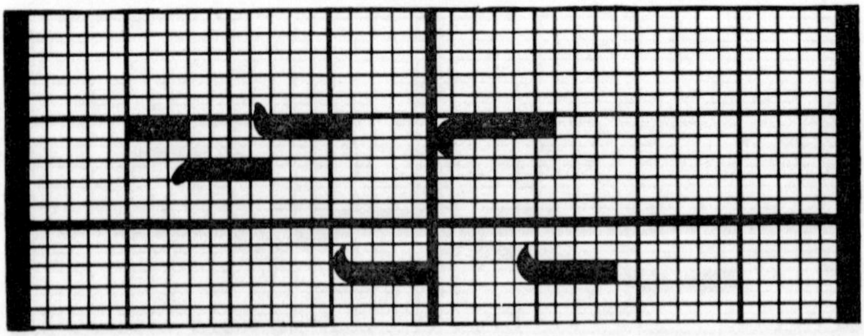

Fig. 4-12. The segment drive signal for segment g, at the driver transistor emitter, six-digit display. See Table 4-2.

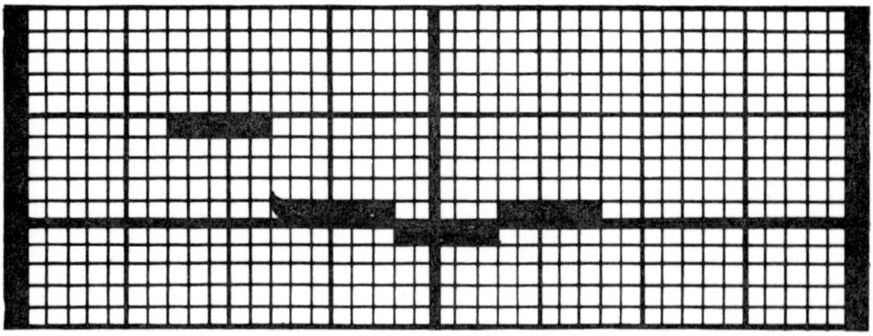

Fig. 4-13. The segment signal at the LED. Fast switching transients are evident in the original; this output point must be used with caution. See Table 4-2.

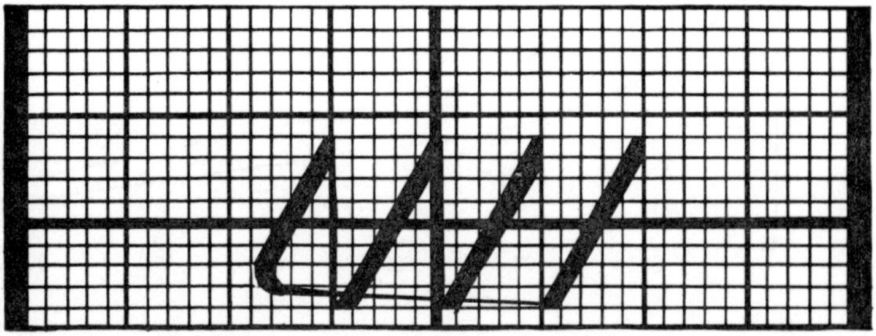

Fig. 4-14. Free-running scan signal, at the Sync-Out terminal. Amplitude 1 V p-p. The wave shape is essentially independent of scan frequency.

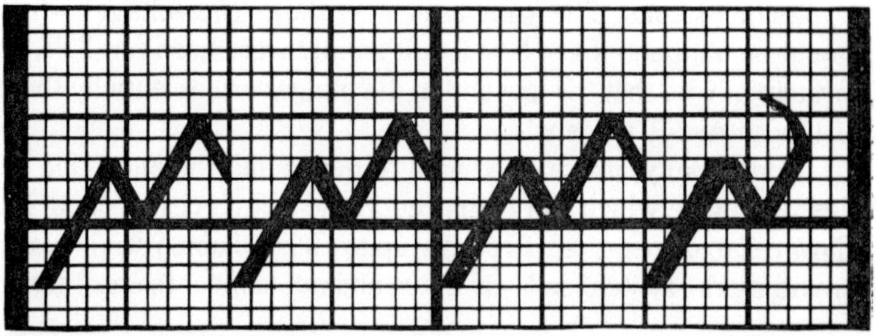

Fig. 4-15. Scan frequency output, sync mode, sync frequency greater than scan frequency.

put points, as a function of the circuit resistance. Note that this is for series resistance only and does not include any LED diode drop. These currents should be compared with those predicted by the equations given in Table 3-4.

During the process of checkout, it is a good idea to record the various voltage levels of the different pins. For example, Table 4-2 gives the digit and segment drive voltages for a supply voltage of

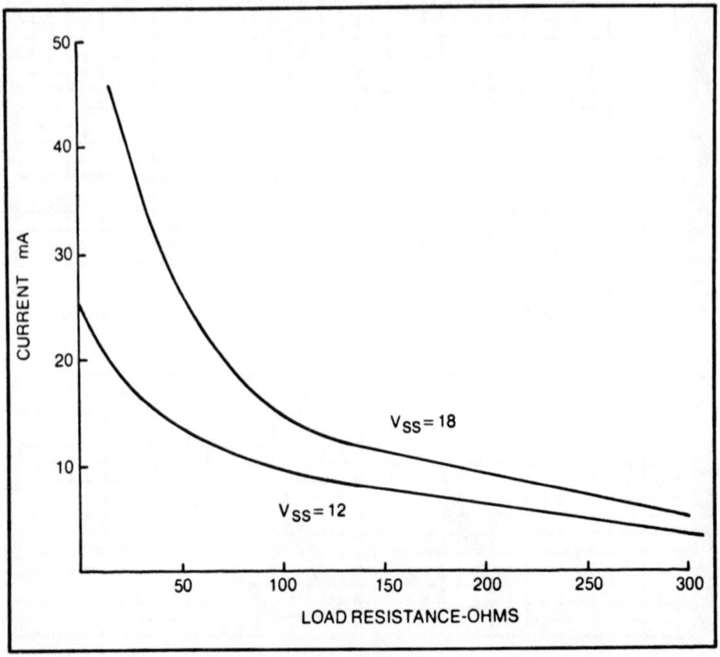


Fig. 4-16. Drive current available as a function of load resistance. This curve does not include the effect of LED internal drop.

9 volts or 14 volts. The same table also shows the voltages which would be obtained if a +5, −9 volt supply were used for TTL interfacing. Note that the low and high voltages are essentially those of the normal TTL level.

Additionally, it is a good idea to determine the allowable series resistance which can be placed in the switch circuits and still allow reliable operation. Typically up to 100,000 ohms can be used, providing the switch leads are short and not subject to stray signal

**Table 4-2. Voltage Levels of Experimenter Clock.**

| Location | Stage | Voltage Vss = 14, Vdd = O | Voltage Vss = + 5, Vdd = −9 |
|---|---|---|---|
| Digit Output | Lo | 8.4 | −0.6 |
| | H1 | 11.2 | + 2.2 |
| Segment Output | Lo | 8.4 | −0.6 |
| | Hi | 13.0 | + 4.0 |
| Segment Transistor Emitter | Lo | 8.4 | −0.6 |
| | Hi | 13.3 | + 4.3 |

pickup. With long leads connected, reliable operation will be secured only with lower resistance.

Finally, it is also a good idea to determine the maximum frequencies of operation. For the input, at constant amplitude, satisfactory operation should be secured from the range of dc up to several hundred kilohertz, possibly approaching one megahertz. The scan oscillator will probably have an upper frequency limit of one to two hundred kilohertz, and should operate down to frequencies of fractions of a cycle per second. For these low frequencies, high quality capacitors must be used, however.

## SIGNAL INJECTION

One of the characteristics of most clock chips is that signals can be injected at the various leads to produce results not compatible with normal clock operation. For example, a signal can be fed to pin 11, the output enable, to strobe the output digits. Figure 4-17 shows the digit drive with this strobing signal being a square-wave. For usual external use, for example in feeding a computer, the strobe should be synchronous with the digit drive and the digit position. Note that an identical signal fed into the 4/6 digit selector will serve as a strobe for the seconds and 10 seconds digits only, an action which is sometimes useful.

The slow set, fast set, hold and 50/60 Hz pins can also be strobed. There are some restrictions on the flexibility of these; for example, if the hold is in the off position and fast set is strobed, the fast set is functional.

Use of these techniques of strobing opens a very wide range of experimental use of clock chips. For example, it is possible to convert the linear accumulator, the h, m, and s counters, into a

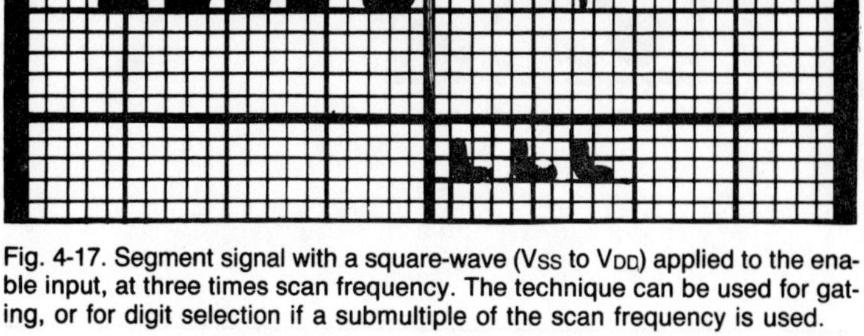

Fig. 4-17. Segment signal with a square-wave ($V_{SS}$ to $V_{DD}$) applied to the enable input, at three times scan frequency. The technique can be used for gating, or for digit selection if a submultiple of the scan frequency is used.

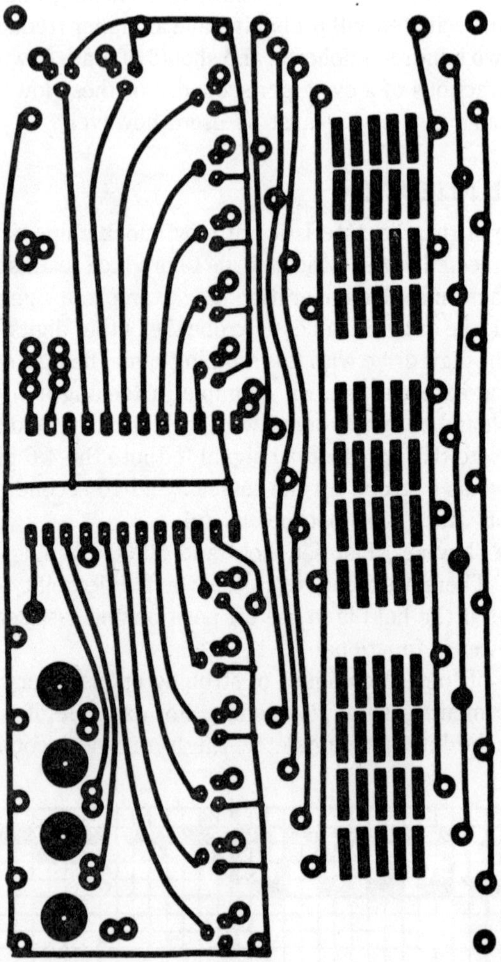

COMPONENT SIDE

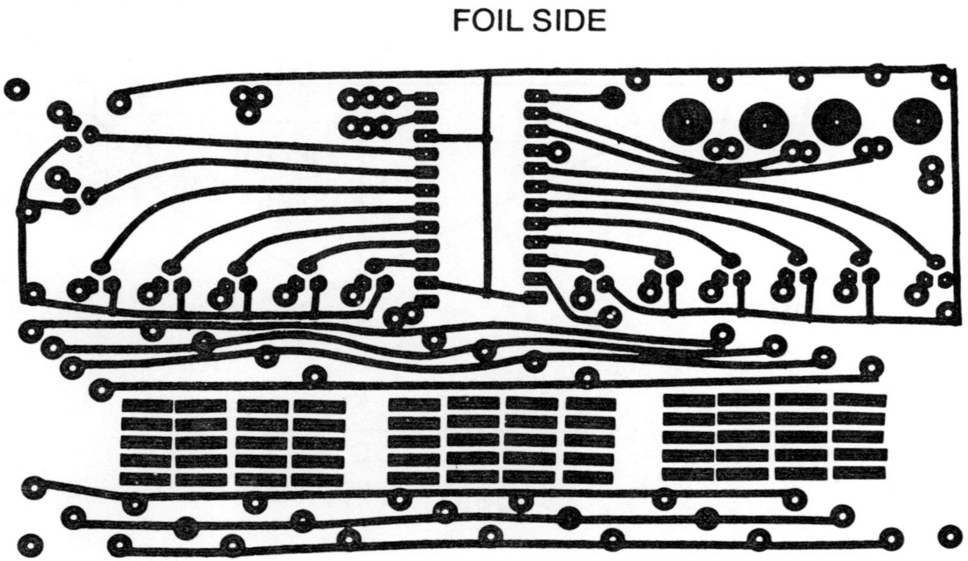


Fig. 4-18. Printed circuit layout for a 6-digit clock, using FND70 displays. The circuit is the same as Fig. 4-7.

COMPONENT SIDE

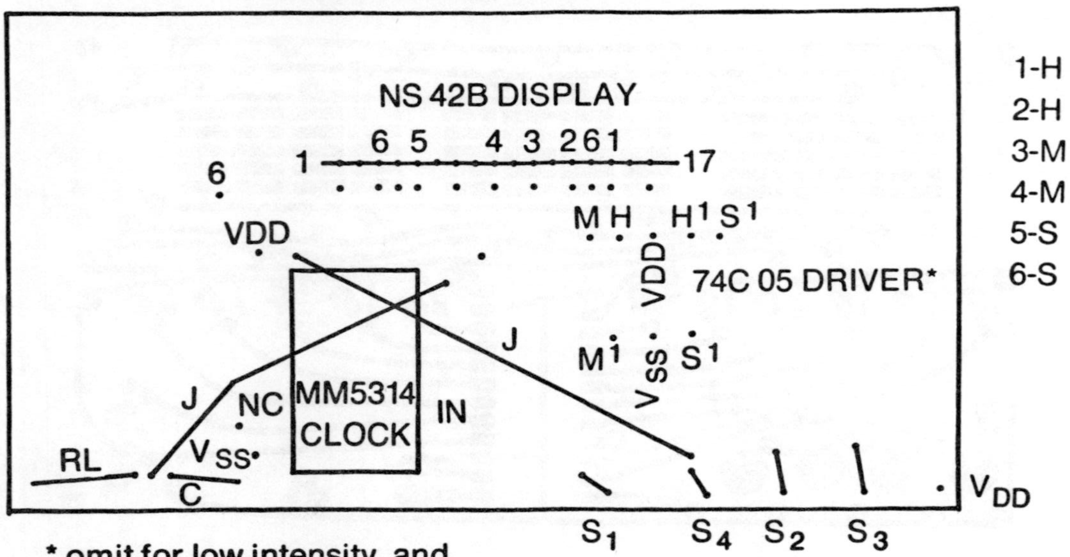


* omit for low intensity, and connect display 1-6 directly to chip leads

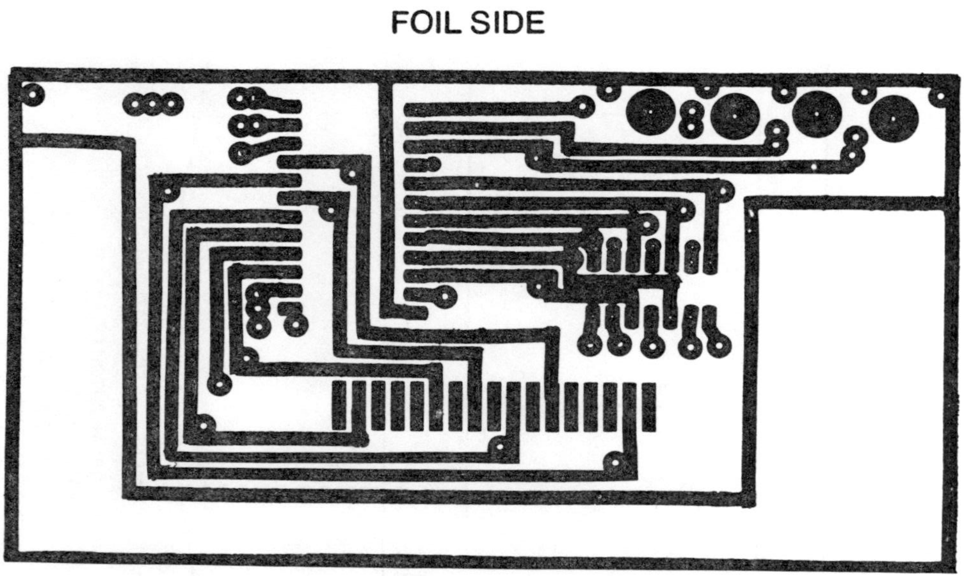


Fig. 4-19. Printed circuit layout for a 6-digit clock using a NS-428 calculator display strip. No driver is needed for Vss = 12 to 17 volts, but space is available if a brighter display is desired. Controls should be arranged to suit use; normally the enable, $12^h/24^h$, 50/60 Hz and 4/6 digit-select terminals will not be used.

quasi-nonlinear device, which runs at one rate for a period of time, and then at another rate. By suitable switching it is possible to make the clock run fast for a period of time, then run slow at an equal or different rate, etc. See Chapter 10 for some possibilities for these techniques.

## VARIATIONS IN DESIGN

While the experimenter's board is extremely useful and most instructive, it is not really a good layout if timekeeping is the goal. For that, a smaller sized board with simpler operation is indicated.

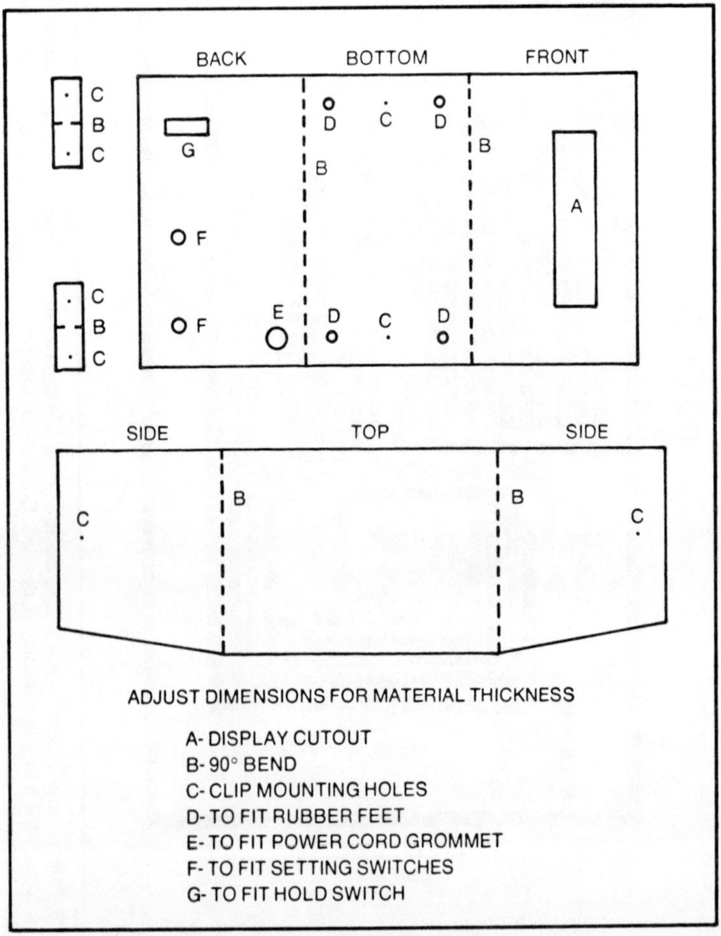


Fig. 4-20. Cabinet design for a small clock. This basic design can be used with the boards of Figs. 4-20 and 4-21, and was used for the kit clock of Fig. 1-4.

For example, the decision as to whether to use four or six digits, or 12 or 24 hour format, could be made prior to the wiring, eliminating many switches. Figure 4-18 shows a small board, intended for clock service. The operating conditions for this are identical to the experimental board, and the wiring should proceed in the same way. However, the unnecessary switches should be replaced by jumpers.

For some clock purposes, such as desk clocks where the vision distance will be small, miniature displays can be used. Figure 4-19 shows a suitable layout, using a six-digit display of the type used in the least expensive hand calculators. Since four to five milliamperes of drive current is ample for segments no segment driver is needed. At most, a digit driver may be needed. In this layout, this is done by an integrated circuit, also of the type used in hand calculators.

## CASE OR CABINET CHOICE

No cabinet is shown for either the experimental board or for clocks built from the small board. The major reason for this is that there is room for considerable ingenuity here, and each builder will want to choose his own mounting. Very good cabinets are available from kit manufacturers, and are generally sold independently of the electronics package. The additional sources of cabinet material are many; for example, various household objects, plastics, and even aluminum extrusion intended for other purposes. A little thought should come up with interesting or novel cabinetry.

For general electronic use, in the laboratory or around the radio shack, the cabinet of Fig. 4-20 serves nicely and is very inexpensive.

# Chapter 5

# Decorator Digital Clocks

Commercial digital clock designers have standardized on a design which is highly utilitarian, but is not really adapted or adaptable to decorator use. A set of glowing red numbers covered by a transparent filter is very visible, but rather difficult to blend in with decor of various types.

The clock described in this chapter is intended to provide a decorators' alternate to this glowing numerical display. It is a design which is not now available commercially; it must be built, proving that the home constructor can do things not being undertaken by commercial enterprise.

## THE CONCEPT

A decorator digital clock was conceived as a bridge between conventional clock design and presentation, and the new all-electronic digital techniques. Basically the timekeeping functions are completely electronic. On the other hand, the presentation functions, while using electronic techniques, are intended to give a time presentation which is immediately understandable as being a standard clock face.

To give this clock face presentation, two concentric circles of LED indicators are used, as shown in Fig. 5-1. As is standard practice, the inner circle is for hours, corresponding to the indication of the short hand on clock faces. The outer circle gives the minutes

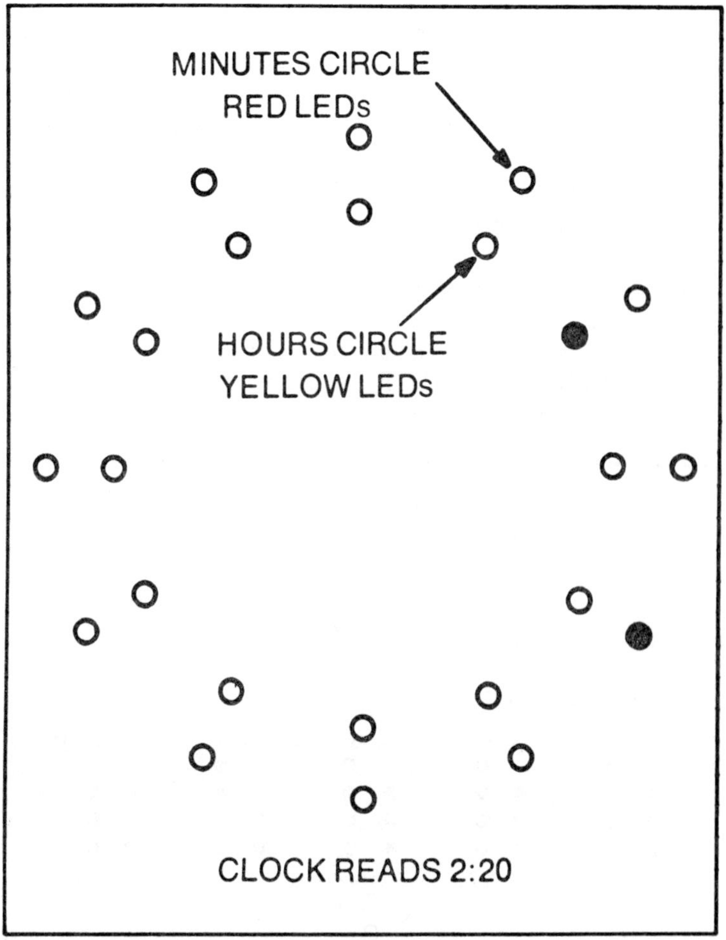


Fig. 5-1. Basic timekeeping principle. The LED position corresponds to the hours marks on a conventional clock and is read the same way.

indication, corresponding to the indication of the long hands of standard clock faces.

To avoid excessive complexity, each of these concentric circles is formed from 12 light emitting diodes. Thus, there are LED indications for each hour, but only one for each five minutes. This is not as strange a presentation as may seem at first glance—if you examine a number of electric clock faces, you will find many with no marks between the hour indications, that is, with no indications of minutes positions. In addition, you will find others which are very tricky since they have four marks between each five minutes instead of the five which one would expect. This common alarm face is really set up to indicate the 15-minute points of the hand provided to show the alarm setting, and has nothing to do with minute indication. This is a clockmaker's version of "Sneaky Pete strikes again!"

To provide a way of reading the exact minute, an auxiliary bank of LEDs is used. This is composed of two vertical rows of three LEDs in one row and four in the other. As shown in Fig. 5-2, the electronics are arranged so that one LED is on, then two, then three, then four. The exact minute is obtained by first determining the five minutes position, then adding to this the number of these supplements LEDs which are lit. For example, the illustration in Fig. 5-2 shows a time of 22 minutes.

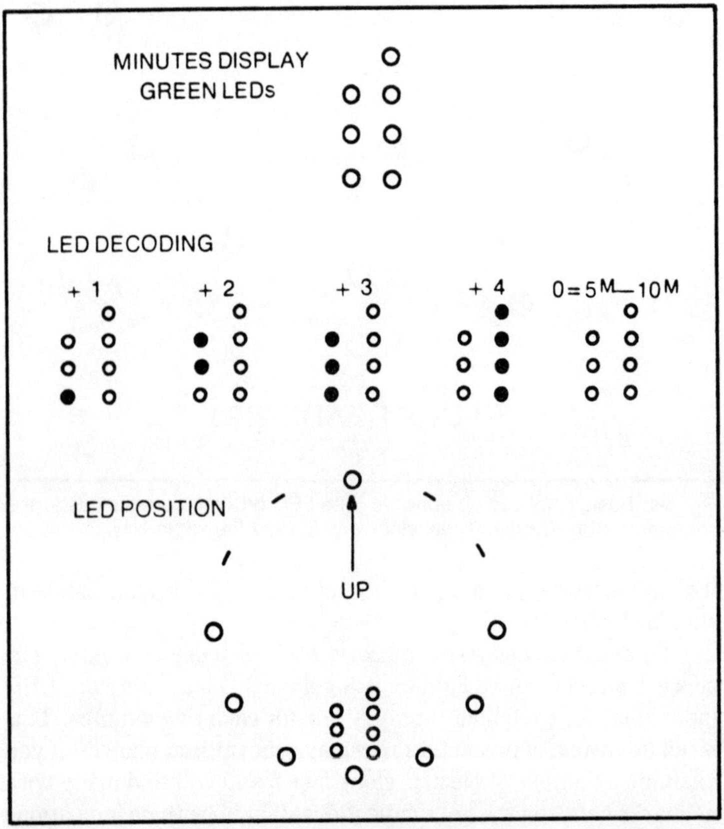


Fig. 5-2. The minutes interpolation scheme. Seven LEDs are used to give indication of 0, 1, 2, 3, and 4. These values are added to the minutes reading of Fig. 5-2.

Perhaps surprisingly, it takes almost no time to learn this simple code. Most times, when one glances at a clock, one is not really interested in the exact time, but on the time remaining to do something or before something must be done. Usually, the time is only needed to an accuracy of five minutes or so. In these instances, only the LEDs in the five minutes circle are scanned. When more accurate time indication is needed it appears that the average person looks at the five minute position first: he then decides whether the one minute time should also be used. A few practice readings will give complete training.

Of the many possible ways to arrange this additional bank of LEDs, the one chosen gives a vertical array, which was used for an additional reason. It gives a way of orienting the clock face at night, when the general outline cannot be seen. It also gives a reference for reading the LED positions. Unlike most decorator clocks (in fact, most clocks with hands), this decorator digital clock is just as usable during the night hours as it is during the daytime.

There are a great range of decorator possibilities which can be worked out using the basic design concept. One, for example, is in color. With present LEDs, four rather intense colors are available. In addition, by using different indicators, the display can be changed to show neon red, fluorescent green, some shades of blue, and so on. It is also possible to set up blinking displays or displays which appear for a time then turn off. Blinking can also be introduced for special purposes, such as an alarm function.

There is also a great range for experiment on display arrangement. For example, the displays can be arranged in a horizontal straight line, or a vertical straight line, or even with the hour in one direction and the minutes in another. The many books on clocks and clock faces can be consulted for ideas of design.

## THEORY OF OPERATION

As can be seen from Fig. 5-3, the first order block diagram of the decorator digital clock is just the same as for many other digital electronic clocks. That is, there is an accumulator or counter which determines and stores the hours and minutes reading, plus a display which shows the reading so accumulated.

However, the decorator digital clock design differs in many details, partly as a result of the method of display chosen, and partly as a result of "designer's choice" in the details of the counting circuit. These differences are illustrated by the block diagram in Fig.

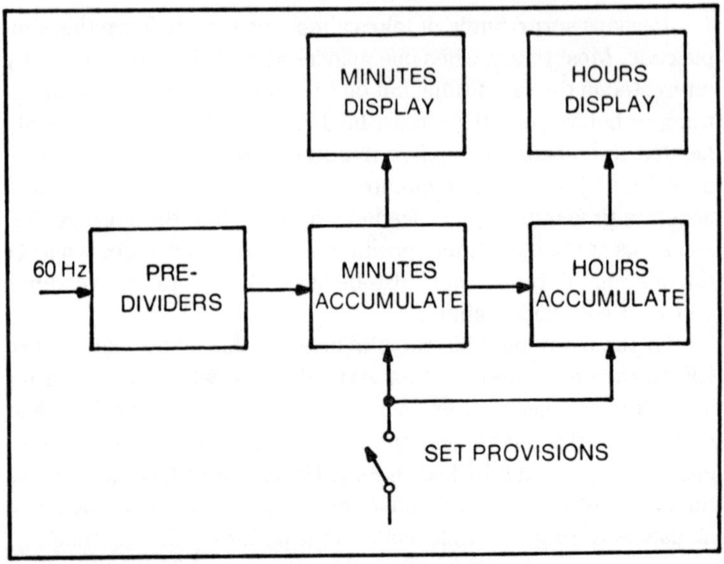


**Fig. 5-3. Block diagram of the entire clock. It is essentially the same as a 4-digit chip in performance.**

5-4. Starting at the left, the 60 Hz supply line frequency is first divided by a factor of 3600, to obtain minutes. Designer's choice entered here; since there was no plan for using a seconds display with the decorator digital concept, the usual initial division by 60 to obtain seconds, then a further division by 60 to obtain minutes was not used. Instead, the dividers were arranged in a 15, 16, 15 bank, as shown in Fig. 5-5, to give the overall divide-by-3600 action. The output of the first divider is at a rate of 4 cycles per second, (4 hertz) and the output of the next two dividers gives the basic one-per-minute for the remainder of the clock. The output of the first divider could be divided to 1 cycle per second (1 hertz) if desired. (See Chapter 7 for an idea on using such an output.)

The divide by 15 stages are made from 7493 hexadecimal dividers, using a diode AND gate to detect the 15 count, as shown in Fig. 5-6. The gate signal is fed to one of the reset-to-zero inputs of the counter. The other reset-to-zero input is provided with a normally closed switch which can be used to reset the scale of 3600 counter to zero on demand, and which also serves as a HOLD for the subsequent stages.

The divide by 16 counter uses a straight hexadecimal connection, with input to A in and output at D. The same reset scheme is used.

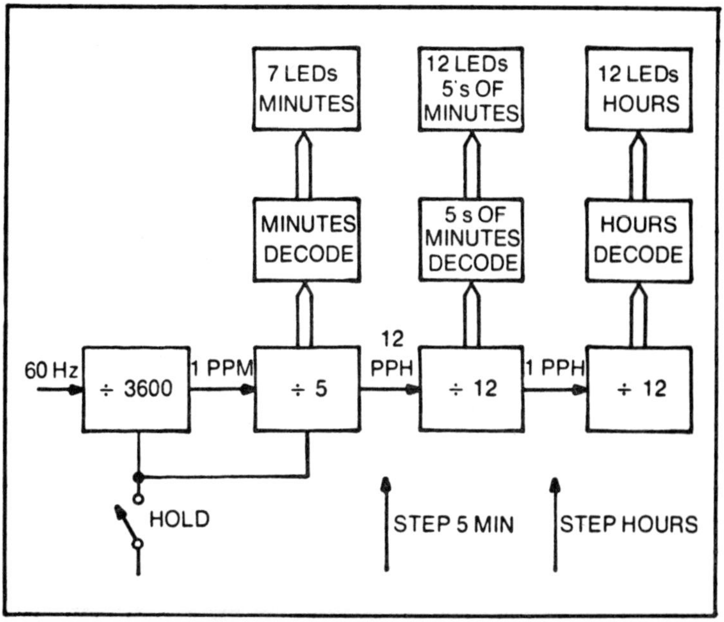


Fig. 5-4. Expansion of the block diagram to show the divider stages and LED feeds. The "set" points are indicated.

If 50 Hz clock operation is desired, the dividers must be changed to a ratio of 3000:1. A suitable chain for this is shown in Fig. 5-7. No additional packages are needed, since there is an unused 2:1 divider in the subsequent minutes divider chain: see the discussion below.

The main display requires not a one-per-minute rate, but a one-per-5-minute rate. The auxiliary display, on the other hand, requires the one-per-minute rate, but one which is decoded to give indication of one, two, three, four and off. Both of these actions are accomplished in a single divider of a scale of 5:1. The final output of this divider is the one-per-five-minute rate which drives the main minute display.

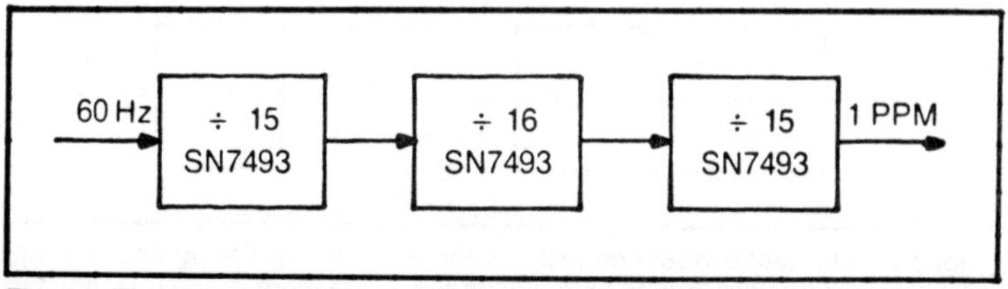


Fig. 5-5. The form of the divide-by-3600 circuit. Type 7493 TTL dividers are used.

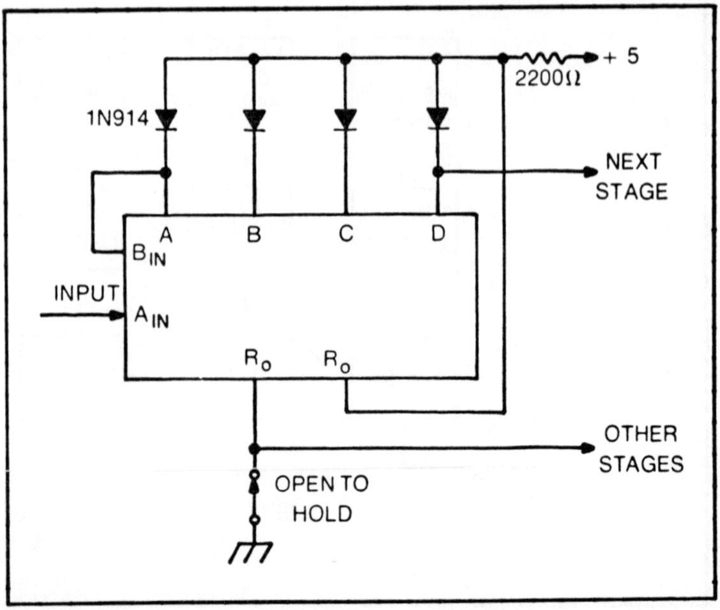


Fig. 5-6. The connections of the divide-by-15 stages. The output is taken from D, which goes low at the 15th pulse.

The divide by five counter which is used to give the minutes indication, is normally in the low state, and goes high when active. At the counting speed required here, the high level can supply a considerable amount of current, easily the 10 to 20 mA needed by representative LEDs.

It will be recalled that the divide by five counter uses a binary coded scale, with indications of one, two, and four. The indication for three is obtained with the one and two count both high. Since it is desired that the count be in the continuous sequence of one, two, three, four, a small diode decoder is used, as shown in Fig. 5-8. As can be seen, the first count will cause the bottom diode of

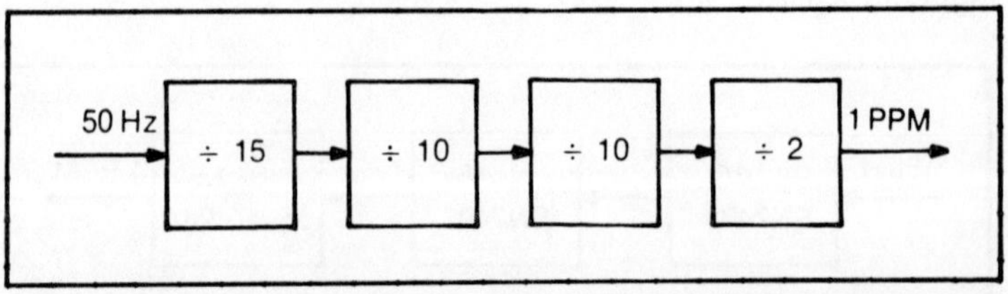


Fig. 5-7. Alternate divider connections for 50 Hz input. Wiring and package changes are needed, but no extra packages, since there is an unused binary stage available.

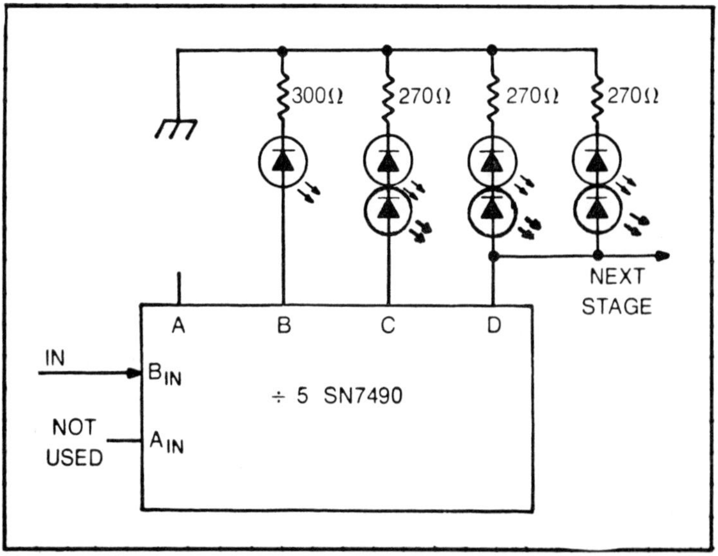


**Fig. 5-8. Connections for the minutes stage, including LED connections. The B and C LEDs are in one column, the D units in another.**

the left column to light. On the second count, the two top diodes in this column appear. For the third count, all three will light. At the fourth count all diodes in this column are extinguished and the four in the right most column light up. At the end of the fourth count all diodes are extinguished; this occurs at the same time as the shift from one five minutes position to the next. All diodes are off for one minute, and then the cycle resumes.

Note that the A section of the divider is not used. It is available for other purposes, such as the 50 Hz divider just mentioned. With some rearrangement of the first divider chain, this unused section could give a one pulse-per-second signal.

The one-per-five-minute pulse of the main minutes display must be divided by 12 to give the one-per-hour-rate. This is accomplished by a divide by 12 counter, a 7492. Coupled to this is a decoder, actually a one-of-sixteen binary decoder. Only 12 of the outputs of this are used. This decoder drives the "five-minutes" LEDs of the basic minutes display.

The outputs of the one-of-sixteen decoders are normally at the high level, and go low when selected by the decoding action. Accordingly, the LEDs must be wired in a current sink arrangement. When selected, the TTL output circuits can sink a considerable amount of current, up to some 50 mA. However, for most operating conditions and most types of LEDs, only 10 to 20 mA of current are needed. Therefore, the output is arranged to provide a single series resistor, to permit setting the LED current to the desired value. The connections, including the output circuit of the decoder, are shown in Fig. 5-9. Table 5-1 gives the "Truth Table" of decoded indication and pin connections which correspond to the five minutes positions, which are numbered sequentially from 1 to 12.

The output of the divide by 12 counter is at the one per hour rate needed for hours accumulation. Since conventional clock construction concepts are being followed, the hours need only be accumulated up to 12, then repeat. This is accomplished in a second divide-by-twelve circuit. Again, the output of this is fed to a one-

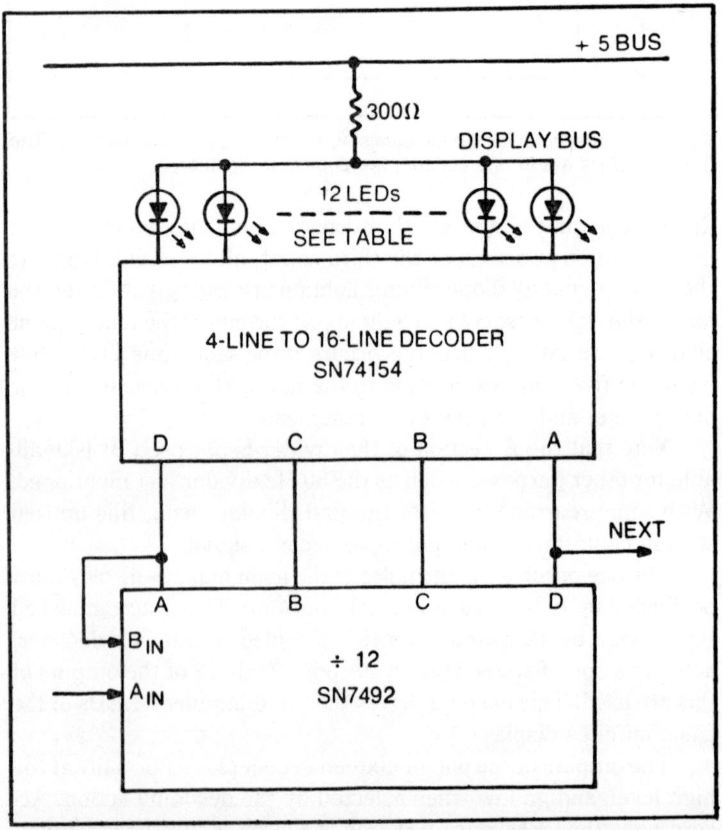


Fig. 5-9. Connections for the $5_S$ of minutes and hours stages. The binary outputs of the 7492 stage are fed to individual lines by 74154 decoders. Not all of the decoder outputs are used, some being inactive.

Table 5-1. Truth Table for 74154 Output.

| Time | | 7492 Output<br>74154 Input | | | | 74154 Output | |
|---|---|---|---|---|---|---|---|
| Hr | Min | D | C | B | A | Value | On Pin |
| 0 | 0 | L | L | L | L | 0 | 1 |
| 1 | 5 | H | L | L | L | 8 | 9 |
| 2 | 10 | L | H | L | L | 4 | 5 |
| 3 | 15 | H | H | L | L | 12 | 14 |
| 4 | 20 | L | L | H | L | 2 | 3 |
| 5 | 25 | H | L | H | L | 10 | 11 |
| 6 | 30 | L | H | H | L | 6 | 7 |
| 7 | 35 | H | H | H | L | 14 | 16 |
| 8 | 40 | L | L | L | H | 1 | 2 |
| 9 | 45 | H | L | L | H | 9 | 10 |
| 10 | 50 | L | H | L | H | 5 | 6 |
| 11 | 55 | H | H | L | H | 13 | 15 |

Valid for Connections

| 7492 | 74154 |
|---|---|
| A | D |
| B | C |
| C | B |
| D | A |

or-sixteen decoder, to provide the twelve different hour signals. The connections of this unit are the same as in Fig. 5-9, and the truth table of Table 5-1 applies. Note that the truth table must be redone if the input arrangement is changed for any reason. No indication of A.M. or P.M. is provided in the prototype unit. However, if desired, the unused divide-by-two counter available in the five-minutes package could be used to drive an LED A.M.-P.M. indicator circuit.

In addition to these functions of dividing and presentation, provisions for setting the clock to the correct time are needed. Most of the commercial chips use a fast advance, obtained by shorting out one of the divider sections, usually the first 60:1 divider for slow set, or this plus the following 60:1 divider for fast set. For the decorator digital clock design, it was deemed simpler (and also easier to operate) to provide a press-to-advance-one unit, plus a hold function. The hold is provided by using the preset to zero capability of the counter chain in the prescaler and minutes units. Open-

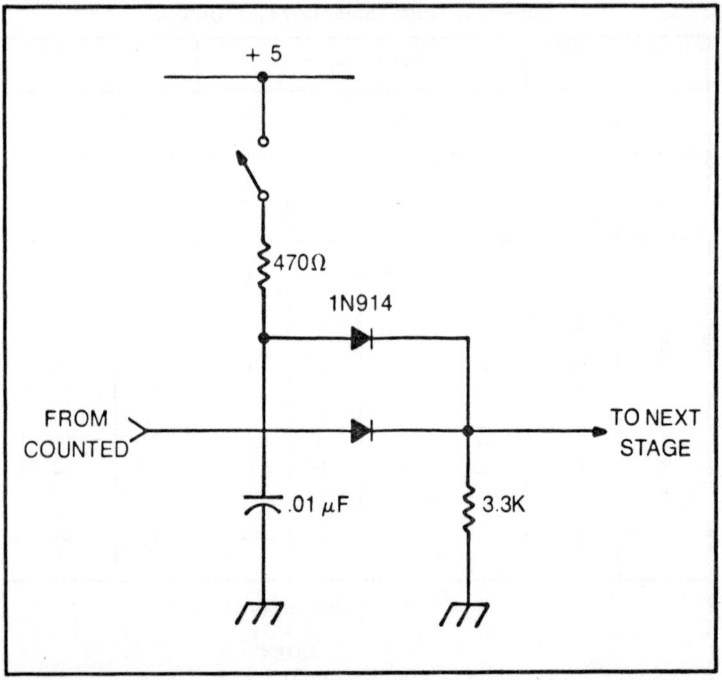


Fig. 5-10. The OR circuit used at the input of the 5S of minutes and hours counters to step the count when setting. The RC circuit following the switch is a debounce circuit.

ing this switch sets the counter to zero and stops the signal feed-to later stages. A separate circuit provides for advancing the five minutes counter by one step per switch operation. A similar step circuit is provided for the hours chain. These stepping switches are connected to the divider circuit by a simple diode OR gate, as shown in Fig. 5-10. Note that there is a capacitor connected at the switch circuit. The purpose of this is to eliminate the effect of "bounce" of the switch, which can cause the count to step an indeterminate number of times at each switch closing. The capacitor slows the rise time, to give only a single step per switch closing.

This method of setting seems to be far, far simpler than that used by the conventional clock chip, and it seems odd that it is not more widely employed.

## THE COMPLETE CIRCUIT

The complete circuit of the prototype decorator digital clock is shown in Fig. 5-11. It follows directly from the block diagram and individual circuits shown before. Note again the method of utilizing four AND diodes to force the divide-by-sixteen counters to reset on fifteen. Note also the OR circuits which are used in the hours and five minutes setting circuits, and the diode arrangement for the minutes indication.

There are really no critical circuits in this design. The closest to this is in the hours and five minutes settings, as has already been mentioned. The remainder of the circuits should operate perfectly if no mistakes in assembly have been made.

## CONSTRUCTION

Printed circuit board layout for the clock is shown in Fig. 5-12, and parts placement in Fig. 5-13. The entire assembly is on a piece of printed circuit board eight inches square. The parts are arranged so that the six o'clock-twelve o'clock line lies along one of the diagonals of the board, and the three o'clock-nine o'clock line along the other. The top area of the board includes the prescalers, i.e., the 3600:1 dividers, and the minutes divider and decoder. The five-minutes and hours dividers are just below the center line, and their associated decoders are near the bottom.

This layout does have one wiring difference from the original. Two return circuits are provided for the five-minutes and hours LED circles, instead of the single return bus of the original. This eliminates a "blink" of the hours LED as the five minutes indication changes.

The location of PC packages, decoders and switches within the LED circle was made to be symmetrical, although this did give a few awkward connection problems which would probably be corrected by moving a few of the parts. No attempt was made to use printed circuit interconnections between the decoders and the LEDs themselves. Instead, these connections are made by wire jumpers, on the reverse side of the circuit board, generally arranged in a circle lying along the LED bank.

The prototype was intended to be used with a separate 5-volt regulated power supply. A suitable supply is shown in Fig. 5-14, one part showing a design based on a standard 9-volt "battery eliminator" package plus a 5-volt regulator, the second based on a small transformer plus discrete component rectifiers and regulators. Either of these will work perfectly well. The power supply can be made as a separate package if desired, or is small enough to be placed in a case holding the basic clock board.

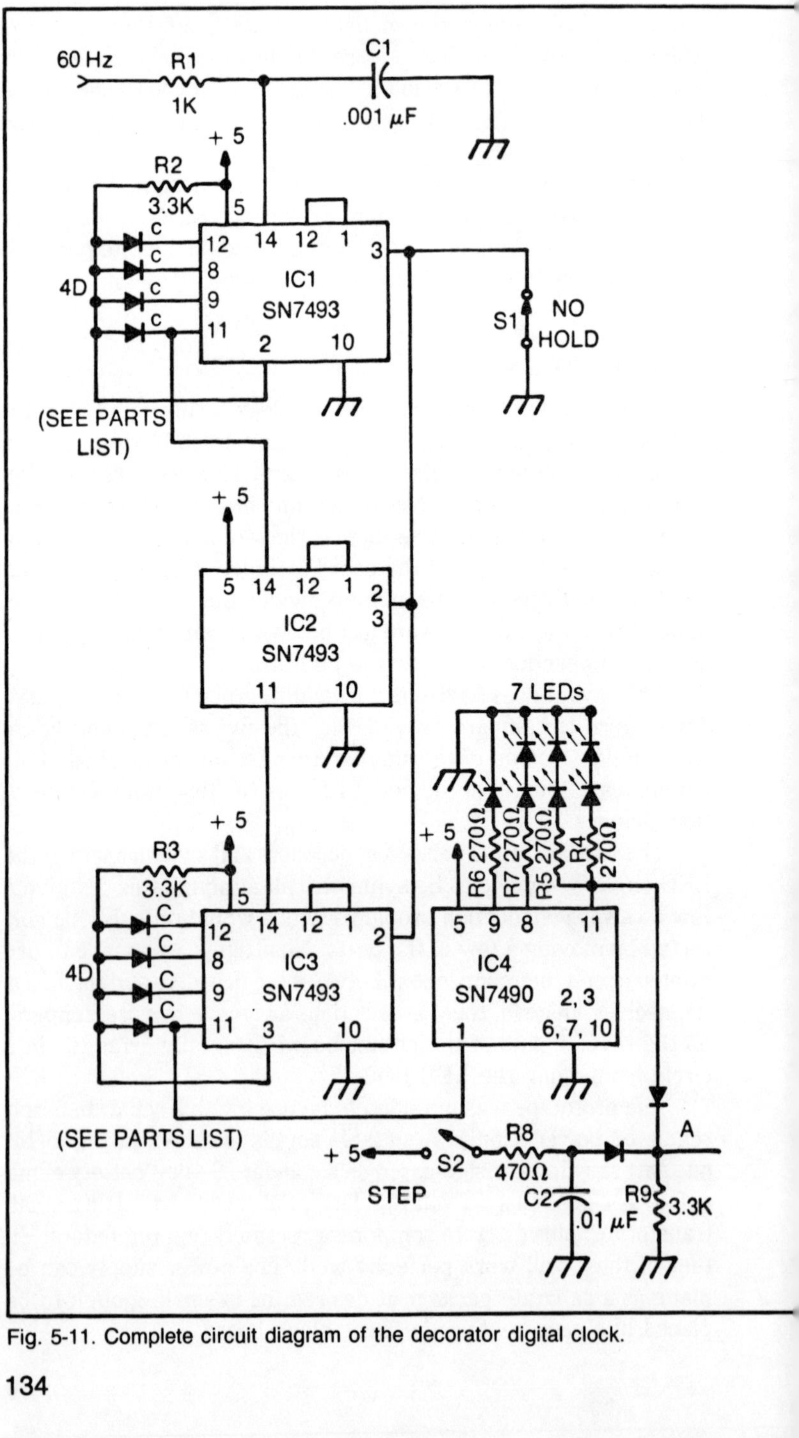


Fig. 5-11. Complete circuit diagram of the decorator digital clock.

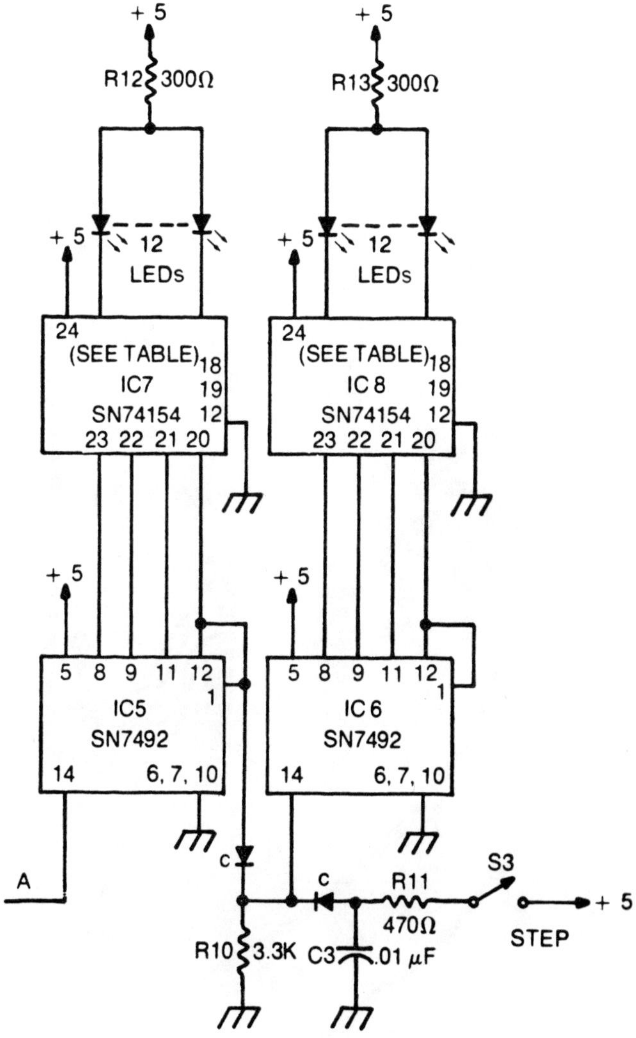
+ 5
R12 300Ω
+ 5
R13 300Ω
+ 5
12
LEDs
+ 5
12
LEDs
24
(SEE TABLE)
18
IC7
19
SN74154
12
23 22 21 20
24
(SEE TABLE)
18
IC 8
19
SN74154
12
23 22 21 20
+ 5
+ 5
5 8 9 11 12
1
IC5
SN7492
14
6, 7, 10
5 8 9 11 12
1
IC 6
SN7492
14
6, 7, 10
c
A
c
R11
470Ω
S3
+ 5
STEP
R10 3.3K
C3 .01 μF

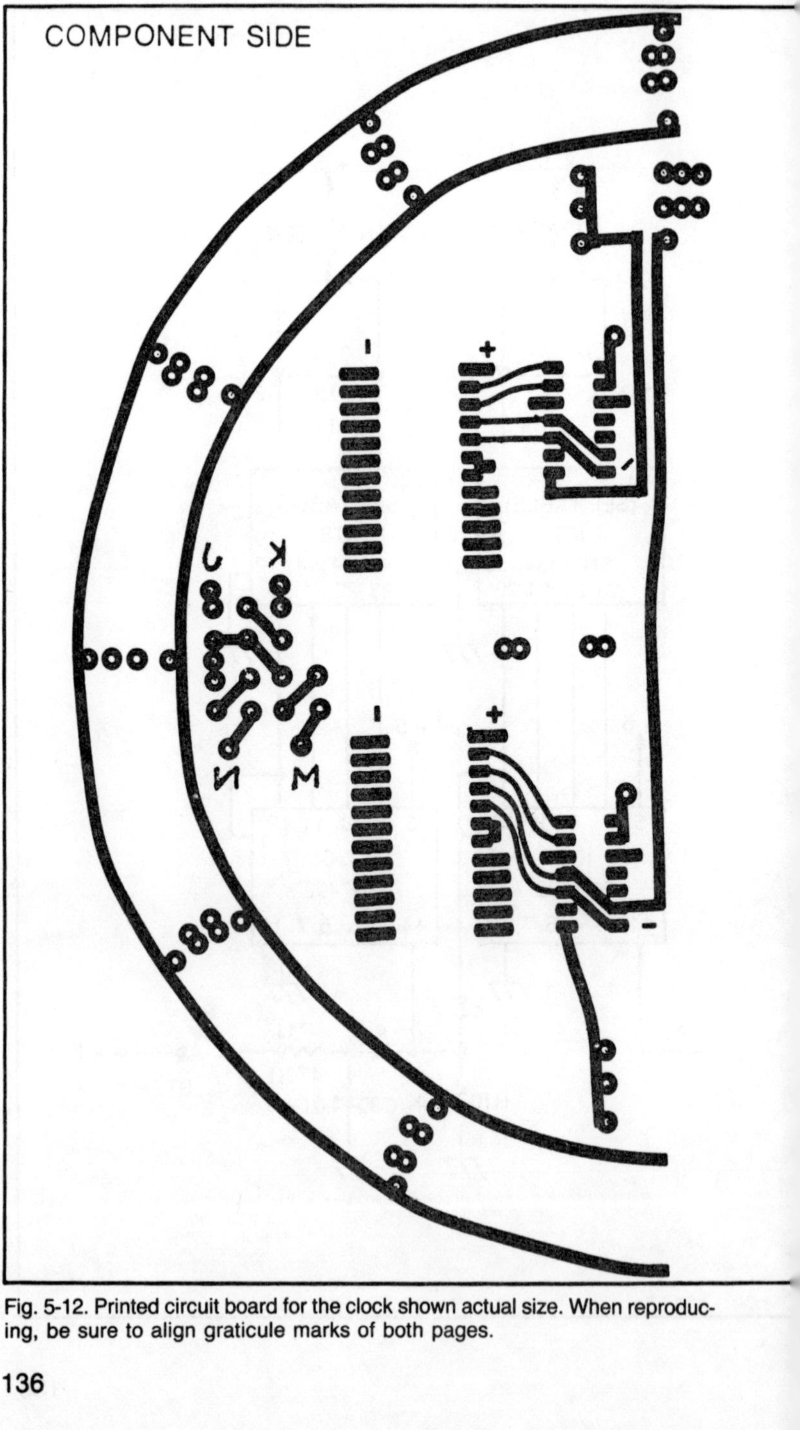


Fig. 5-12. Printed circuit board for the clock shown actual size. When reproducing, be sure to align graticule marks of both pages.

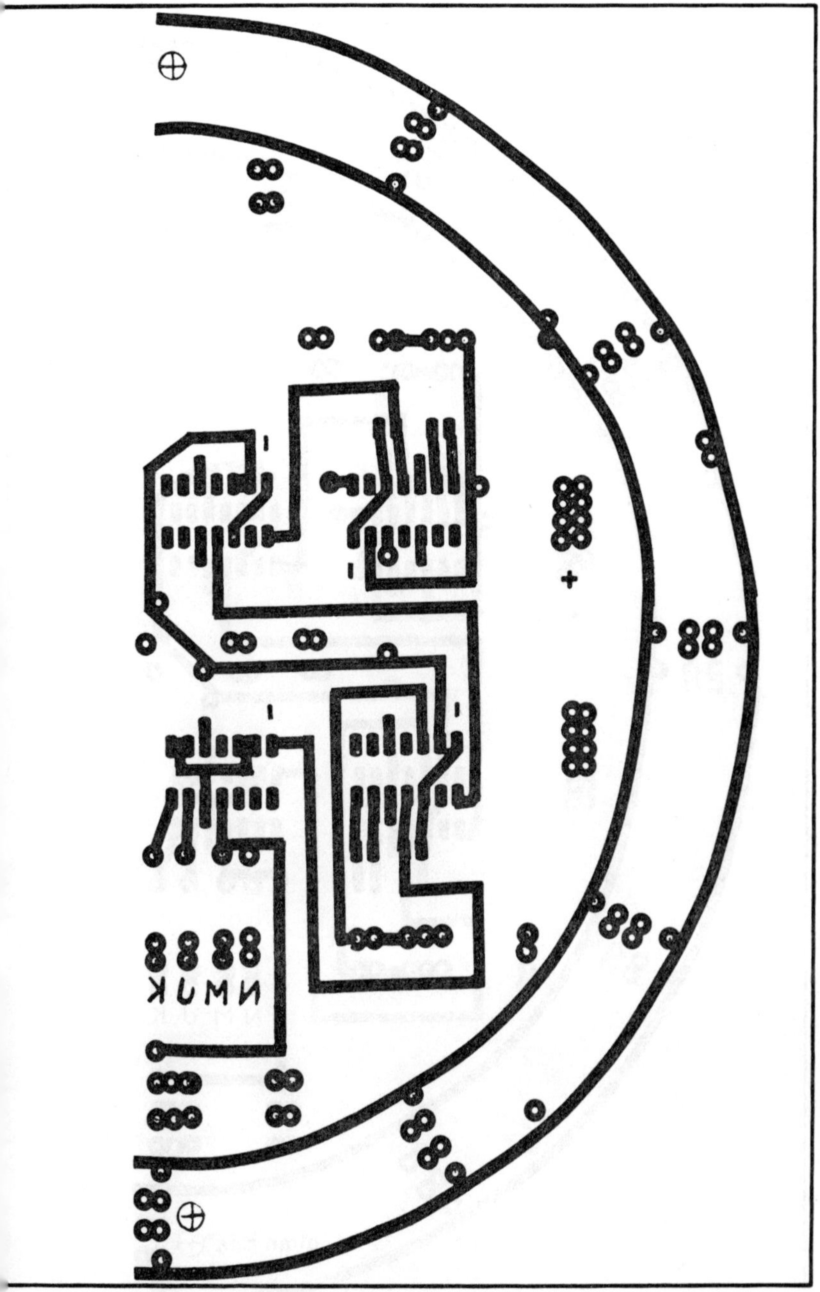

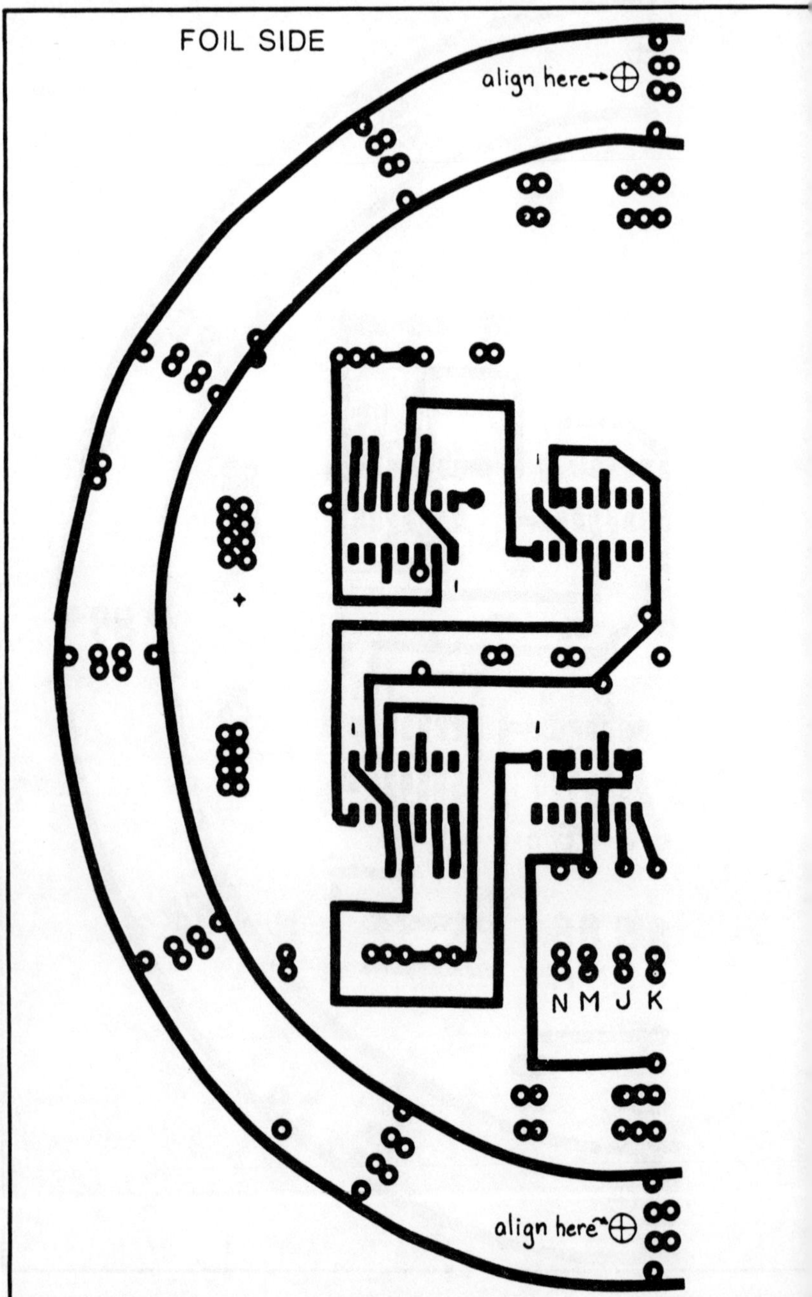

Fig. 5-12. Continued.

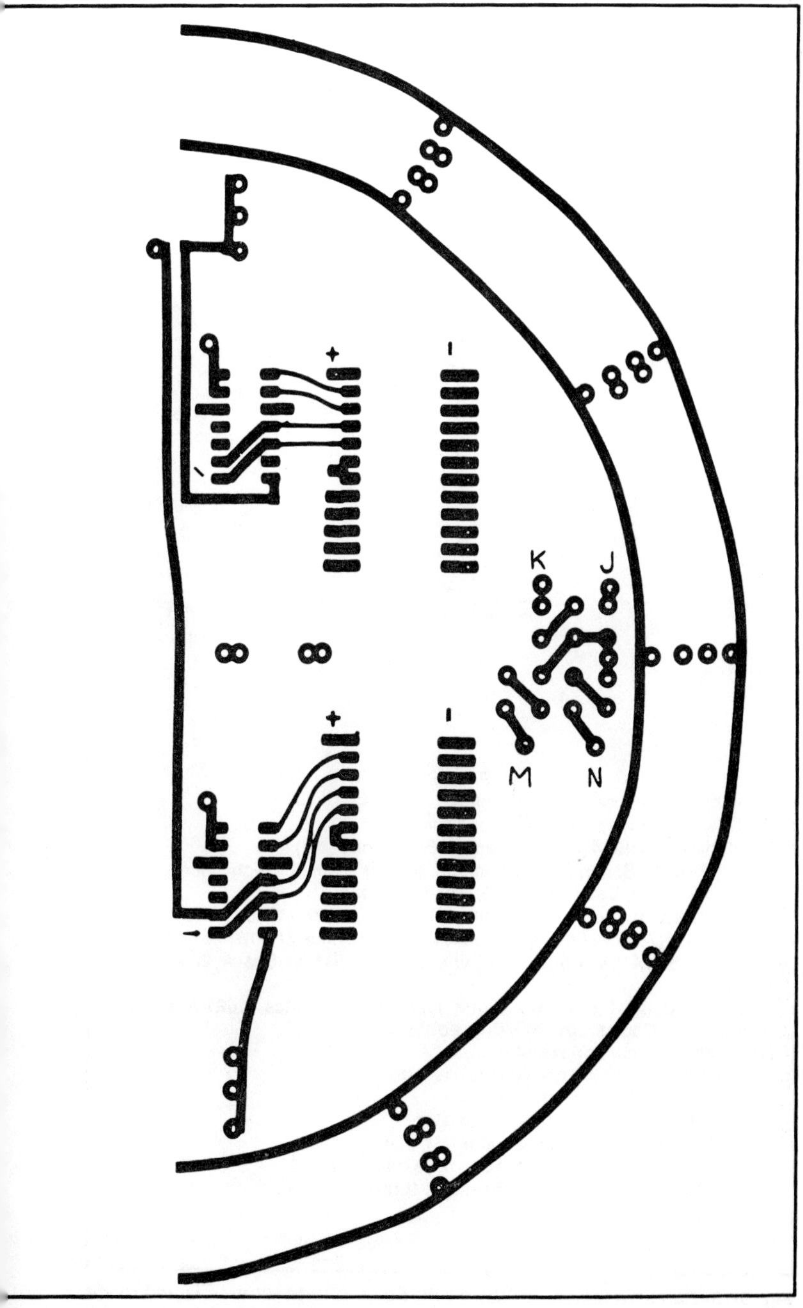
K
J
M
N
+
+

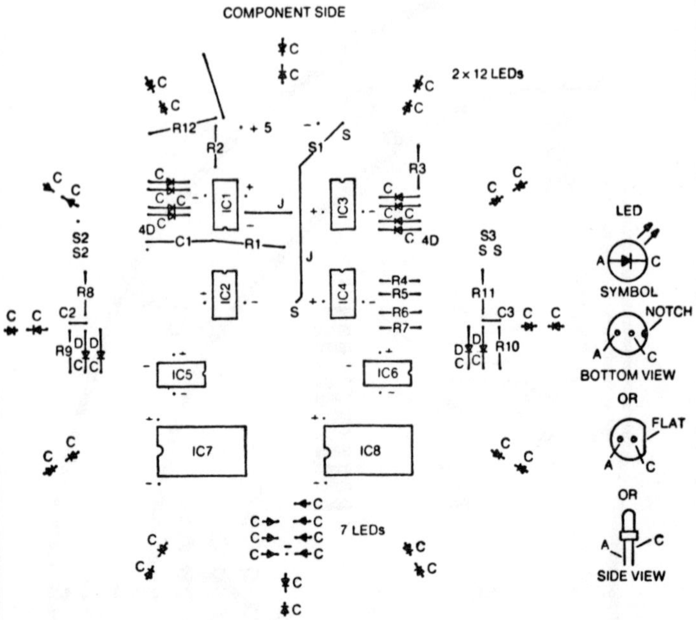


## PARTS LIST

C1—.001 $\mu$F disc ceramic capacitor
C2, C3—.01 $\mu$F disc ceramic capacitor
D—all diodes 1N914 (Radio Shack 276-1144 or equal)
IC1, IC2, IC3—SN7493 hexadecimal divider (James Electronics, 1021-A Howard Avenue, San Carlos, CA. 94070) or equal
IC4—SN7490 divide by five counter (Radio Shack 276-1808 or equal)
IC5, IC6—SN7492 divide by 12 counter (Radio Shack 276-1819 or equal)
IC7, IC8—SN74154 4-line to 16-line decoder (Radio Shack 276-1834 or equal)
LED—MV50 light emitting diode (James Electronics, 1021-A Howard Avenue, San Carlos, CA. 94070) or equal
R1—1000 ohm carbon resistor
R2, R3, R9, R10—3300 ohm carbon resistor
R4, R5, R7—270 ohm carbon resistor
R6, R12, R13—300 ohm carbon resistor
R8, R11—470 ohm carbon resistor
All resistors 1/4 watt or more, 10% more tolerance
S1, S2, S3— switch, single-pole single-throw

Fig. 5-13. Parts placement for the clock. Be sure to observe LED and diode polarities. All components are mounted on the nonfoil side of the board.

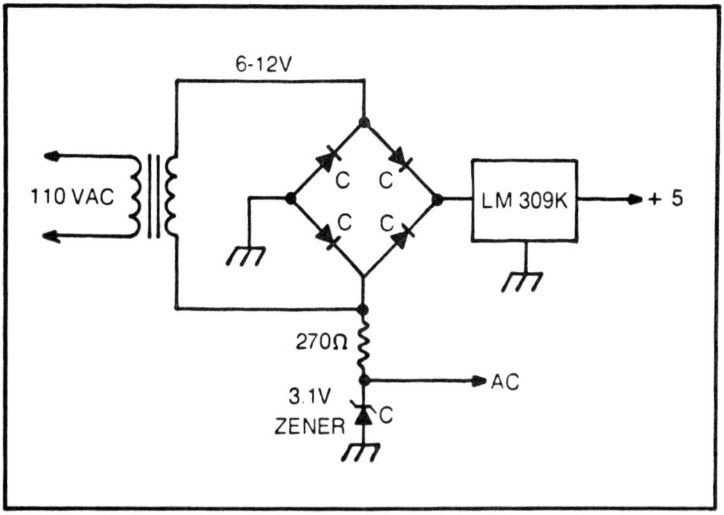


Fig. 5-14. Power supply for the clock. The transformer can be a type which is built into the wall plug. The remaining components can be mounted in the clock board cabinet or base.

## DECORATOR DESIGN

One of the features for this type of clock design is the very wide range of decorator arrangements which are possible. For example, the entire assembly can be covered by a piece of translucent plastic, so that the hours and minutes LEDs produce a small glowing spot on the plastic surface. Alternately, colored and transparent plastic rings can be used with various designs inside and outside of the rings. The ingenuity and the artistic sense of the experimenter are the best guides for variation on the general theme.

Since the prototype was intended both to work out ideas and also to serve as a conversation piece, its construction has been left visible. This was done by leaving the integrated circuit packages and other components visible within the circle of the hours and minutes LEDs.

## CHECKS AND ADJUSTMENTS

There are really no adjustments necessary on this clock. However, because of the opportunity for mistakes or accidents such as solder bridges, a careful check of proper divider and decoder action is recommended. This is most conveniently done by substituting the output of an audio oscillator for the 60 Hz source. By setting the audio oscillator to a high frequency, the divider action can be speeded up, permitting rapid check of their performance.

To start the check, it is convenient to work at a frequency of about 600 Hz. With this applied to the 60 Hz input, an oscilloscope probe can be used to show divider action. An example of this check is shown in Fig. 5-15. Setting the oscilloscope to give 15 input pulses should give one pulse at the output of the first divider. The second divider can be checked for the correct two-eight division mode, and the third for the fifteen-to-one mode.

At this point in the circuit, events are now occurring at the rate of 10 per minute. For a further check it is convenient to increase the oscillator rate to perhaps as high as 60 kHz. Division by five and then successive divisions by two and six should be obtained. Decoder action can now be checked, which may be simpler if the oscillator rate is reduced. Watch especially for skipped lamps, and lamps which do not show uniform brightness level. With surplus LEDs, a few faulty or low output units must be expected, although any one which does not light may have been installed backwards.

After the decoder and display actions have been checked, the setting circuits should be examined. It should be possible to stop the display and then to restart it without disturbing the hours and five-minutes counters. The counter for minutes, however, will reset to zero each time the HOLD is operated. The advance switches should cause an advance of one step each actuation when the HOLD circuit has been activated. If it has not, the advance circuit will still function, provided that the divider input at the particular instant is not already high. When operating from the 60 Hz source, this is no particular drawback since the step can be visually observed.

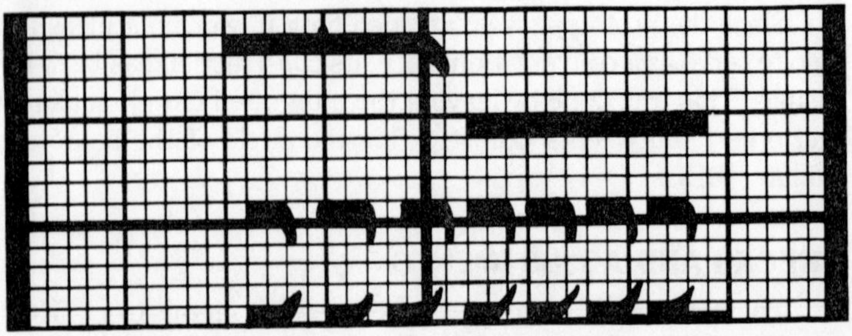

Fig. 5-15. Checking divider action at 50 kHz input to the clock. Lower oscilloscope trace is the A divider output, upper trace the D output. There are seven output pulses in one cycle, then eight in the next, as required for 15:1 division. They can all be seen with the lower trace synchronized to the A input, but it is easy to miscount.

When all circuits have been checked out and are operating properly, it is a good idea to give a complete check for a period of several hours, using a conventional electric or digital electric clock alongside for a check. It is easy to make a mistake in the fifteen-to-one or twelve-to-one count down circuit; for example, if a solder bridge exists and is not corrected, the counts can be off one unit, which is not readily noticed. A check for several hours is a good way of ensuring that this type of error is detected.

As a final step, mount the board in the way you have chosen, and give it a final check. An interesting little test is to place the unit in an unobtrusive location, and start it at the correct time. Then see how long it takes the family to work out the technique of decoding.

## SINGLE ROW DISPLAY

Two color LEDs are now available. Basically, these are two single color LEDs in parallel, but with reversed polarity. When current is passed in one direction, one color appears. The second color appears when the current is reversed.

A dozen of these in a circle allows design of a single circle clock. The electronics would be somewhat more involved than for the double row type. The complication lies in providing for both colors at the same point. A way of handling this is to multiplex the hour and five minute signals, easily done by using the 60 Hz line as a driver. However, unlike the common multiplex, the current must be reversed. Suggestion—use a centertapped transformer with the two color LED connected between the center lead and ground. Connect the outer leads of the transformer to open-collector high voltage TTL drivers, one for the hours and one for the five minute counters. Red and green will be normal, except that when both must be illuminated, the colors will combine, and appear as yellow.

With some care, the two LEDs of the two row clock can be set into a single row. One method of mounting is to use twelve of the clear faceted jewels used to house "pilot lights" in radio circuits. Most of these have a shiny tube of sufficient size to house two good sized LEDs. The shiny tube helps give a bright display, and the jewel facets make it visible over a very wide angle.

## MICROCOMPUTER DRIVEN DECORATOR CLOCK

Experimenters interested in timekeeping with computer techniques will find the decorator clock described in the magazine EDN for May 12, 1982 of interest. The article gives a specific design, using a MC6800 processor, latches to hold time, plus provisions for time setting. A machine language program to be placed in ROM is included.

# Chapter 6

# A Precision Timer

Clocks and timers are close relatives in that they both measure time. They differ in that the timer usually does not have an indicating element, and in the fact that the timer is intended to give a signal or action after a definite and known time interval. Clocks and timers are sometimes combined, as in the familiar clock-timer of the kitchen oven, or even as in alarm clocks. Some timers do have provisions for indicating the fraction of time elapsed, or the time remaining, but generally clocks and timers are two separate devices.

Timers may be built to work on the basis of analog computation, or on the basis of digital computation. The familiar photographic darkroom timer (of the electronic type) is an example of an analog timer, the analog being secured by determination of the time required to charge a capacitor in an RC circuit. The oven timer is actually digital, the dial being gear coupled to the electric motor which integrates the 60 Hz power line pulses to obtain angle versus time relations in digital form.

The precision timer described here is intended to give a wide range of time intervals of very high accuracy. As designed, the accuracy is determined by the power line to which the timer is connected and, on the average, will be as good as the master time standards of the country. This accuracy is obtained directly, since the timer operates by counting the cycles which have elapsed since timing was initiated. For short time periods, however, the timer accuracy will be less, on the order of one part in ten thousand, since power line frequency is allowed to vary slightly under conditions of varying loads. The timer may be operated from a separate precision frequency source and will have the same accuracy as that source.

The timer is intended to give periods ranging from fractions of a second to several hours, with a resolution of 8 binary bits, over one part in one hundred; the resolution may be extended if desired. The basic range of the timer is in seconds, with a total range of 2 to 255 seconds. This basic range may be multiplied by factors of 0.1, 10, or 100, to give a total range of 0.2 seconds to 25,500 seconds. Any value within the range can be selected in unit steps.

In addition, the timer may be set to operate on a scale of minutes rather than seconds, the range now being from 2 to 255 minutes in steps of one minute. Likewise the range may be set to operate in cycles of a basic supply frequency; for the normal 60 Hz line, in steps of 2 to 255 cycles of 1/60th second each. The total range covered is from 1/30th of a second to over four hours.

To extend the usefulness of operation, the timer is provided with a separately set OFF cycle. It may be set to operate in a one-shot mode, or may be set to cycle continuously. The one-shot mode timing commences from a switch closing, or an external signal, and continues until the preset time clocks out. In the cycling mode, the duration of the ON cycle and the OFF cycle can be independently set, but with the limitation that the multiplying factor established by the time base applies to both the ON and OFF cycle. The total range of the repeated cycle thus extends from 1/15 second to over eight hours.

The timer may be wired to give positive voltage output corresponding to either the OFF or the ON interval, separate positive voltages on two leads for the ON and OFF interval, or may be wired to give pulses whose spacing corresponds to the preset time interval.

## THEORY OF OPERATION

The operation of the timer can be evaluated by following the block diagram in Fig. 6-1. The basic reference frequency, normally 60 Hz, is fed to a time base, a chain of dividers. This gives the basic rate of seconds and the multiples. The time base may also be set to give a count rate of minutes.

The selected rate is fed to a second counter chain. Connected to this chain are a pair of switch banks, one for the ON cycle duration, and one for the OFF duration. When the accumulated count

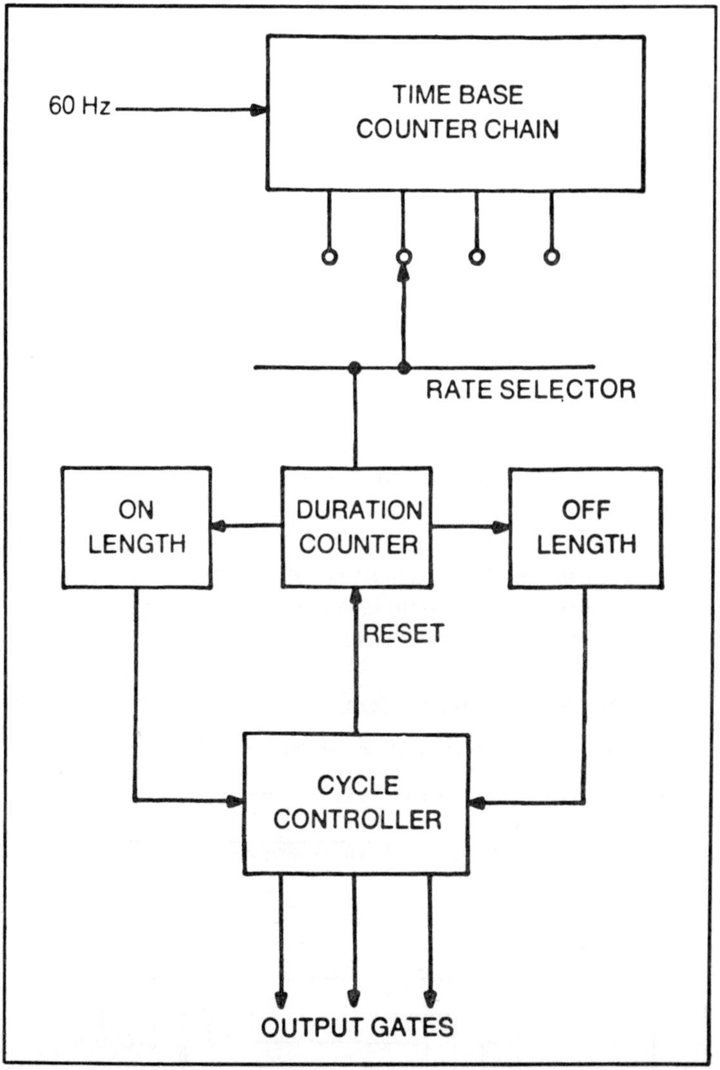


Fig. 6-1. Block diagram of the timer. The duration of the OFF and ON periods are determined by the setting of binary-coded switches which are independently set.

reaches the value set on the switches, a reset signal is generated, setting the counter to zero, and transferring from one cycle to another. The action then repeats, giving repeated ON and OFF cycles, each of precisely controlled duration. Single cycle operation is obtained by interrupting the cycling. Various outputs may be taken from the duration generator as needed. Outputs may also be taken from the two sets of counters.

## THE TIME BASE CHAIN

The time base chain is shown in Fig. 6-2. It is basically intended for operation on the 60 Hz power line. The input frequency is first divided by 6 and then by 10, to give the basic 1 Hz rate. This is further divided, to give multiples of 10 and 100. The input, or the output of the first divider can also be used, to give multiples of 1/10th and 1/60th of a second. In addition, the connections of the last stage can be changed, to give a divide-by-6 counter. This gives a time base of minutes. The desired rate is selected by a two-pole, six-position switch, one pole making the change from seconds to minutes base.

The counter connections are the normal ones for the type indicated, except for the last stage. As shown in Fig. 6-3, this uses a pair of diode AND gates to force the counter to reset on a count of 4 + 2, or 6, to give the minutes count. This gate is shorted out for all other settings; the counter output runs at the 1/1000 second rate for all settings except the minutes setting.

## THE COUNT-DIGIT SET CHAIN

The output of the time base, at the selected rate, is fed to a

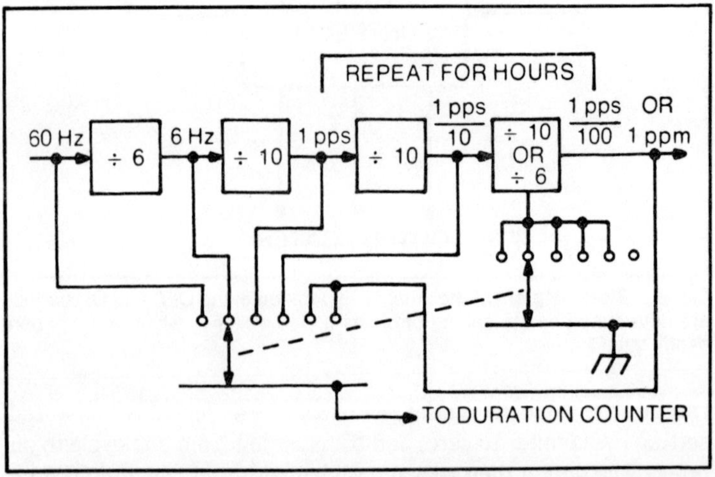


Fig. 6-2. The time base chain for the counter. The nominal rate is one pulse-per-second, with multiples of 1/10, 10 and 100. Rates of one pulse-per-minute and the line frequency are also available.

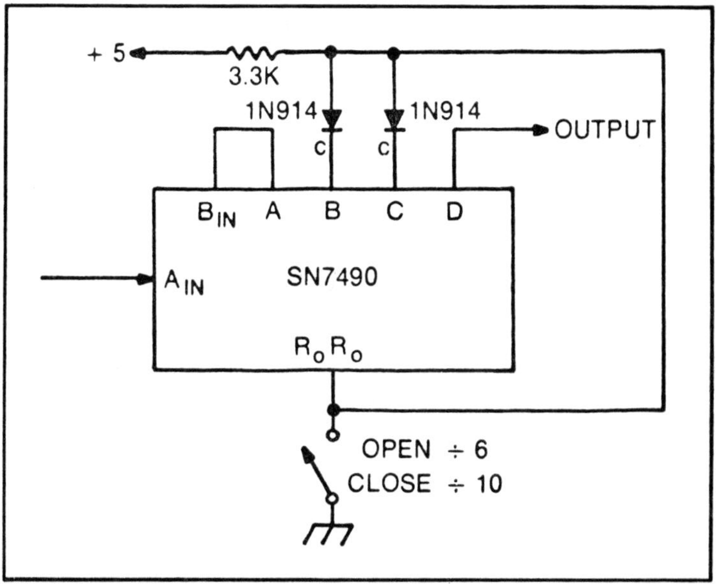


Fig. 6-3. Connection of a 7490 counter for division by 10 or 6, as required for the ×100 or minutes scale.

pair of cascade hexadecimal counters, giving a count capability of 2 to 255 units. As shown in Fig. 6-4, any of the eight outputs of these two gates can be fed to an 8-input NAND gate by a bank of switches. The setting of these switches gives the length of the count period. The switches are labeled in binary, the one connected to the fastest changing counter being labeled 1, the next 2, the next 4, and so on to 128. The switches are thrown in combination, the sum of the switch label values giving the count duration. For example, suppose the A and D switches of each counter are thrown. The sum is equal to 1 + 8 + 16 + 128, so the duration setting is 153 units of the time base setting. If the time base is on 1/10th second, the set duration is 15.3 seconds.

When the accumulated count reaches the value set on the switches, the output of the 8-NAND gate goes from the high to low state. This action occurs because all of the 8-gate inputs which are not connected are automatically high for the TTL logic family. However, until the preset count is reached, one or more of the switch-connected inputs are held low, so the 8-gate output is high. It remains in this state until all of the switch connected inputs go high.

If a simple one-cycle counter were wanted, this would be the complete circuit. However, to permit repeat cycle operation, with independent control of OFF and ON duration, a second 8-NAND gate is added and a second switch bank. These switches are also connected to the eight outputs of the hexadecimal counters. Switch labeling and operation are just the same, except this bank controls the off cycle duration.

## OFF-ON CYCLE CONTROLLER

The OFF-ON cycle controller has two major functions. One is to reset the two hexadecimal counters to zero at the end of each on or off cycle. The second function is to switch from the ON bank of switches to the OFF bank of switches at the end of the on cycle, and vice versa. As shown in Fig. 6-5, the transfer from one switch bank to the other is accomplished by a 2:1 divider, which changes

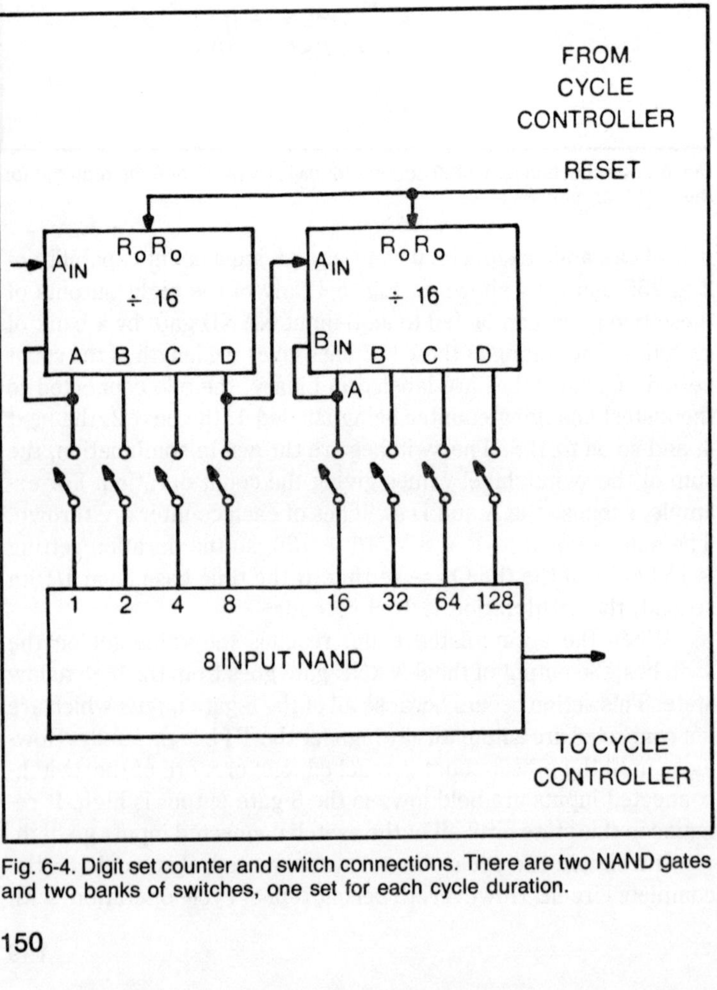


Fig. 6-4. Digit set counter and switch connections. There are two NAND gates and two banks of switches, one set for each cycle duration.

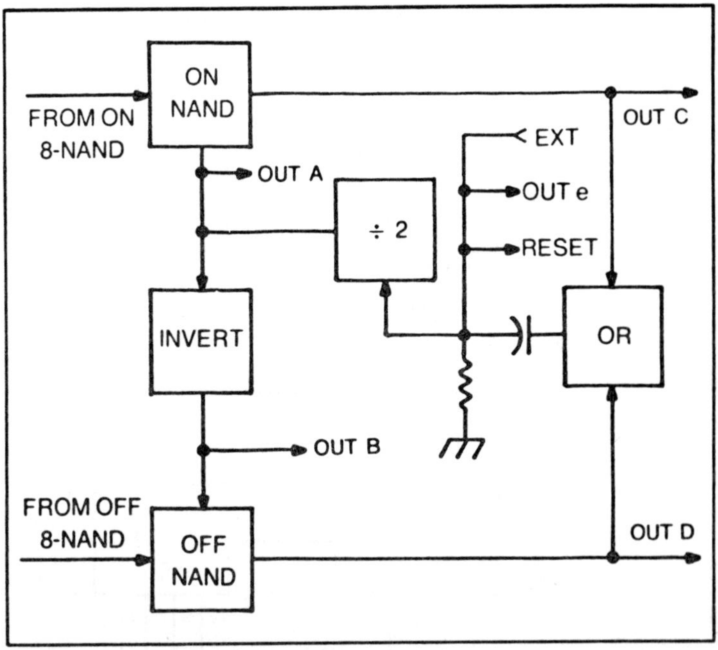


Fig. 6-5. The cycle selector, which changes from one switch bank to the other each cycle. Several output points can be used.

state at each reset pulse. The divider condition and its inverse are fed to two NAND gates, one for each of the 8-gate outputs. The outputs of these two NAND gates feeds a NOR gate, which generates the trigger for the 2:1 counter, as well as the reset pulse.

To see the action of this circuit, suppose that the output of the 2:1 counter is high. When the ON 8-gate output goes low, the ON NAND output goes high. The output of the OR goes high, triggering the binary divider after the short delay introduced by the differentiator, and also resetting the two hexadecimal dividers in the digit set chain. The 2:1 divider output is now low, so the ON NAND is inactive and the OFF NAND will not respond during the ON period even though the OFF cycle duration is to be shorter than the ON period; however, the OFF 8-gate does produce one or more extra pulses at its output.

There is another action available. Suppose that the ON cycle switches are set to give some interval, and all OFF cycle switches are open. The controller then rests in the OFF state. Closing one of the OFF switches momentarily starts the ON cycle. When this times out, the controller returns to the OFF state and again rests

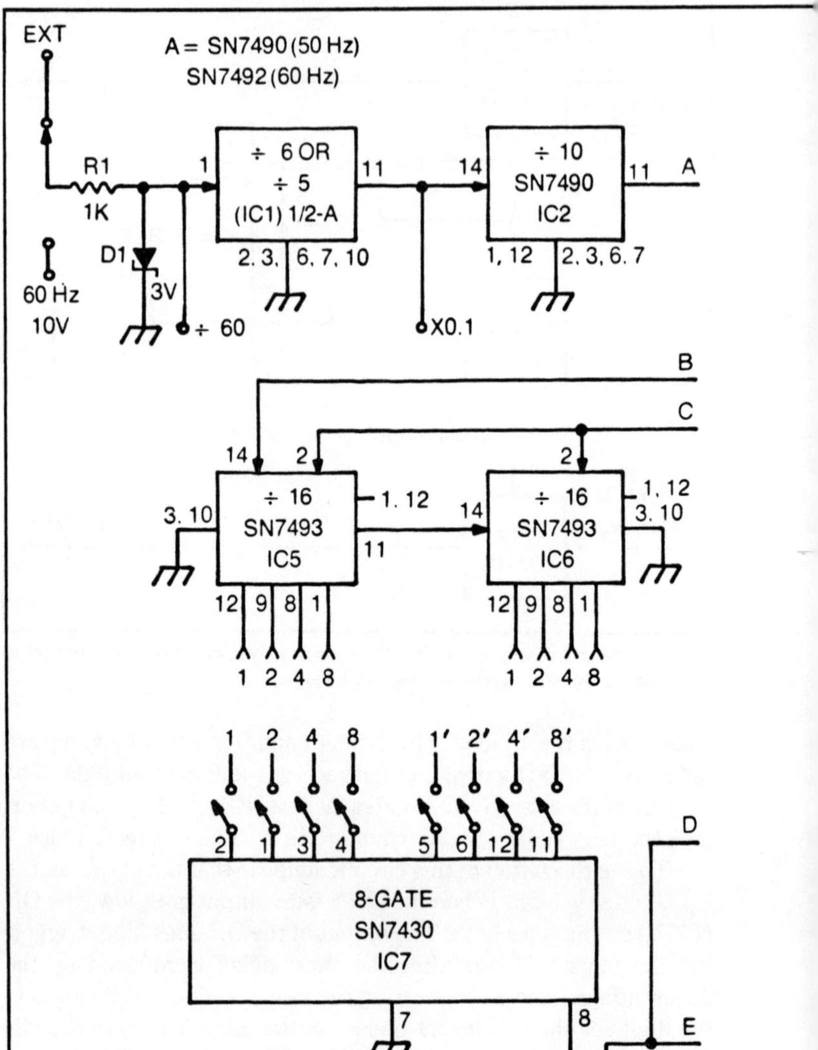


Fig. 6-6. Complete circuit diagram of the precision counter. In wiring, all points marked 1 are connected together, etc.

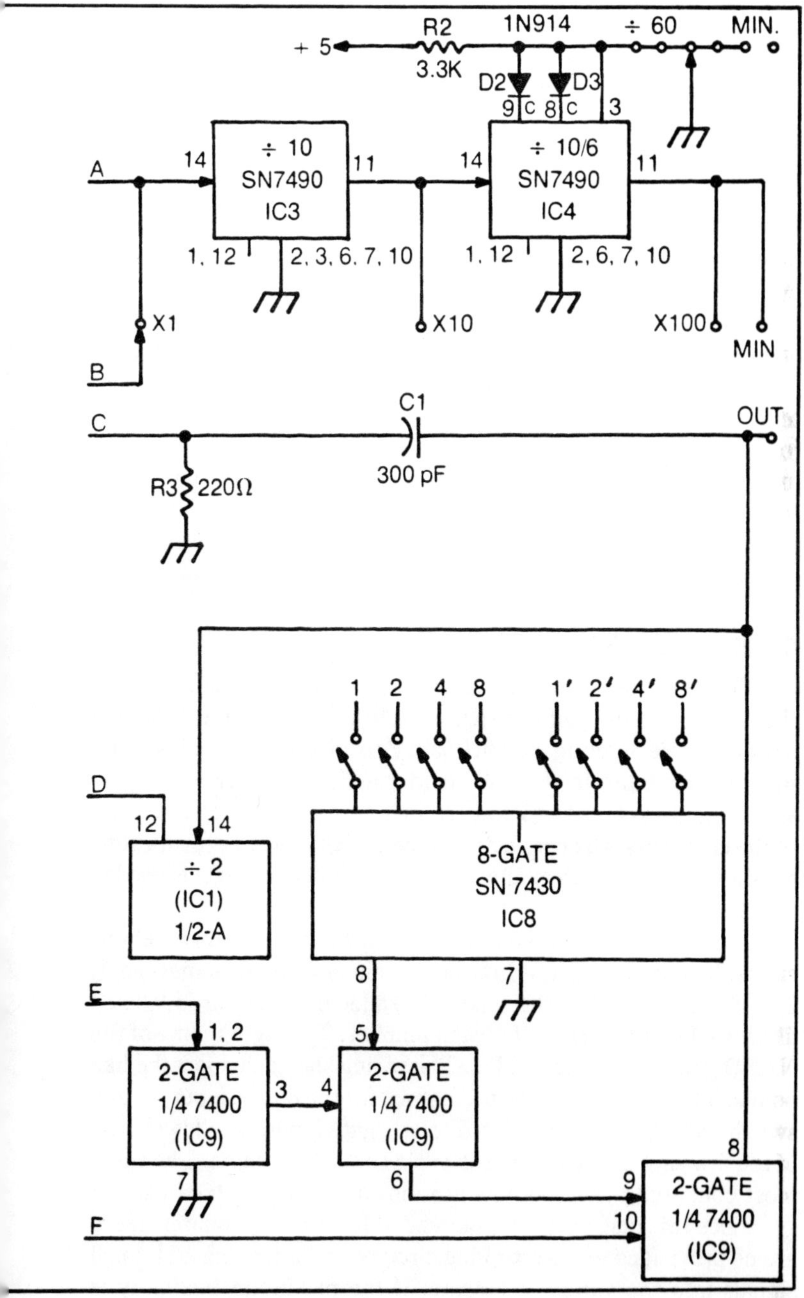
+ 5
R2
3.3K
1N914
÷ 60
MIN.
D2
D3
9 C
8 C
3
A
14
÷ 10
SN7490
IC3
11
14
÷ 10/6
SN7490
IC4
11
1, 12
2, 3, 6, 7, 10
1, 12
2, 6, 7, 10
X1
X10
X100
MIN
B
C
C1
300 pF
OUT
R3
220Ω
1 2 4 8
1′ 2′ 4′ 8′
D
12
14
÷ 2
(IC1)
1/2-A
8-GATE
SN 7430
IC8
8
7
E
1, 2
5
2-GATE
1/4 7400
(IC9)
3
4
2-GATE
1/4 7400
(IC9)
7
6
8
9
10
F
2-GATE
1/4 7400
(IC9)

there. This one-shot action can be used to generate either an ON or an OFF cycle. The action can also be initiated by external signals, for example, a positive pulse fed to the 2:1 counter.

## TIMER OUTPUTS

Several output points can be used, as required by external circuitry. The normal output is labeled as OUT A in Fig. 6-5. This point goes high during the ON cycle and is low during the OFF cycle. Its inverse is also available, and is labeled OUT B.

OUT C and OUT D provide alternate output points. And finally, OUT E provides a pulse signal at the end of each of the cycles.

If desired, outputs can also be taken from the hexadecimal counters, or from the 8-NAND gates. Various pulse chains can be formed, some continuous, some interrupted, and some corresponding to the end of a cycle. The interrupted pulse chains have durations and interrupt periods determined by differences in count switch setting. Additionally, of course, the time base outputs can be used externally.

## COMPLETE CIRCUIT

The complete circuit is shown in Fig. 6-6. The time base is shown across the top of this figure. Note that the input is provided with a simple limiting circuit, partly to give a reasonably good square-wave input for reliable divider action, and partly to serve as the input protective circuit. A switch is provided at the input so the counter can be connected to the 60 Hz source or to an external signal generator. A minimum of approximately 4 volts is needed for reliable operation.

The central part of the drawing is devoted to the digit set circuitry. The output of the first hexadecimal counter is marked 1, 2, 4, 8. The output of the second hexadecimal counter is labeled likewise, but with primes for each number. The eight inputs of the NAND gate are also so labeled. The wiring should connect the like points. There is no order or preference for the 8-gate input, so any switch output can be connected to any gate input. The lower part of the drawing is devoted to the cycle control elements. The three most commonly used outputs are indicated on this drawing.

Since this instrument was intended for use as a counter, there are no provisions for interrupting circuits, or for feeding additional signals into the various elements. If the maximum flexibility is desired, circuit interruption can be provided by jumpers.

## NOTES ON THE CIRCUITS

While the basic circuits shown can be used for any logic family, the particular circuit connections are intended for TTL units. For these, an unconnected input automatically assumes the high state. Use of this fact is made in setting up the switch circuits, with simple circuit opening switches being used. In other logic families this simple technique does not work, since unused gates must be connected to ground, or to the supply. For these, the switches must be of the single-pole, double-throw type. Actually, because suitable double-pole, double-throw switches are available in surplus at low cost, it is easy to change the circuit to connect the unused gates to ground or to a positive voltage. The positive voltage is correct for the TTL logic family, and would give slightly greater noise immunity than the simple circuit opening arrangement used. Also, if desired, the second pole of the switches could be taken to external circuitry, for example, to a remote indicator.

Since the prototype was intended for use only with TTL logic elements, there is no output interface circuitry as such. If other logic elements are to be used, or if the timer is to control power circuits, interfaces will be required. They may take the form of drivers, level changers, high voltage elements and so on. Some of the possible forms will be described later.

No protection is provided for the output circuits. Also there is no protection for the 5-volt supply. Series diodes could be installed to provide reverse voltage protection, or one of the "on board" voltage regulators of the '309 family could be installed, if these protections are needed.

## CRITICAL ELEMENTS

There is one circuit which is fairly critical, and one which may be a source of trouble. The critical element is the reset chain of the count accumulator dividers, which are capacity coupled in the prototype. If the shunt resistor in this circuit is too low in value, the reset is not reliable or may not occur at all; if the resistance is too high, the counters may remain in the reset condition. There is a reasonable range of resistance which can be used, with the specified value being approximately in the center. However, if trouble is encountered, due to the characteristics of a particular counter, a change in resistor value may be needed. Problems in this circuit would be increased if there were leakage in the coupling capacitor.

These problems could be avoided if a TTL inverter were in-

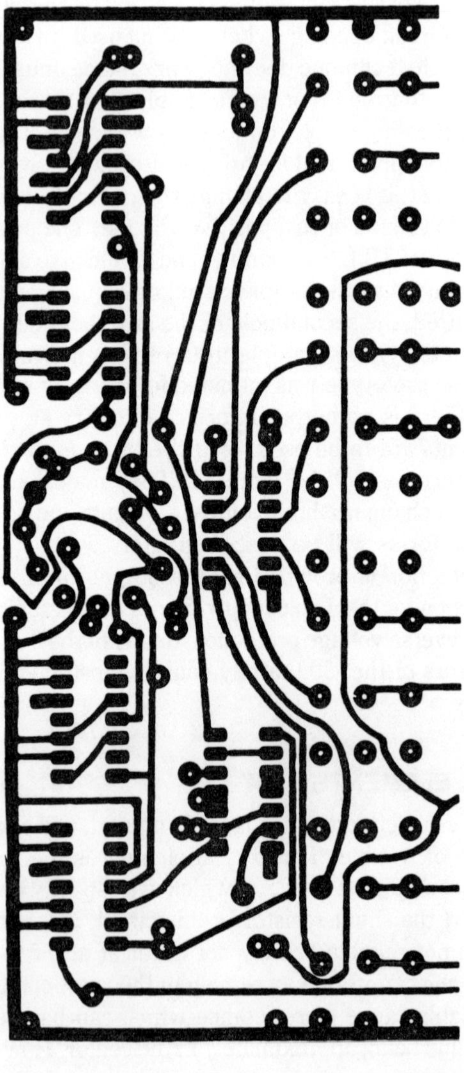

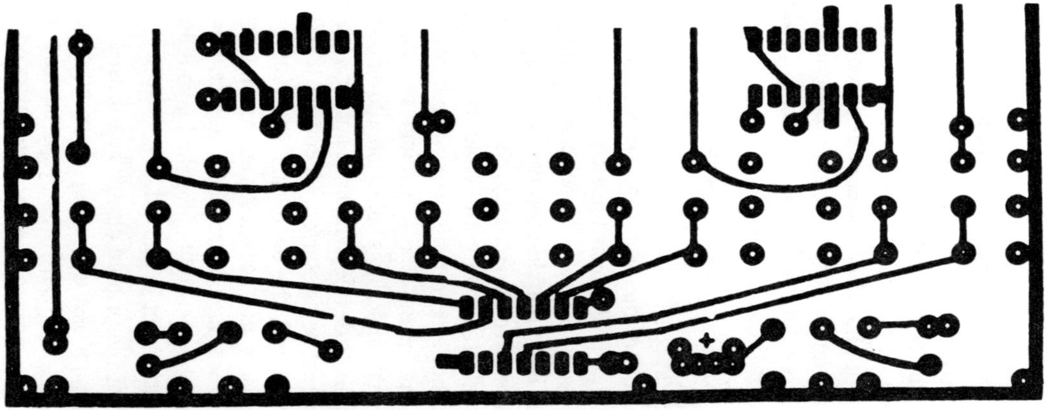

Fig. 6-7. Printed circuit board layout for the counter. The prototype board was prepared from a positive photosensitized board.

stalled in place of the simple differentiator. There is ample room on the circuit board for this. The remaining sections of this inverter could be used as output buffers, if desired.

One source of trouble which might be encountered is common to all counters of the reset family, and arises if reset delays are large. Slivers of counts can then appear in the output, generating false action in subsequent stages. It can also occur with counters having internal preset, a well known phenomenon. If this trouble is encountered, the solution is to replace gates and counters.

There should be no problem with cross talk between the off and on cycles. If it should be encountered, look for such factors as resistance in common leads, capacity coupling between leads, or poor voltage regulation. Signal tracing with an oscilloscope should find any of these problems.

## CONSTRUCTION

All components of the prototype are mounted on a single circuit board, 4 1/2 × 5 3/8 inches in size. The major factor in controlling board size is the type of switches employed; this size is adequate for the large double-pole, double-throw rocker switches widely available in surplus. Actually, the size of the printed circuit board was also selected to be compatible with the standard instruments of the author's preceding book. The board requires a case size just twice that of the 3 × 4 × 5 inch minibox used for most instruments of that book. This selection allows compatibility with the other instruments developed for the author's laboratory, including a new series just being started. Of course, if suitable miniature switches can be found, the board size can readily be reduced to one-half of its present size, so that the entire timer could be mounted in a single 3 × 4 × 5 inch chassis box. In fact, it is recommended that the switches be secured before final layout and board preparation of the instrument is started.

The layout of the printed circuit board used is shown in Fig. 6-7, with the major feature being the mounting holes for the switches. The original board was prepared from photosensitized board, printed from a transparency. Equally good results could be obtained from the ink resist type of transfer, and satisfactory results could be obtained with hand layout, although the large number of 10 pins makes this a tedious operation.

The placement of the DIP packages, and other components is evident from the drawing of Fig. 6-8. This also shows the location of the input and output points as used in the prototype.

As supplied, all the switches did have ears. To save space, these were removed from all but four of the switches. The remaining ears become the mounting pads used to hold the printed circuit board assembly to the mounting case.

The layout of the mounting case is shown in Fig. 6-9. This layout is predicated on the use of a case which is identical in size to two of the 3 × 4 × 5 inch minibox chassis. While these double-size boxes were readily available at one time, the writer was unable to find any locally and so made a box of sheet metal, using a small metal brake. The box was made in the form of two U channels, one for the front panel and two ends and the second for the back and two sides. Any other convenient construction can, of course, be used.

## CHECKS AND ADJUSTMENTS

There are no adjustments as such in the timer. However, because of the opportunity for the usual problem of poor soldered joints or solder bridges on the integrated circuits, and the effect of these on the accuracy of the timer, careful checks of the timing chain is needed. The most convenient way to do this is with a fairly high frequency source, since this allows use of an oscilloscope to observe the dividing action. The suggested method of checking is to start with the time base. Check first the initial divide by 6 counter, ensuring that this is actually resetting on the 6 count. Each successive counter can be checked easily by increasing the input frequency to give the same output rate, counting the number of input cycles to each counter stage for one output cycle of that stage. The last stage of the counting chain should be observed for the correct count of 10 and count of 6 action as required for the minutes conversion. Improper action in this counter is probably due to a diode being connected in the reverse direction, or to an open or shorted diode.

When the timing chain has been checked, the signal progression through the period counters should be observed with all time selector switches set to the on position, which allows the two counters to work to their full capacity. There should be an output from each of the eight output points: these are in the 1, 2, 4, 8 sequence. When this is observed, the switch setting should be changed and the timing action again checked. Note first that there is no output for stages beyond the highest switch setting; for example, if one

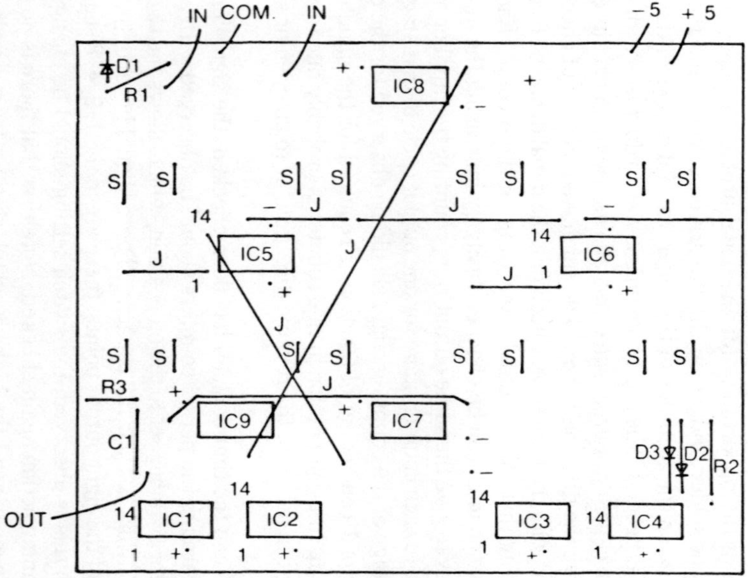


PARTS LIST

C1—300 pF disc ceramic capacitor
D1—3-volt Zener diode. 1N4372 (ADVA Electronics. Box 4181K. Woodside. CA 94062) or equal
D2 D3—1N914 diode (Radio Shack 276-1144 or equal)
IC1—SN492 divide by 12 counter (Radio Shack 276-1819 or equal) for operation with 60 Hz input. or for 50 Hz input. SN7490 decade counter (Radio Shack 276-1808 or equal) See text
IC2 IC3. IC4—SN7490 decade counter (Radio Shack 276-1808 or equal)
IC5. IC6—SN7493 4-bit binary counter (James Electronics. 1021-A Howard Ave . San Carlos. CA 94070) or equal
IC7. IC8—SN7430 8-input NAND gate (Lafayette 32-06810V or equal)
IC9—SN7400 quadruple 2-input NAND gate (Radio Shack 276-1801 or equal)
S—switch. single-pole single-throw

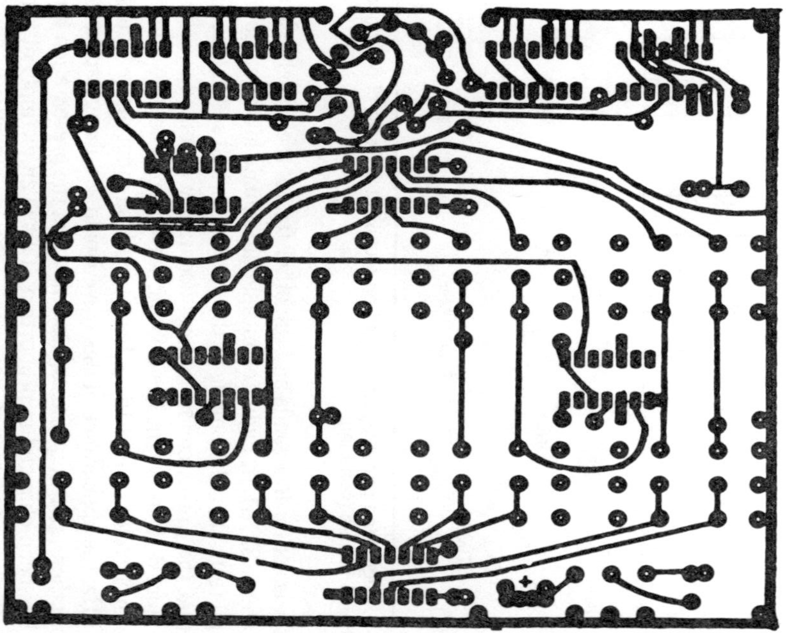

Fig. 6-8. Component placement for the counter. Connect + and – points to the buses.

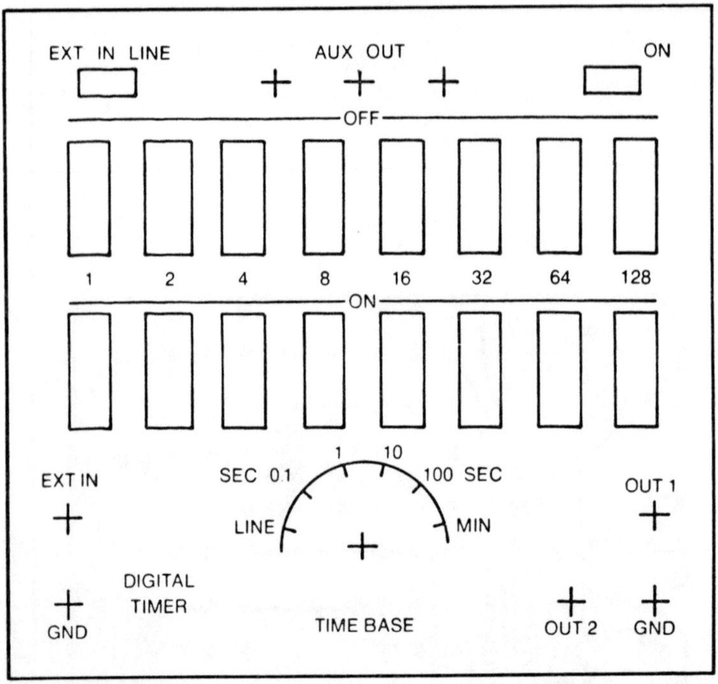


Fig. 6-9. Suggested panel layout, based on a cabinet the size of two boxes 2 × 3 × 5 inches (nominal).

counter is set for 8 seconds and the other for 16, the output corresponding to 32, 64, or higher counts should remain low at all times. Enough tests should be made of the various switch settings to ensure that the preset count is being accurately obtained.

Figure 6-10 shows the basic output wave form when the counter is set for 8 cycles on and 16 cycles off, with an input in the vicinity of 50 kHz; the appearance of this wave would not change appreciably for other input frequencies, assuming that the presentation time base has been changed in proportion. This is an example of time determination, and also of frequency division. Figure 6-11 shows output of one of the 8-gates with unequal time setting. This output point is occasionally useful in generating a class of digital words. Note that the number of digits present is affected by the setting of both switch banks.

If these and their equivalent presentations for other switch settings are not obtained, it is suggested that the reset bus be grounded, which will cause the counter chain to run continuously, and cause one of the 8-gates to give a steady stream of outputs.

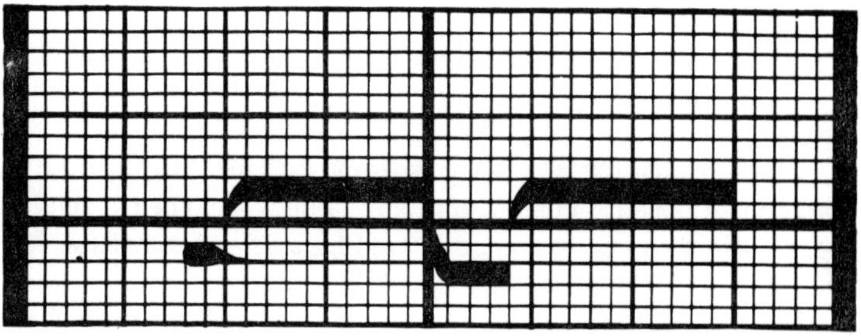

Fig. 6-10. Use of the counter at 50 kHz as a variable rate frequency divider. The setting is for 8 cycles on, 2 off. Division up to about 25 MHz is possible.

Removing the ground and applying a short voltage step should transfer from one NAND gate to the other. With techniques such as these, it should be no great problem to get the counter operating properly even if trouble is encountered.

## OUTPUT DRIVERS

When used with TTL or most of the MOS logic families, the timer needs no interface elements, since the voltage levels and current capabilities are compatible with these families. As noted, outputs may be taken from any point in the timer. Incidentally, signals may also be fed into the timer, if this is needed to secure a desired action.

With other devices, an interface element is usually necessary. Actually, this may be desirable even with TTL elements, for increased drive, for isolation, or to secure better noise immunity.

Several of the possible interface elements are shown in Fig. 6-12. The first shows inverting and noninverting buffers. These use high-voltage open-collector devices and can be used with logic

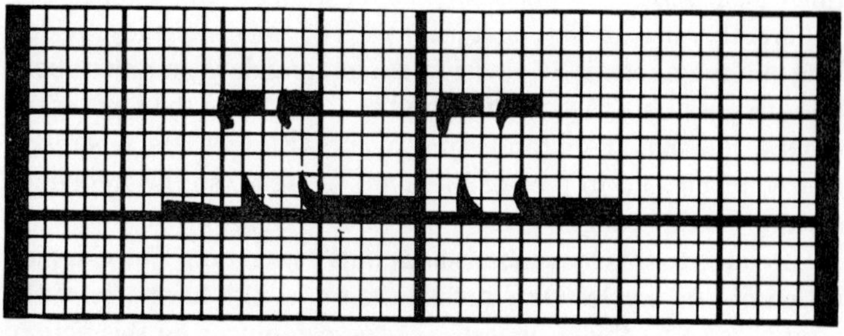

Fig. 6-11. Output of the OFF-NAND gate with switches set for 8 on, 4 off. Various patterns can be developed, simulating digital words.

families operating at much higher voltages than the TTL types. Other packages besides those indicated may be used, for example, the 5413 Schmitt trigger for waveshaping, or the 5433 for higher current.

Simple transistor buffers are shown in Figs. 6-12(B) and 6-12(C), with connections for current sourcing or sinking. The capacity of these buffers can be extended by adding a power transistor connected as a Darlington pair.

Where real power must be handled, it is difficult to improve on the capability of a relay. Figure 6-12(D) shows a possible method of use, with a 5406 IC as a driver. This relay would need to be of a fairly sensitive type, operating on 12 volts at about 25 mA. Surge protection as shown is a good safety feature, although the IC specified is a high voltage type. Relays may be cascaded if still more power must be handled, the relay delays being compensated for, of course.

An alternate to the relays is a family of controlled latching semiconductor devices. Figure 6-12(E) shows the use of a silicon-controlled switch (SCS) to control the power applied to a load. Many devices of this general family are available—for applications see the literature on their use.

## USE OF THE TIMER

The obvious use of this instrument is to secure definite time intervals. Periods up to a third of a day, at minute intervals, are available, and shorter periods in smaller increments can be set. The OFF or ON can be set, up to capability limit, or OFF and ON periods can alternate.

If fed with frequencies other than 60 Hz, the timing action can be speeded up or slowed down. For example, with 600 Hz input, timer scales are multiplied by a factor of 1/10, and so on.

Another use for the counter is as a precision frequency divider. The minimum ratio is 2:1 division. All integral division ratios above this are possible, 3:1, 4:1, 99:1, and so forth, up to the limit of the timers. The maximum ratio is just over 500:1 divisions. Operation of frequencies up to tens of MHz are possible, so the frequency divider is wide range.

Related to this operation is harmonic generation. The chipping action of "m" cycles OFF caused the (m + n)th harmonic of the time base to be the strongest. See the references on this action.

The timer can be used to develop a series of pulse trains. Some

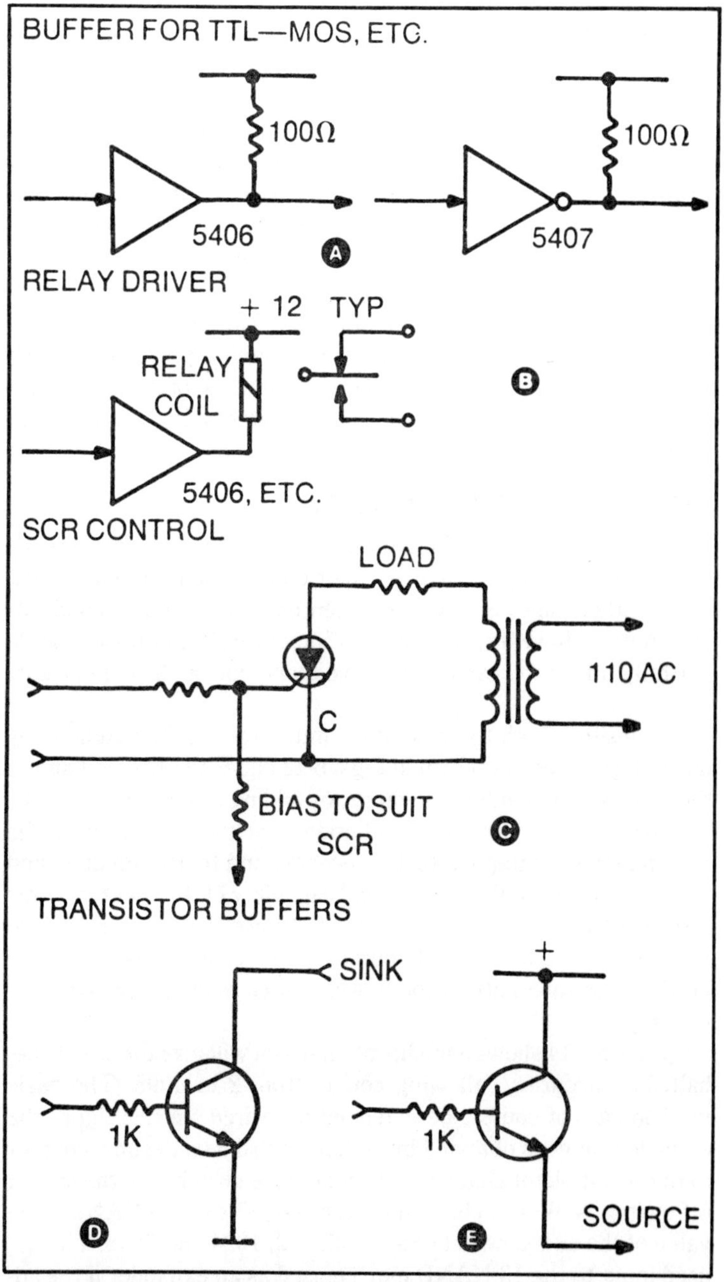


Fig. 6-12. Output drivers for interfacing several types of external circuits. Buffers are not needed if the load is no greater than a few TTL inputs.

are available at the outputs of the hexadecimal counters. Others appear at the output of the 8-gate. Many of the useful wave trains occur when there is an interval in the switch settings; for example, when the 4-16-64 positions are set but none of the others. Use an oscilloscope to study the formation of these wave trains. Some of these pulse trains can be used as digital words for some purposes. Various other digital words can be formed by using AND gates, fed from two or more of the counters.

The prototype counter has no provision for presentation of the accumulated count or time interval. The outputs of the two count accumulators are, of course, binary and could be presented directly by adding LEDs and a set of drivers, or could be presented in various forms by using some of the decoder-display combinations available.

## EXTENDING TIMER CAPABILITY

While the timer has high precision, that of the basic reference source, it has only two and one-half digit resolution. The range, on the other hand, is large, from tenths of a second to hundreds of minutes. However, the combination of range and resolution do not fully utilize the capability of the technique or the components used.

If desired, both the resolution and range can be extended by using the techniques shown in Figs. 6-13 and 6-14. The first shows the method of extending the range, by adding more counters to the time base chain. With the two additional packages shown here, the multipliers are extended to $10^4$ seconds, and to $10^2$ minutes, and hours are made available. The total capability of the counter is now well over two million seconds, over 555 hours, that is, over 23 days. At these long durations, of course, timer reliability becomes important; the reliability of the power source may be the dominant factor in this.

Figure 6-14 shows a modification which will give three and one-half digit accuracy, allowing counts from 2 to 4095. The basic method can, of course, be extended if desired. In this figure the extended range is obtained by adding a third hexadecimal counter to the accumulator chain, adding four more switches to the on and off switch bank and changing the 8-NAND to a 13-NAND. The value of the added switches are 256, 512, 1024 and 2048. The unused input to the 13-NAND may be used as an expander if the addition of more digits is needed, or it may be connected to the plus bus through a small resistor.

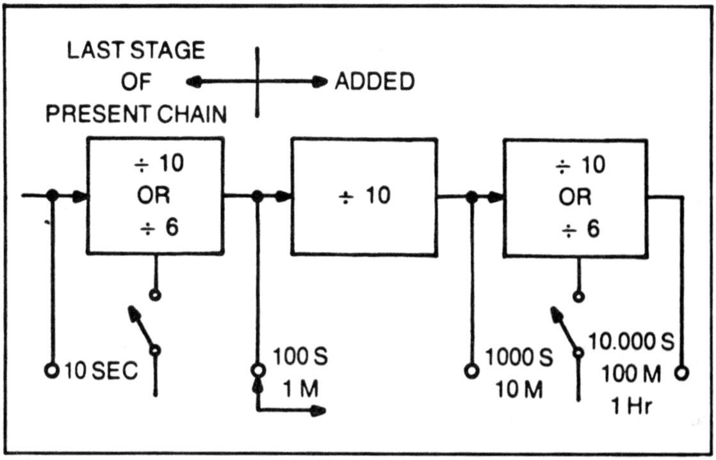


Fig. 6-13. Extension of the divider chain for range extension. With two added dividers, the range can be extended to 255 hours OFF or ON.

Of course, these extensions may be used with other reference frequency inputs, to give greater time resolution at shorter intervals. For example, operation from a 1 kHz source would permit resolving one second to an accuracy of one millisecond.

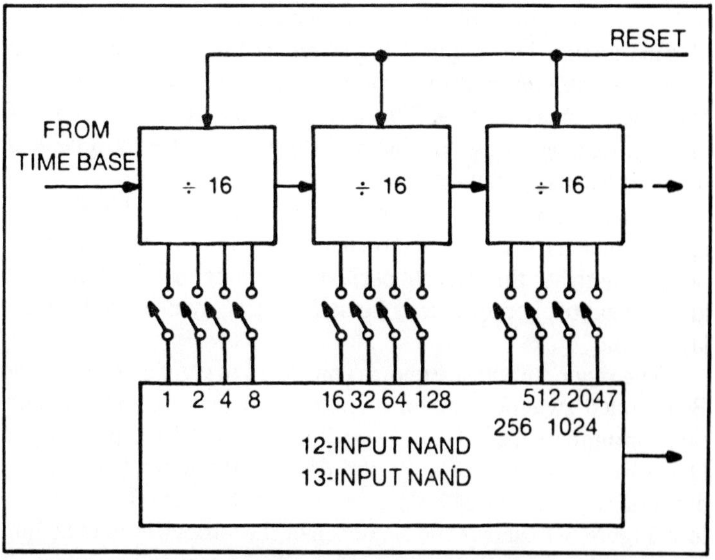


Fig. 6-14. Extension of resolution by adding another 4 digits of switching, and changing to a 12 or 13 input NAND. Time can be set to over 4000 seconds, by one second intervals.

Various other methods of using the device will occur as its capabilities become appreciated.

## COMMERCIAL COUNTERS AND TIMERS

The digital timer continues to increase in importance as more and more control, measuring, and computing functions are based on digital rather than analog techniques. As a result, most manufacturers of integrated circuits have added one or more specialized types of counters to their standard line. The following covers the major points of two such timer chips from one manufacturer and illustrates the capabilities which are now available.

## INTERSIL ICM7240 FAMILY

The Intersil ICM7240 family of timers/counters is basically a programmable eight-stage counter, plus control circuits and a built in RC oscillator (clock generator). Members of the family count in binary, in binary coded decimal, or in seconds, minutes and hours. Maximum division ratios are 256, 100 or 60, respectively, for the three types. Multiple periods are obtained by cascading chips.

As for all MOS devices, the maximum divide rate varies with supply voltage. It is about 5 MHz at 5 volts. However, if full control is to be secured, the count rate must be limited to about 100 kHz maximum. This requirement is imposed by the need for simultaneous signals for control purposes. This signal is easily obtained from the on-chip oscillator. This has reasonable stability, about 250 ppm per degree Centigrade, but an external oscillator must be used for high accuracy applications.

The counter is very easy to use in moderate resolution applications. The basic circuit is shown in Fig. 6-15. Two external components activate the internal oscillator. Timing may be started by an external trigger, or it may be self-starting by use of a time delay circuit.

The timer outputs are shown connected to eight switches. Two BCD coded switches, or, for a specific application, solid connections or jumpers may also be used. These output circuits are Open Drain FETs, so the output is the AND of the individual counter outputs, which is low as long as any one of the counter outputs are low. Figure 6-16 shows the sequence of the various individual outputs, starting from the trigger pulse which resets all counters. A great variety of pulse patterns can be developed by changing switch settings. A few of these are shown in Fig. 6-16. These and others

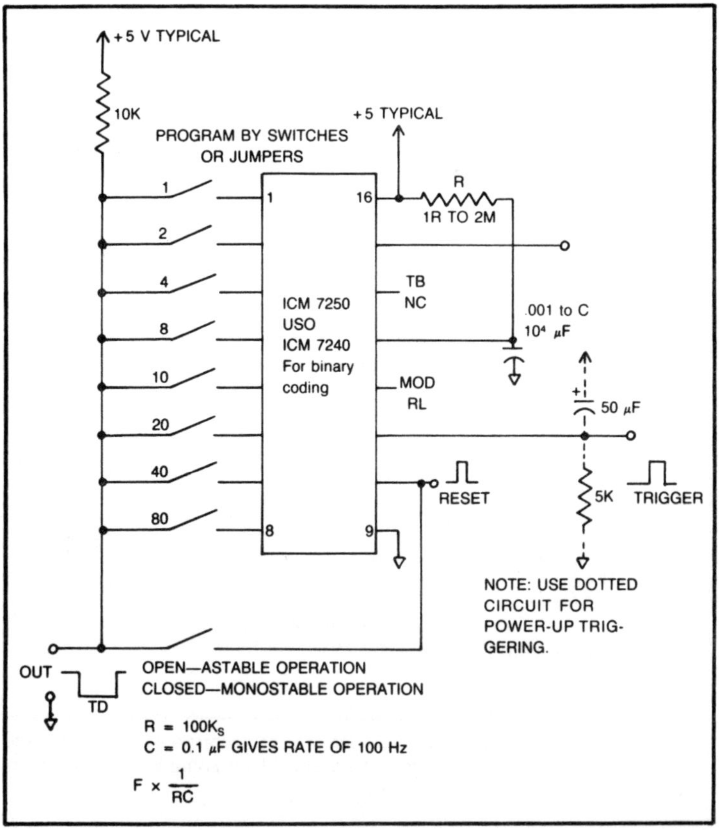


Fig. 6-15. A simple timer based on the ICM 7250 special chip. Basic time setting is derived from two banks of four counters each. In the 7250 the four counters are connected to give a reset on ten, so the maximum count is 99. The 7240 can count to 255, coded in HEX. The output is the AND of closed switches. See specification sheet for other capabilities of this chip.

can be developed by considering the AND function of the output whose switches are closed. There may be a single pulse repeated at intervals, two or more in a chain, or alternating long and short period groups. Alternately, with BCD switching, there may be only a single pulse given by the binary sum of digits.

The application note available gives several variations on this basic circuit, including laboratory and computer use. The complete list of applications covers the range from microseconds to days.

## INTERSIL 7045/A TIMER

The Intersil 7045 precision decade counter is a much more com-

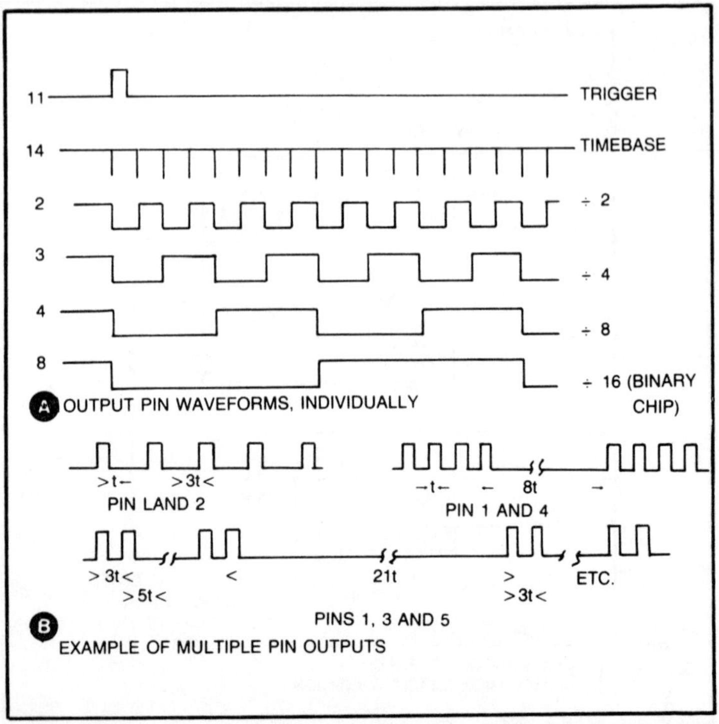


Fig. 6-16. This figure shows the formation of output signals by the AND of the individual timer stages. See Figs. 6-10 and 6-11 for some waveform photos of this type of counter.

plex unit, since it includes an oscillator, frequency divider, multiplexer, decoder, and digit/segment drivers on a single chip. Some of the possible applications are precision timers, a four-mode stopwatch, and a 24-hour clock.

The basic block diagram of the unit is shown in Fig. 6-17. The multiplexer and decoder functions relate only to display generation. Actual counting and control is handled by the low frequency counter and latches, as instructed by the controller. For good resolution and accuracy with low crystal cost, a 6.5536 MHz crystal plus a prescaler divider is used.

The application of this chip to a clock is illustrated in Fig. 6-18. The connection to pin 14 functions as a power-on reset. Start and stop operation is possible, valuable in setting to an exact time. A useful feature is that the minutes advance and hours advance are separate, and not combined, as in the usual clock chip.

Modifications in the external connections convert the chip to

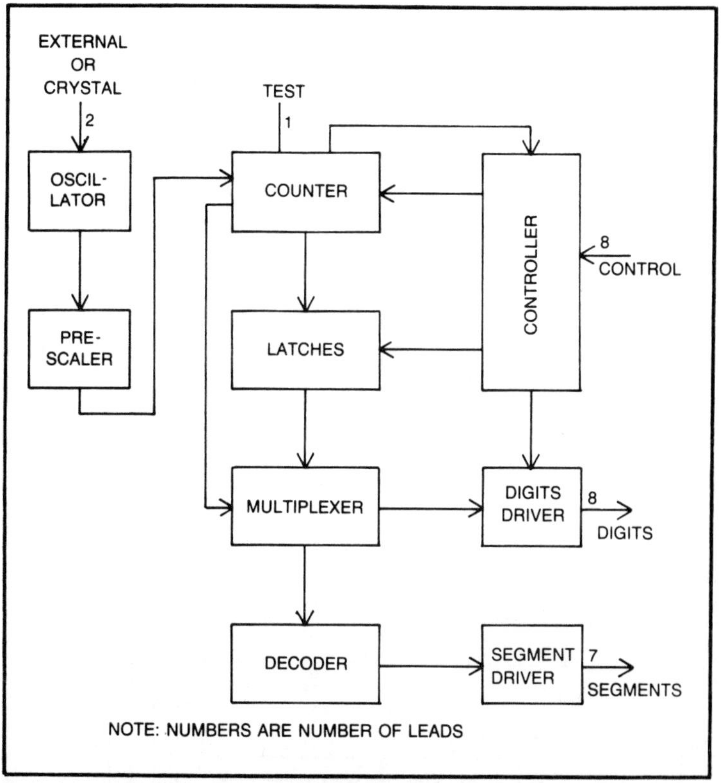


Fig. 6-17. Block diagram of the ICM7045 Timer chip, a design of good flexibility. The heart of this lies in the freedom built into the controller.

a four-mode stop watch. These are shown in Fig. 6-19. Note that an eight digit display is used, covering the range from tens of hours to one-hundredths of seconds. This wide range capability is one of the features of good digital timers.

The control circuits include a four position selector for the mode, a pair of circuit opening switches to start or stop timing and another to cut off the display for battery conservation, and a momentary contact reset switch. The start/stop circuit can be easily connected for remote or automatic operation.

The standard mode uses one actuation of the start switch to activate time accumulation and a second actuation to terminate it, as in normal stop watches. A third actuation causes a reset—then a restart. The separate reset function can be used at any time.

The sequential mode is intended for use in events of more than one part. Timing starts with the first actuation of the start switch.

The second actuation stops and holds the display, but the internal counters instantly reset and start counting again. This second actuation would normally be the end of a lap, whose duration can then be read out. This can be repeated for each lap. Total time is the sum of lap times.

The split mode is similar, but does not reset the counters at switch actuation. The first actuation starts both counter and display. The second actuation stops only the display. The third actuation causes the display to jump to total time from start, and so on. At the last actuation, the total time is read.

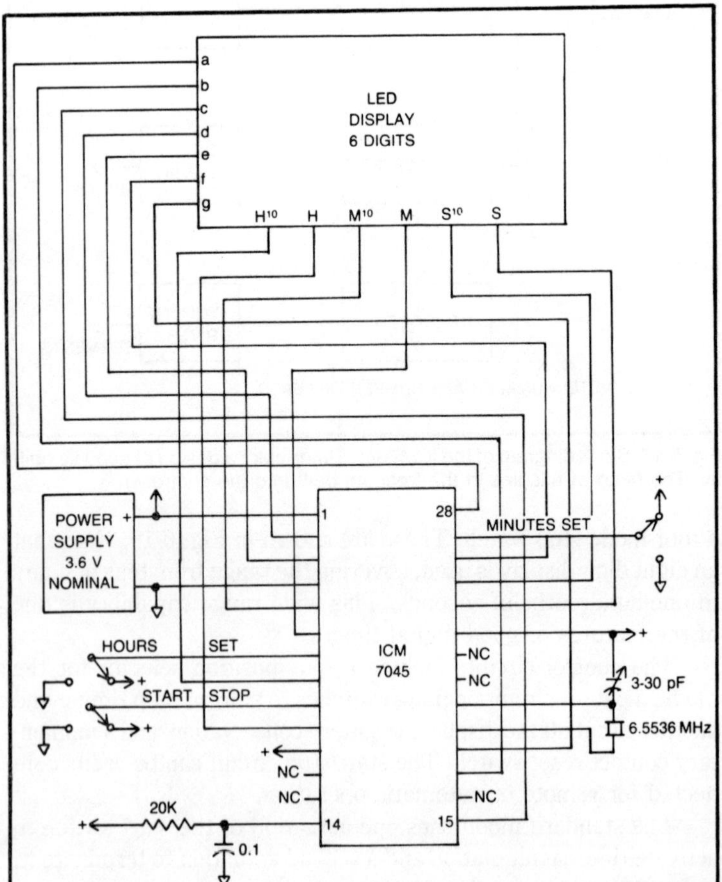


Fig. 6-18. Clock based on the ICM7045 Timer chip. Note that hours and minutes can be set independently, a departure from usual clock design. This basic design is also usable for the precision clocks of Chapter 8.

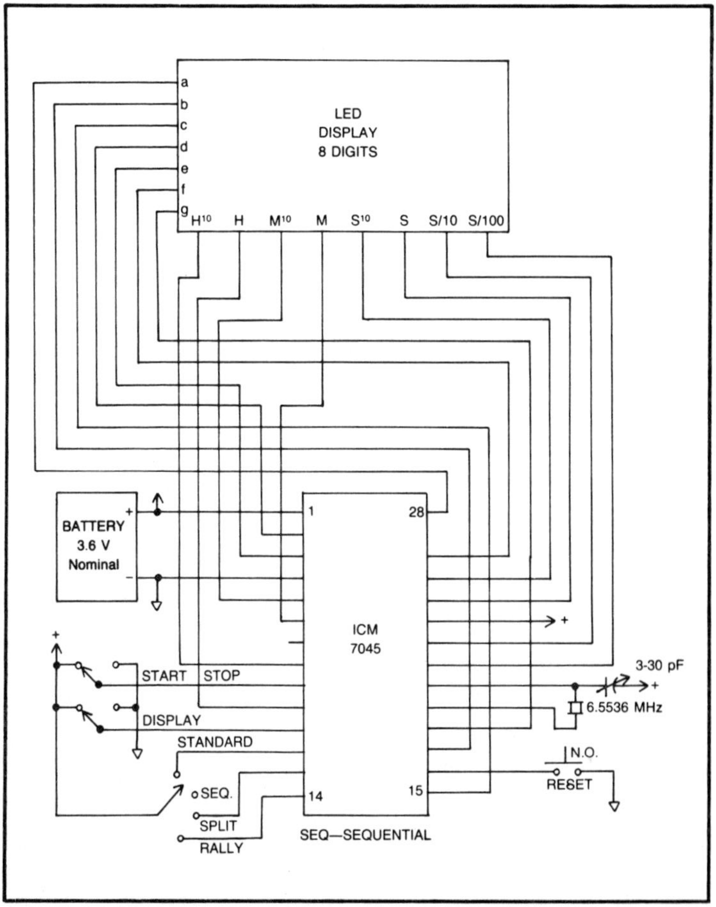


Fig. 6-19. A four-function stopwatch based on the ICM7045 Timer chip. See text for the principles of the four functions. The display can be turned off to conserve battery power. This can also be done with the clock of Fig. 6-18.

The rally mode is intended for events with an interval between laps. The sequence must start with reset. The start/stop switch is actuated at the start, and again at the end of each lap. Lap times are determined by subtraction.

There is automatic switching to minimize the chance of error due to accidental reset. Display off is possible in the standard and rally modes, to save power, but not in other modes.

Stop watches using this chip or an equivalent are available in kit and in assembled form. Prices are such that it is not attractive to attempt "construction from scratch" unless a nonstandard mode or a special application is planned. An example of this would be a special timer such as a giant display for school athletic events—a good way to earn the gratitude of students and coaches. Even there, a kit might be a good starting point.

# Chapter 7

# Some Specialized Clock Attachments

The basic purpose of this chapter is to show some attachments to basic clocks, additions which may be added for experiment or to make a different appearing clock, or to adapt the clock signals for other than display purposes. The first attachment is to change from the normal four-digit or six-digit display to a two-digit display, or if desired, even to a single-digit. Descriptions of methods of securing commonly needed signals follow, including a method of generating a one pulse-per-second signal from chips which do not internally provide this output; the method can be extended to develop an output of one pulse-per-unit time for any of the chip time signals. Also shown is a method of developing a nonmultiplex pulse at each digit time. Associated with this are methods of conversion from seven segments to BCD output and from BCD to seven segments. The seven segment-BCD conversion also provides decimal output. Finally, some attachments to produce audible or flashing light signals are described.

## TWO-DIGIT DISPLAY

The largest percentage of clock chips use multiplex display signals, with segment signals fed to the cathode of the display. Demultiplexing is done by combination of this set of connections and the timing of the segment and digit signals.

Additionally, it is possible to multiplex the display digits themselves, that is, to allow them to be used for more than one purpose—such as displaying hours and minutes to minutes and seconds, but this is not an automatic function.

Offhand, it would seem that a single chip display could be obtained by simply leaving its cathode on at all times, and feeding the segment signals to it in multiplex form. The multiplex oscillator would have to be operated at a very low rate, say one cycle each five or ten seconds, but the display would sequentially present each digit. The problem with this simple approach is that most clocks perform the multiplexing operation backwards: the seconds digit appears first, then the 10 second, then the minutes digit, and so on. Too much mental addition and sorting is needed to make this a practical method.

In any event, a two-digit display seems to be a better approach. It can show both digits of the hours, then both digits of the minutes, as shown in Fig. 7-1. In most applications, this is all that is needed.

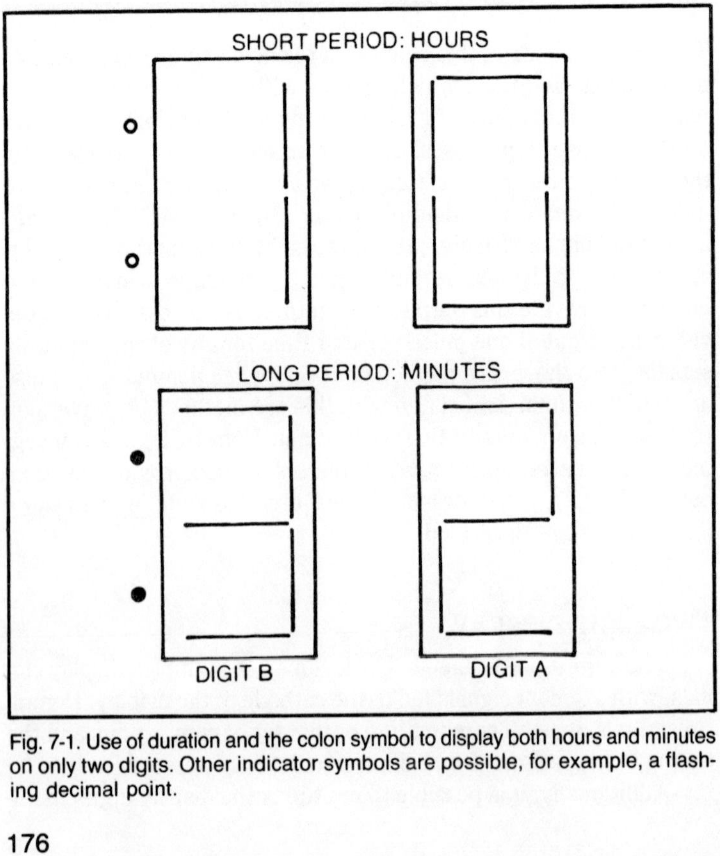


Fig. 7-1. Use of duration and the colon symbol to display both hours and minutes on only two digits. Other indicator symbols are possible, for example, a flashing decimal point.

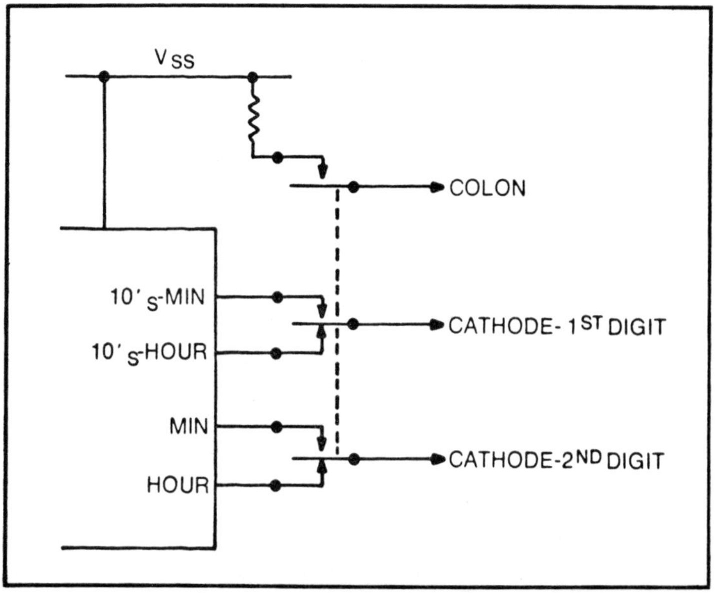


**Fig. 7-2. Basic switching principle for a two-digit display. For one digit, a single-pole four-position switch would be needed.**

If necessary, additional circuitry can be used to provide switching from the hours-minutes display to the minutes-seconds display.

Some method of telling whether the display is on the hours or on the minutes position seems necessary. Two methods are readily possible. One is to use a different time duration for the two display periods. In most uses of a clock, the minutes information is the most important, since one glances at the clock to see if there is time remaining to complete some task. Therefore, it would seem desirable that the minutes display be on for a longer period than the hours display. A usable choice might be two seconds for the hours display, then five seconds for the minutes display, then repeat. These periods can, of course, be adjusted for individual preference.

A second method is to use a symbol to indicate the difference. The standard symbol for separation of hours and minutes is a colon, so this might be adopted as the standard. The hours display could show the number only, and the minutes display could show the two numbers, preceded by a colon. The appearance of the display for the successive presentations would be as shown in Fig. 2-1. As shown in this figure, the time 10:32 would appear first as the numeral 10, then as the display :32. Much of the time there would be a separate indication, in that only one figure would appear in the hours timing. Of course, these two methods of time duration and symbol addition can be combined, and would seem to eliminate all problems.

All that is necessary to realize this display multiplex is a double-pole, double-throw switch, arranged as shown in Fig. 7-2. This switches the digit cathodes from the "tens of hours" and "hours" signals. The time duration of presentation is determined by switch dwell and is independently controllable. The schematic also indicates use of a ganged single-pole switch for presentation of the colon signals. In many cases, however, this colon signal can be developed directly from the switch drive signal.

The electronic form of one pole of this switch is shown in Fig. 7-3. This is the regular drive circuit of a discrete component demul-

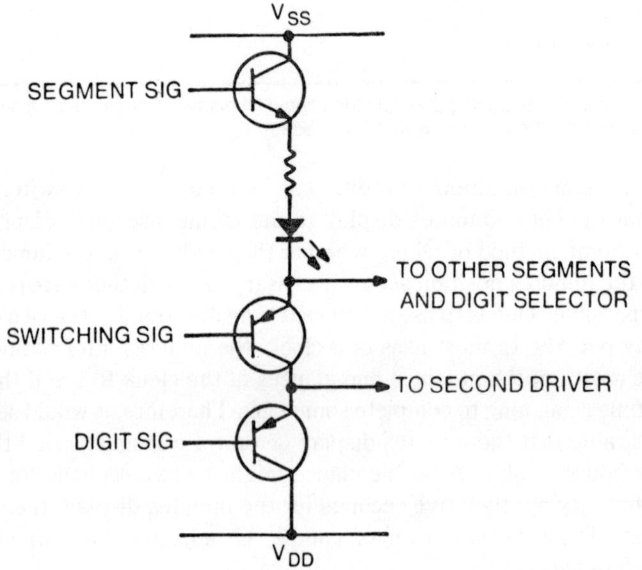


| SELECTOR LEVEL | DIGIT TIME 1 | 2 | 3 | 4 | DIGIT PRESENTATION A | B |
|---|---|---|---|---|---|---|
| HI | HR | 10 HR | - | - | HR | 10 HR |
| LO | - | - | M | 10 M | M | 10 M |

Fig. 7-3. Use of a transistor AND circuit to form one pole of a switch. The complete selection action is also shown.

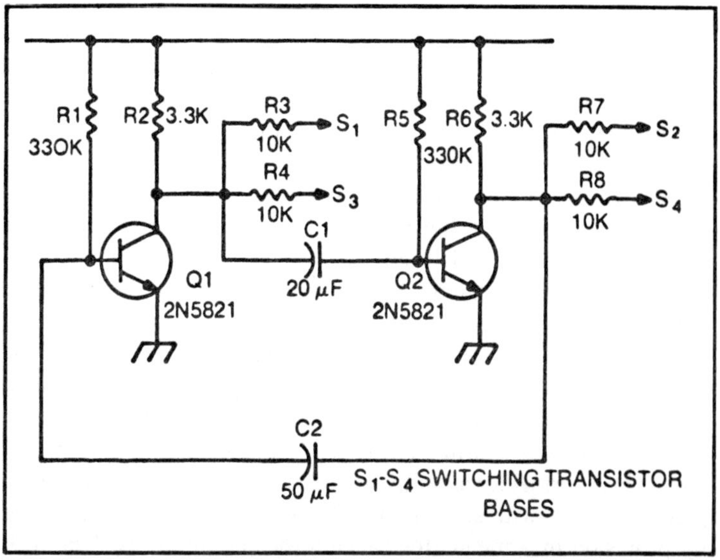


Fig. 7-4. A long period multivibrator to generate the digit selection signals. A third or a fourth 10K resistor, or both, could be added to drive colon LEDs.

tiplexing scheme for LEDs, modified by adding a series transistor in the cathode return line of each time digit. This, together with the regular digit-drive transistor functions as an AND circuit; both the added "digit select signal" and the regular "digit drive signal" must be present for the cathode circuit to be energized. The resulting switching action is shown by the truth table given by the same figure.

The digit-drive signal can be developed from a multivibrator. It does not need to operate in synchronism with the drive signal, since the eye cannot follow the multiplexing signal speeds. A free running, low speed circuit is perfectly satisfactory. A suitable circuit is shown in Fig. 7-4. It is straightforward in design, but there are two points which must be remembered in selecting components for it. First, the transistors must drive the next stage, and also provide an ON signal to two additional transistors. The transistors must thus be capable of handling appreciable current; and their collector resistances must be kept relatively low. In addition, relatively large capacitors are needed to obtain time constants on the order of five seconds, and so the capacitors must be low leakage; tantalum electrolytics would be the preferable choice.

If colon drive is also desired, it can be provided by another isolating resistor connected directly to the appropriate selector.

However, by using the little circuit of Fig. 7-5, the demand for additional drive can be avoided. This connection avoids a potential source of trouble, since it will be remembered that most LEDs have a very low reverse voltage capability. In this circuit the reverse voltage can be kept small.

A printed circuit board layout for this combination of multiplexer and display is shown in Fig. 7-6, with the parts placement shown in Fig. 7-7. This board is arranged to plug into the experimental clock of Chapter 4 in place of the six-digit display shown in that chapter. The board provides for differentiation of hours and minutes by time presentation duration, that is, by selection of the capacitor size, or separately by indication by use of a colon added to the displays by a pair of small LEDs, or by both means.

The basic wave form developed within the unit is shown in Fig. 7-8. This is normal for a multivibrator of this type. Note that if some pains are taken with stability, by choice of component quality, and by provision of a means for varying the timing period length, it would be possible to secure synchronous operation by feeding the number one digit signal to one of the multivibrator bases.

This general idea could be extended to a single-digit display, by using a four-stage ring counter in place of a two-stage flip-flop, as shown in Fig. 7-9. It could also be used to present two digits of a six-digit display by using a three-stage ring counter.

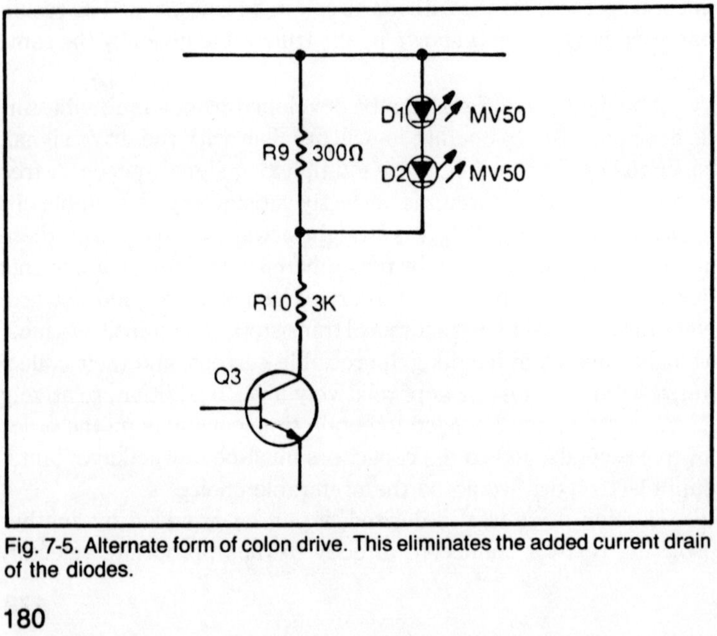


Fig. 7-5. Alternate form of colon drive. This eliminates the added current drain of the diodes.

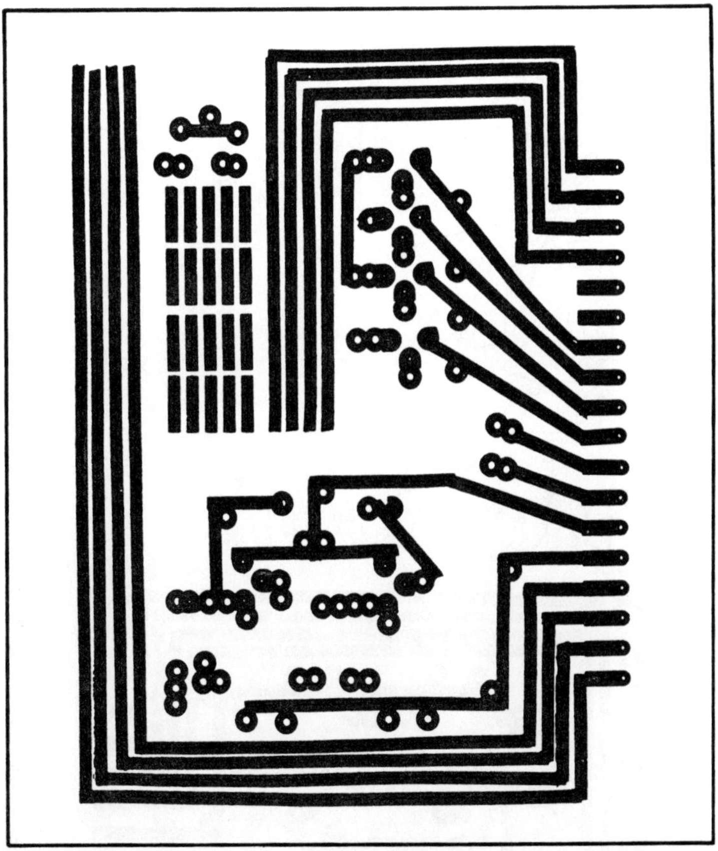

Fig. 7-6. Printed circuit board for the two-digit display. The board is interchangeable with the display shown in Fig. 4-1. The + and – leads must be added to the clock of this figure. Foil side, full size.

For another method of developing a single-digit display, see Fig. 7-10. One reference shows a method of developing synchronous switching, useful in some applications.

The multivibrator of Fig. 7-4 can be used in another fashion with chips which rearrange the presentation by a switch, such as the MM5316. This design normally presents hours and minutes, but operation of the switch causes it to present minutes and seconds.

If one of the multivibrator output leads is fed to the pin used for this switch (pin 32 of the MM5316), the presentation will switch at the multivibrator rate. Some method of indicating which data is being presented should be used; see the earlier comments on this.

Still another way of using the multivibrator is to blank the dis-

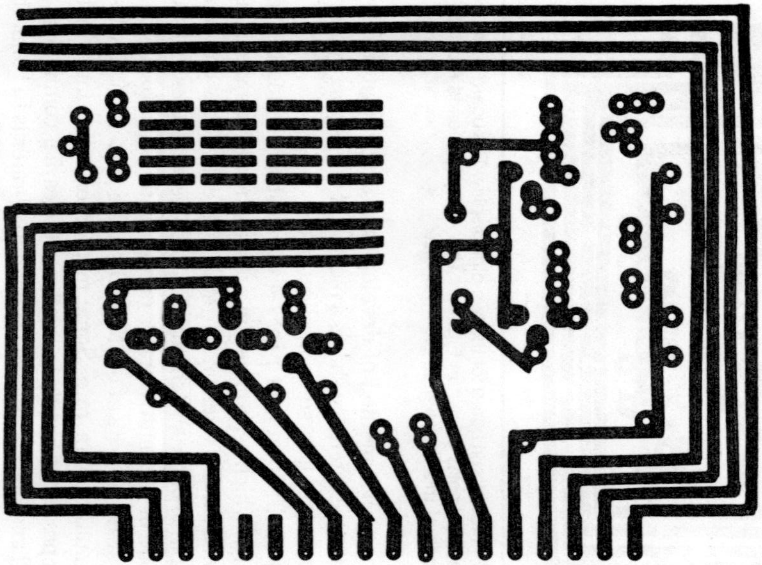

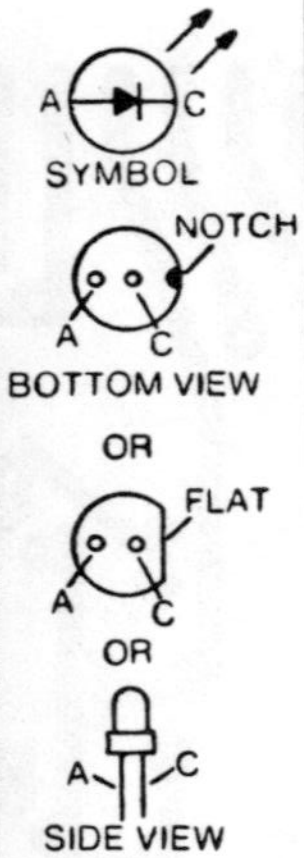


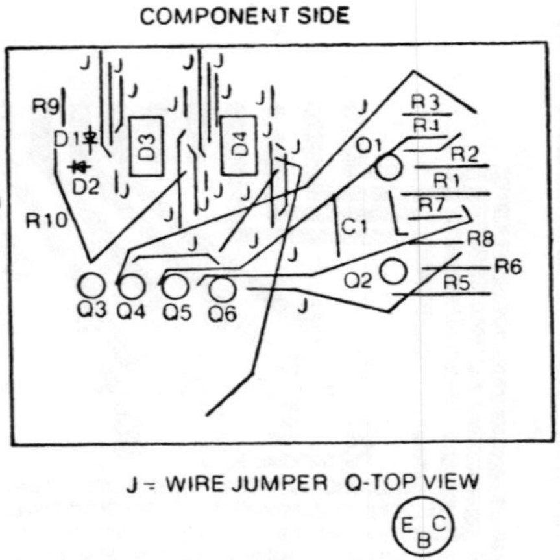


PARTS LIST

C1—20 μF tantalum electrolytic capacitor 25-volt
C2—50 μF tantalum electrolytic capacitor 25-volt
D1 D2—MV50 light emitting diode (James Electronics 1021-A Howard Ave San Carlos CA 94070) or equal
D3 D4—FND70 7-segment single-digit LED readout (James Electronics 1021-A Howard Ave San Carlos CA 94070) or equal
Q1 Q2 Q3 Q4 Q5 Q6—2N5821 transistor (Radio Shack 276-2025 or equal)
R1 R5—330 000 ohm carbon resistor
R2 R6—3300 ohm carbon resistor
R3 R4 R7 R8—10 000 ohm carbon resistor
R9—300 ohm carbon resistor
R10—3000 ohm carbon resistor
All resistors 1/4 watt or more 10% tolerance

Fig. 7-7. Parts placement for Fig. 7-6. Tantalum capacitors are the best for the large value coupling capacitors.

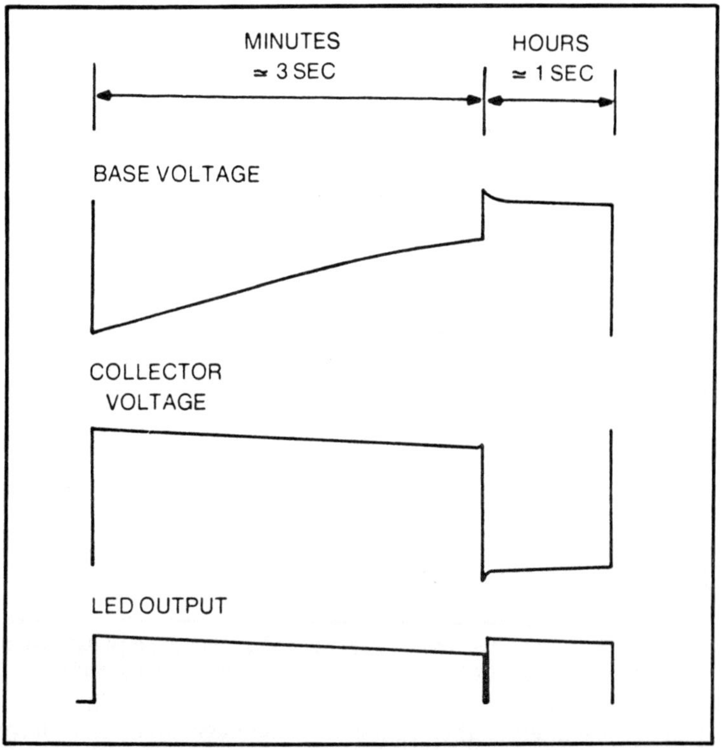


Fig. 7-8. Approximate waveforms for the multivibrator. These are difficult to display on most oscilloscopes, but can be observed with a low damping VTVM. The "droop" of the ON state is very sensitive to capacitor leakage.

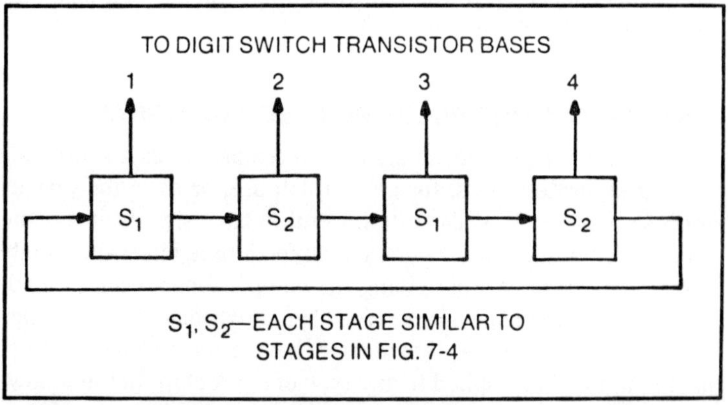


Fig. 7-9. Ring counter to drive switching transistors for a one-digit display. Such displays can also be created by driving the enable circuit with a gate slaved to the scan oscillator.

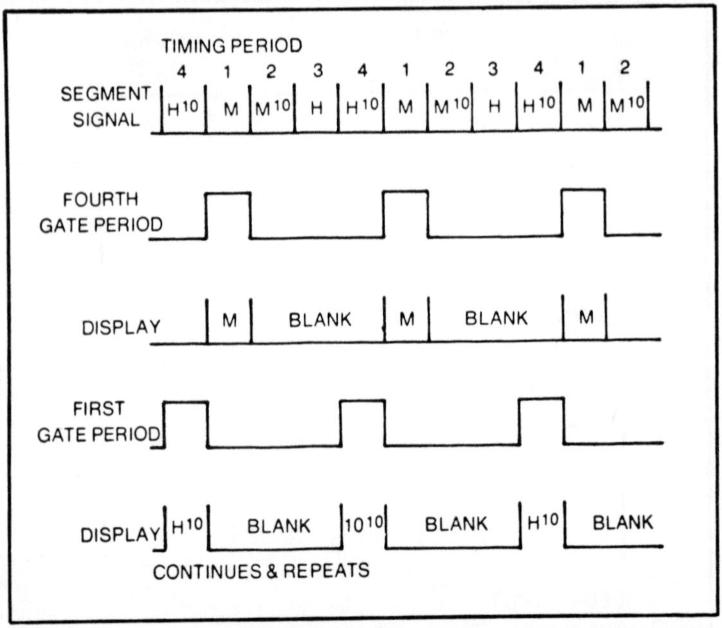


Fig. 7-10. Timing of a one-digit display obtained by blanking digit data selectively. The blanking may be done at the chip output, or at the enable.

play for one interval, then turn it on for another. Many chips have a pin connection for blanking, and connecting it to a multivibrator output will give the blanking. If the pin connection is not present, series transistors connected as in Fig. 7-3 can be used; the bases should be connected to the multivibrator output with a small series resistor for drive equalization.

## OBTAINING A ONE-PULSE-PER-SECOND SIGNAL

A one-pulse-per-second signal is often useful, as a source for driving separate timers, for general lab use, or even for various process controls. It is also often useful to have signals at the one minute, ten minute, one hour, and so forth time intervals. While some chips do have a pin giving the one-pulse-per-second signal, many do not, and none of the chips have provisions for the one per minute, etc., signals. A detector for these signals would be most useful, if it can be applied to any type of clock chip. In the following, the one-pulse-per-second is assumed as the design goal, but it should be remembered that the same technique can be applied to any of the digits which are present in the multiplex display.

Since the display for seconds must have a change each second, the data needed to develop a one-pulse-per-second signal is present in the seven-segment output lines. Actually, because of redundancies in the seven-segment presentation, only two segment lines are needed to secure the information. This is shown in Fig. 7-11, which shows the segment signals plotted against time for segments "a" and "e". Since the digits are continually changing in sequence, the time plot is also a plot of segment amplitude versus digit sequence. Examination of the sequences shows that there is a change in level for segment "a" for all digits except 3, 8, 9 and 0. If both digits are considered, there is a change in level for one or the other, or both, at each digit change. Thus, if an OR of "a" and "e" leading and trailing edges is taken, there will be a pulse each second.

Actually, the information is not quite so clean in the clock output circuit. The scan process must be remembered, at least for most chips. This cuts each of the segment ON periods into a train of signals, causing the time signals to appear as shown in the second half of Fig. 7-11. However, if this scan process is remembered, no great problem arises as a result.

One of the many ways to use this information to secure this desired one pulse-per-second signal is shown in Fig. 7-12. First,

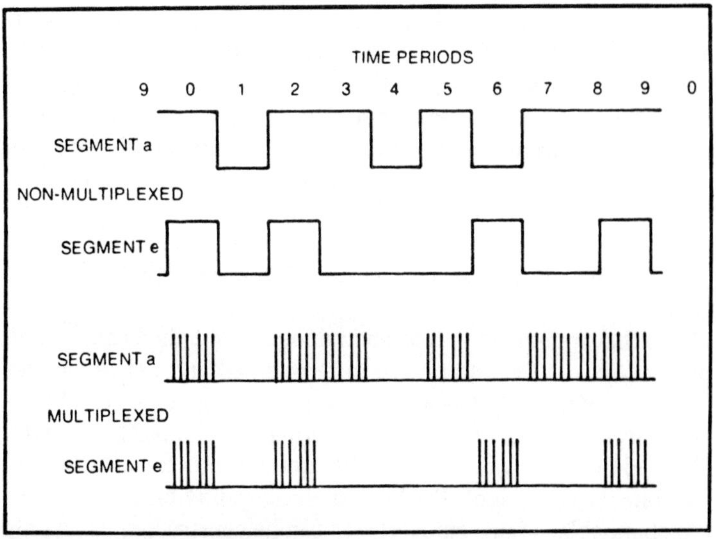


Fig. 7-11. Principle of development of a 1 pps signal from a chip lacking this feature. There is a transition in either segment a or e, or both, each second. However, the circuitry must compensate for the fact that the data is multiplexed, as shown by the lower two graphs.

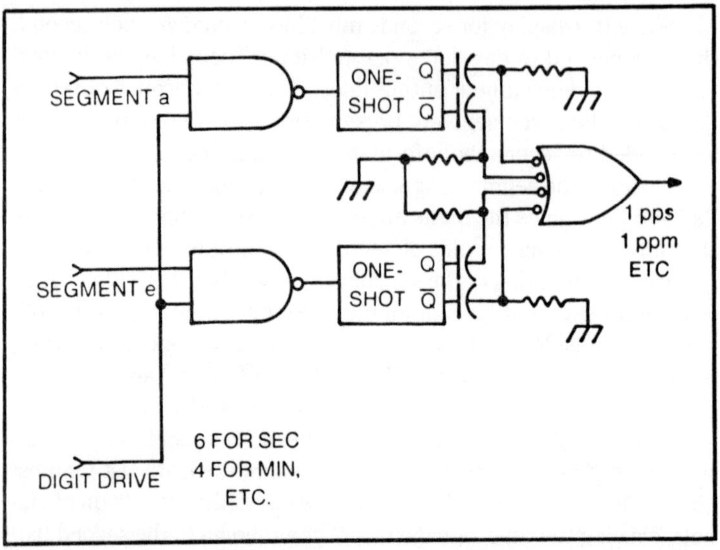


**Fig. 7-12. Possible 1 pps generator. The one-shot period is just longer than the scan interval, the outputs being the top two wave forms of Fig. 7-12. These are differentiated, then clipped and combined to give the 1 pps signal. The circuit can be built with two TTL packages.**

a pair of NAND gates are used to demultiplex the "a" and "e" segment signals at the desired scan time. The output of these NAND gates are just the pattern of pulses shown in the second half of Fig. 7-11. These signals are used to drive a pair of retriggering multivibrators, whose period is set so that it is just larger than the interval between successive scans. The outputs of the one-shot multivibrators produce the waves shown in the upper half of Fig. 7-11, and their inverse, i.e., "a" and "a", etc. There is a small delay in the fall of the segment signal; however, since the scan rate can be high (in the range of kHz), the fall time is negligible in most cases.

In this approach the plus and minus signals, the Q and $\overline{Q}$ outputs of the multivibrators, are differentiated and fed to another NAND gate, operated as a NOR gate. The three gates can be secured from a single chip by using a triple four-input NAND gate. To keep the loading on the segment lines low, this can be a 74LS20 package, or equivalent. Dual retriggering multivibrator packages are obtainable, so this entire circuit can be constructed using only two IC packages.

For chips with nonmultiplexed outputs, the problem is simpler. Often a simple diode OR of differentiated segment "a" and "e" signals is good enough. Also, use a scope to check the chip outputs—there may be a signal superimposed on another one. Remember to watch interface levels as described in Chapter 2. For TTL packages, the interfacing is automatically proper if the chip is powered from a +5, −12 volt supply, with the TTL packages connected between +5 and 0 volts.

## AUDIBLE SIGNALS

Audible or low frequency ac time signals have many uses. The second "ticks" on WWV time signals are an example, as is the hour tone found on many radio stations. Minute ticks can be placed on telephone circuits to indicate the time used. And some people like

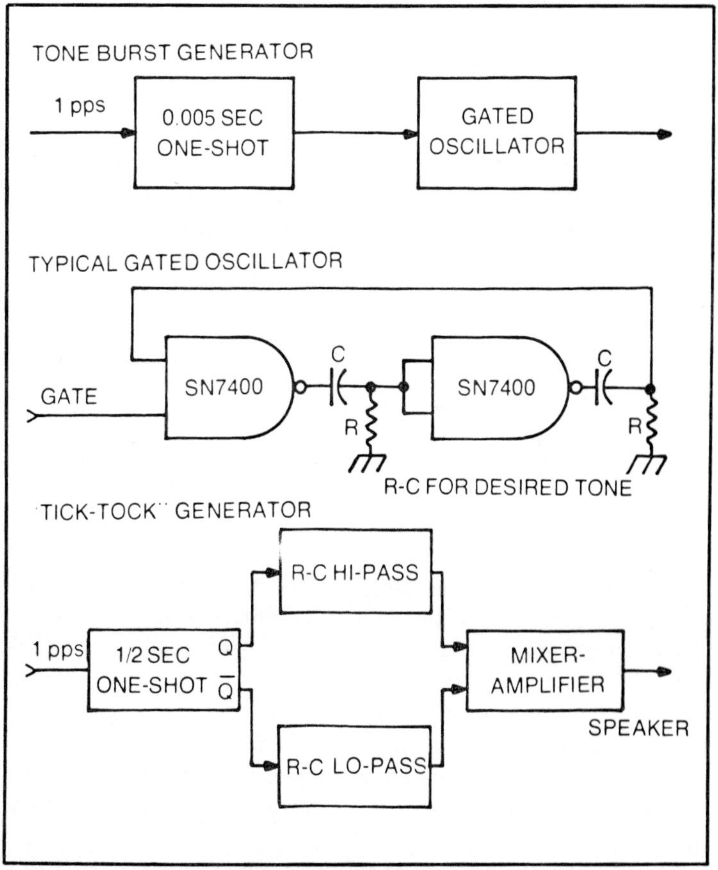


Fig. 7-13. Circuits for audio output for technical or "presence" purposes. Various simulation circuits can be substituted.

the sound of clocks, possibly just the tick of a mechanical clock, or more especially the deep tick-tock of a long-pendulum grandfather clock. Having the one pulse-per-unit time signal available allows easy generation of these audible signals.

For example, Fig. 7-13 shows a technique for generating tone bursts. Here, the one pulse-per-second signals are fed to a one-shot multivibrator to give output pulses of controlled duration. Except for special situations, this duration would typically allow four complete cycles of the tone to be generated, this being about the least that can be reliably recognized. This short gate is used to turn on and off a self-starting oscillator. This can be the multivibrator type as shown in the figure. If tone purity is necessary, a filter circuit can be used in the output of the tone oscillator or a sine-wave gated oscillator can be used.

If the multivibrator ON time of this tick circuit is lengthened to one-half of a second, the output can serve as the basis for a fair approximation of the tick-tock of a grandfather clock. To secure this, one output (Q) of the multivibrator is differentiated, which produces a sharp "tick" sound on the leading edge. Separately, the second output (Q) of the multivibrator is integrated, which produces a low frequency tone, which sounds somewhat like the "tock." These two tones can then be amplified and combined to feed a small speaker, mounted in the clock case. The circuit for these and suitable values are also shown in Fig. 7-13.

An alternate method of developing tick-tock signals is shown in one of the references: this is the original reference for this particular technique.

## STRIKING CLOCKS

Many home owners find the sound of a clock chiming the hours and quarter hours, or even just producing a sound to mark the passage of an hour, a comforting and useful device. With a little interface circuitry, such striking provisions can be added to any clock chip.

Figure 7-14 shows an approach to this design. A scale of 12 counter is used to keep track of the hours, this being easier than attempting a BCD to binary conversion. An OR circuit in the input allows for synchronization of the chime and clock (or demonstration).

Each hour, the reading of this 12-counter is transferred to a presettable counter which can count down to zero. At the same time,

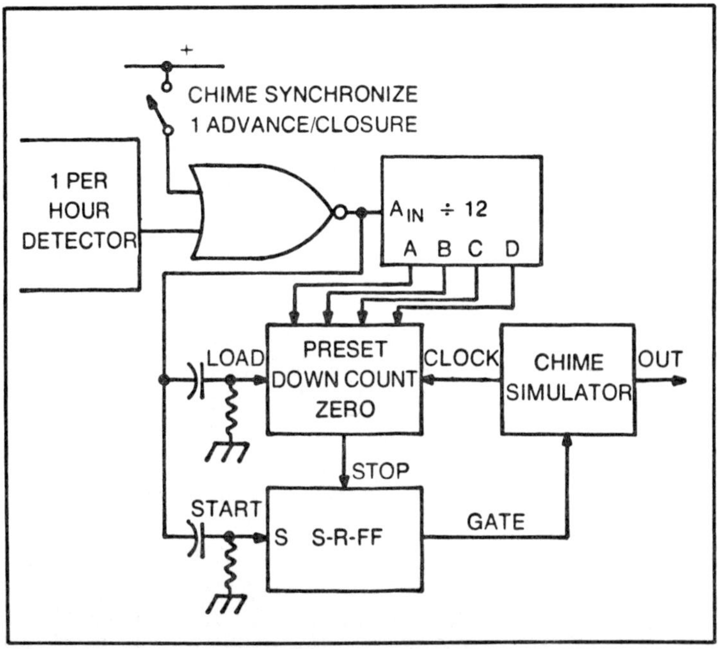


Fig. 7-14. Circuit for a striking clock. Hours information is accumulated, and used to preset a down-counter. The change also starts a chime simulator, which generates a signal approximately every two seconds. Chiming stops when the down-counter reaches zero. The cycle repeats each hour.

a pulse train generator is started. The output of this pulser initiates the chime simulator signal and at the same time steps the down counter one count. The chime is repeated, say each three seconds, until the down counter reaches zero, when the sequence is stopped. The counter is reset at the next hour, repeating the chiming, but with one more chime added. Reset at 12 hours is automatic since a 12-hour counter is used.

If only a tone marking the hours is wanted, the interface circuitry can be greatly simplified. The circuit of Fig. 7-13(A) can be used for such an hours marker if fed from the one per hour pulse. Additional audible signals can be added as desired. And, of course, various audible alarms can be developed from the alarm or snooze functions contained in several clock chips.

## INCREASING CLOCK SIGNAL POWER

It is sometimes necessary or desirable to have a clock function which can be heard over local noise, or which will carry over a considerable distance. The "dinner bell" once was a common example of this.

For a single event, such as a call to dinner, it is hardly worthwhile to go to the trouble of automatic operation. A simple off-on switch is not only cheaper, but it is more reliable. But if several events are to be announced, automatic control is both practical and convenient.

Figure 7-15 shows a possible way of generating such signals. The signal source is the alarm function of a clock chip. In this design, this is used to energize a multivibrator, which in turn drives a relay operated auto horn. The changing level of the sound is much more noticeable than a steady sound would be. In particular, it is much more audible if there is local noise of constant level.

Various niceties can be added, of course. One would be a duration timer, to decrease the "on" period from the ten minutes or so common in clock chips. Another could be an OR circuit, to com-

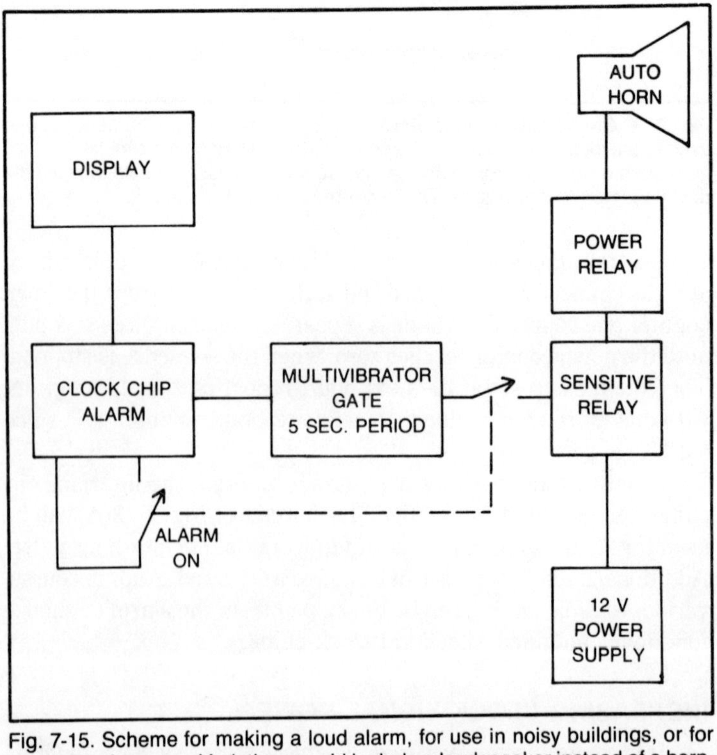


Fig. 7-15. Scheme for making a loud alarm, for use in noisy buildings, or for coverage of an area. Variations could include a loudspeaker instead of a horn, a warbling tone, or others.

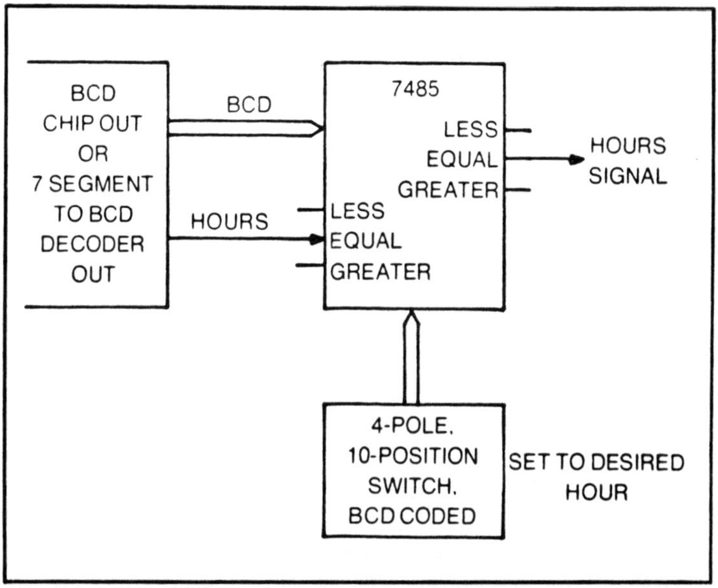


Fig. 7-16. Circuit to give a signal at a selected hour. A 7485 digital comparator is used to compare the BCD output of a clock chip (or data converter) with a preset BCD number, providing a signal that they are equal. Less and greater signals are available. The scheme can be extended to n digits.

bine several signals. However, if many are to be generated, it would seem preferable to adapt a micro-computer-clock chip design, along the lines discussed in Chapter 11.

## DIGIT RECOGNITION

Sometimes it is necessary to have a method of recognition of a particular digit or set of digits. An example would be in the recognition of a particular time in a process, perhaps requiring recognition of as many as six digits of a time presentation. If the clock chip in use has binary coded decimal output available, as does the MM5511, such digit recognition is relatively easy. Figure 7-16 shows the basic technique, using a 7483 magnitude comparator. The magnitude comparison is made between the BCD output of the clock chip, and the BCD coded setting of a switch. If both BCD inputs are identical, the "equals" output goes high. If the chip output is not multiplexed, only the four BCD lines need to be connected to the comparator.

If the chip output is multiplexed and only a single-digit comparison is required (as for identifying a particular hour), the hour digit-drive signal from the chip would be connected to the "equals" input of the comparator, as shown dotted in Fig. 7-16. Since the "equals" input would be high only at the desired digit time, the correct "equals" output could only occur at this time.

These types of digit recognizers are easily cascaded for multiple digit recognition if the input signal is not multiplexed. The output of the lowest level comparator, say the seconds comparator, is fed to the input of the next level comparator, say the tens of seconds, and so on. The last stage so connected is used for the equals output for the entire chain, its equals signal occurring only if all digits correspond exactly to the setting of the entire switch bank.

If chip signals are multiplexed, and only a single time comparison is needed, it is far simpler to use an alarm chip, which has a built-in comparator that functions before the signals are multiplexed. If multiple comparisons must be made, it would be necessary to demultiplex the chip output signals, storing these in an array, and using the stored signals for the comparator.

Recognition of seven-segment output signals can be more difficult, but it can be simplified by using the redundancy of the seven-segment presentation. For numbers from 0 to 9, unique output signals can be generated by the (N) AND of two-, three- or four-segment signals, combined as shown in Table 7-1. This table assumes that both plus and minus versions of the segment signals are available. Still further simplification is possible if the number range does not exceed 0-6, the case for tens of minutes and seconds, or 0-2, the case for tens of hours. The necessary gating can be developed from the data of Fig. 3-3.

An example of such a digit recognizer is shown in Fig. 7-17. This is based on the use of a triple three-input NAND gate, to develop the 0 and 5 second signals. The inputs b, f and g are con-

**Table 7-1. AND Requirements for Recognizing 7-Segment Signals.**

| Decimal Value | AND Inputs |
|---|---|
| 0 | $\bar{f}\,g$ |
| 1 | $a\,\bar{b}\,f$ |
| 2 | $\bar{e}\,f$ |
| 3 | $b\,c\,f\,\bar{g}$ |
| 4 | $b\,c\,\bar{f}$ |
| 5 | $\bar{a}\,b$ (or $b\,\bar{f}$) |
| 6 | $b\,\bar{c}$ |
| 7 | $\bar{a}\,e\,f\,g$ |
| 8 | $\bar{a}\,\bar{e}\,\bar{f}\,\bar{g}$ |
| 9 | $\bar{a}\,\bar{b}\,c\,\bar{g}$ |

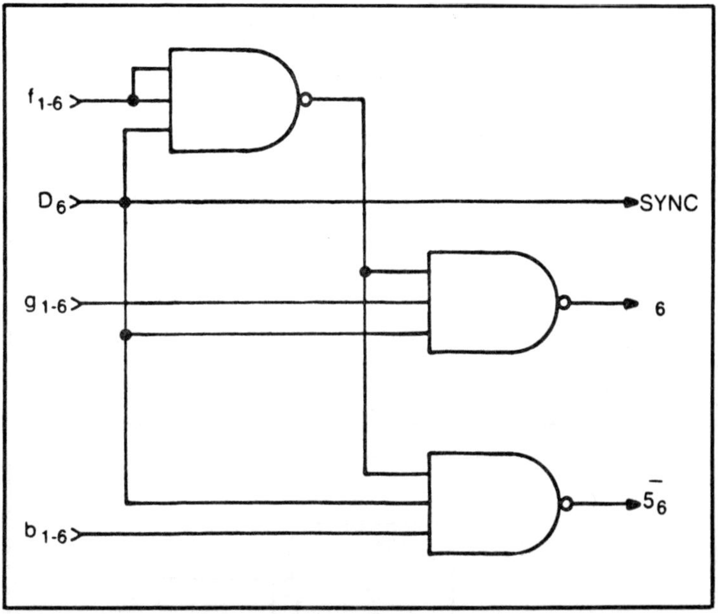


Fig. 7-17. Digit identification by AND gates. Provides a signal when the zero and five readings appear at digit 6 time. Can be extended to give a signal for any digit; see Table 7-1 for the required AND inputs.

nected to the gates, with one input of each gate being fed with the desired digit signal to secure demultiplex output. The output consists of a series of pulses, in this particular case having a duty cycle of 1/6. These pulses last as long as the digit lasts. If the D6 line is left open, or connected to +5 volts through a small resistor, the outputs will be in multiplexed form, giving indication whenever any of the digits in the multiplexed chain reach the 0 or 5 positions.

## SEVEN-SEGMENT-DECIMAL-BCD RECODING

Many times, in adapting clock chip signals for special timing or other purposes, it is necessary to recode the output to another form. The most common recoding problem is from the multiplex seven-segment into multiplexed BCD form. Occasionally there will be a requirement for seven-segment to decimal recoding. Also, at times, it will be desirable to have recognition of all the digits, rather than the one or two easily derived by the preceding technique.

Recoding of BCD to decimal or to seven-segment signals is sufficiently common that an ample number of recoding chips are readily available. Typical chips which are usable are shown in Ta-

**Table 7-2. Chips Available for Data Recoding.**

| Chip Type | Input Form | Output Form |
|---|---|---|
| 7442 | BCD | Decimal |
| 7443 | Excess 3 | Decimal |
| 7444 | Gray | Decimal |
| 7444, 145 | BCD | Decimal |
| 74141 | BCD | Decimal |
| 74154 | 4 Line | 16 Line |
| 74159 | 4 Line | 16 Line |
| 7446 ( 47, 48, 49) | BCD | 7 Segment |
| 74S138 | 3 Line | 8 Line |
| 74147 | Decimal | BCD |
| 74148 | 8 Line | Octal |
| 74280 | 9 Bit | Odd-Even |
| 9309 | 4 Line (dual) | 1 Line |
| 8T04 ( 05, 06) | BCD | 7 Segment |
| 8T50 ('70) | BCD ( + a – f) | 7 Segment |

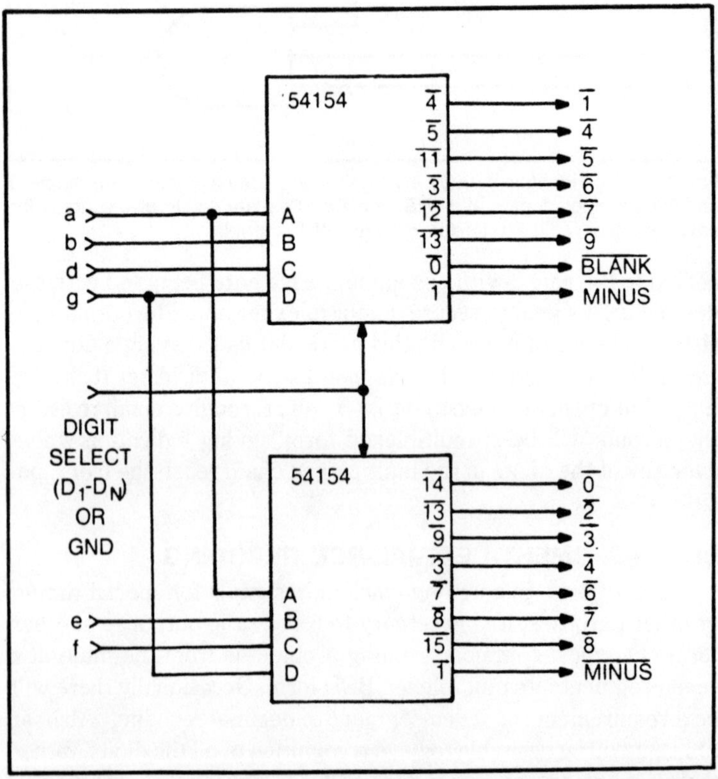


Fig. 7-18. Circuit for completely decoding the 7-segment output. Only five lines are needed as inputs, due to redundancy in the display. These outputs can be fed to a 74147 priority encoder to secure BCD data. Latches at the input or output may be used as needed. Ground the enable line if multiplex data is needed.

ble 7-2. However, recoding of seven-segment signals into BCD is less common and at present only one such chip is known to be on the market. It is not readily available to experimenters, so usually the experimenter must develop his own digit-code converter.

Probably the easiest method of performing this decoding is to first convert the seven-segment information into a digital signal, then, if necessary, to recode this to BCD. Figure 7-18 shows one of several ways of accomplishing the seven-segment to decimal decoding. This uses a pair of 74154 four-to-sixteen-line decoders, feeding one chip from segments abdg and the second chip from segments aefg. This gives a set of unique signals, as shown in the figure. These unique signals are then strapped to a decimal-BCD encoder, for example, type 74147. The 74154 chip also contains gates which can be connected to the digit drive signal to give a demultiplexed output, if this is necessary. Alternately, the multiplexing function can be performed later in the chain, at the decimal level or at the BCD level.

# Chapter 8

# A Clock Chip Frequency-Period Meter

Clocks count. Basically they count seconds, accumulating these in a particular format, the reading. In electronic clocks, the count is really of cycles, which are prescaled to give a count of seconds; the count is then accumulated. It seems a natural idea to attempt use of a clock chip for various counting operations.

Let us look at differences between clocks and the various types of counters. In a clock we have the basic count chain, the accumulator, which is fed by a known frequency and is connected to a display. This is indicated by the upper part of Fig. 8-1. The lower part shows, for comparison, a simple counter or event meter. This also has a count chain and a display. A difference which is immediately noticeable is that the count chain is fed by unknown events rather than a known frequency. Also, normally there is a start-stop circuit that is a means to start counting at some particular indication, and to stop it at another. Start-stop operation is the second major difference—clocks are designed to run continuously, whereas counters are designed to run either intermittently or continuously.

## THE CONCEPT OF A CLOCK CHIP COUNTER

The extension of the basic counter to provide other functions is done by operating on these two differences from clocks. As shown by Fig. 8-2, in a frequency meter the start-stop control is secured from an accurate time base, which generates a unit of time. This

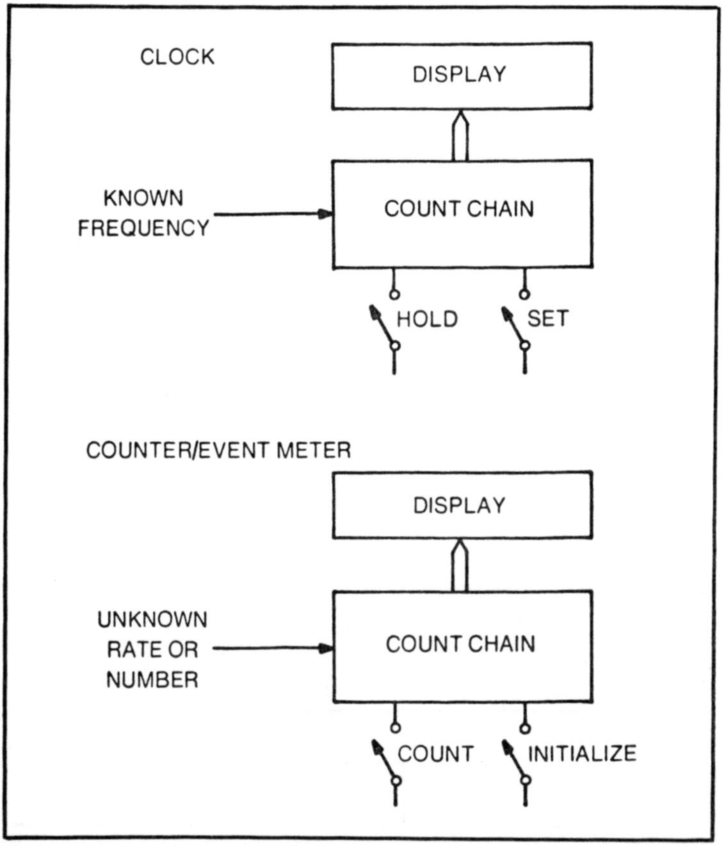


**Fig. 8-1. Comparison of a clock and a counter or event meter. The difference is in input and control.**

turns on, then off, the start-stop control, which is usually called a gate. The number of cycles (or events) of the unknown frequency which occur during this time are accumulated and displayed. They are read as cycles per second (hertz) or a multiple. The alternate nomenclature is events per second, usually called events per unit time.

We should note that this operation really involves an assumption. What is actually read is "cycles in a unit time" or "events in a unit time." We can see the difference by assuming, as in Fig. 8-3, that the input signal stops while the gate is open, say at half of the gate period. Then the display reading at the end of the gate period is just half of the number of cycles that would have been accumulated if the signal had been present the whole period. The

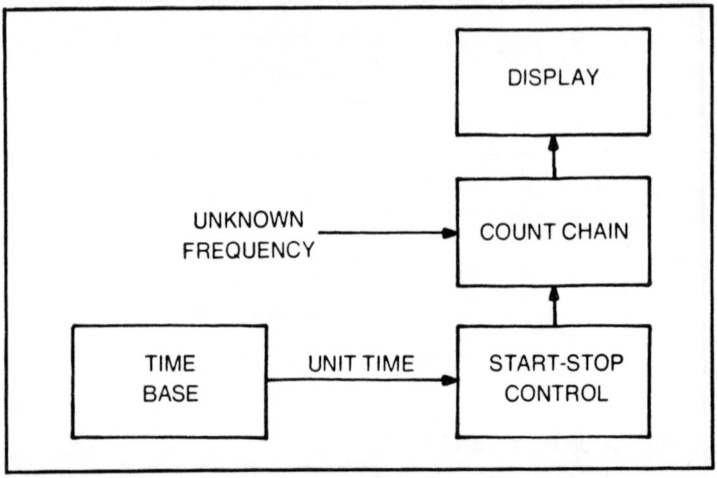


Fig. 8-2. Extension of a clock to provide frequency measurement capability. The known frequency is used as a time base to generate a start-stop gate for the unknown.

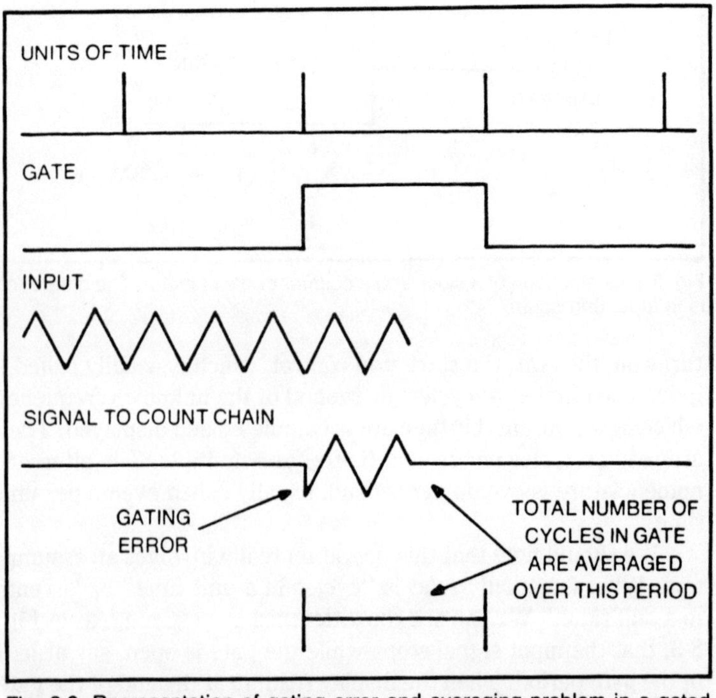


Fig. 8-3. Representation of gating error and averaging problem in a gated counter. With a continuous input of constant frequency, only the gating error problem remains, usually amounting to ± 1 count error.

counter is averaging the input. The reading would also be an average if the unknown had been changing during the count period.

Averaging is characteristic of the entire class of meters which count for a time interval. There is no problem of interpretation if the input is steady and continuous—the average and true values are the same. But with varying frequencies or intermittent signals, the fact that the indication is an average must be kept in mind. The easiest way of telling if there is a difference is to keep an oscilloscope connected to the input of the frequency meter. Any marked change in frequency or chopping of the signal is immediately evident on the oscilloscope display.

This identical problem also exists in the instrument obtained by rearranging the frequency meter components into that of Fig. 8-4. Now, the instrument counts cycles of a known frequency, just as a clock does. However, the unknown is used to control the gate, that is, to determine the starting and stopping of the count accumulator. This is a time-period meter, usually called a period meter. More generally it should be called a time per unit event, or for accuracy, time between two events.

This instrument is especially useful when a low frequency is to be measured. Suppose a 60 Hz signal is to be counted for one second. A counter could read this, but would usually read 59, 60

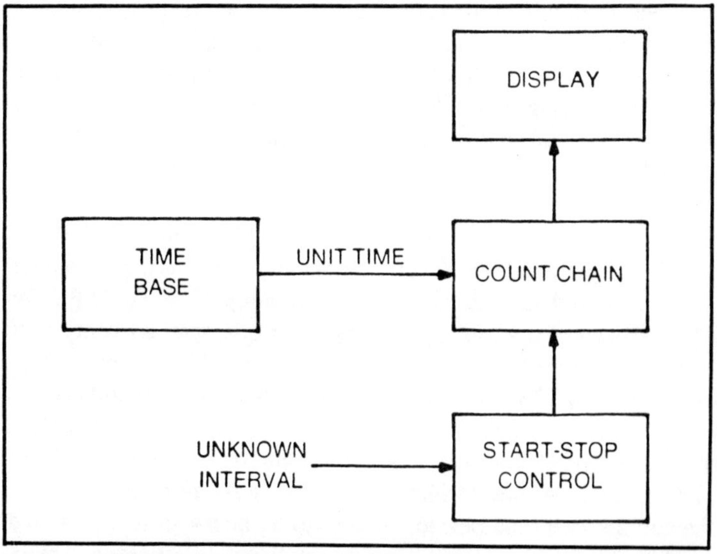


Fig. 8-4. The basic period meter. The time base and unknown are interchanged from their position in a counter.

or 61 hertz. Counters do have some possibility of error, due to the relation between the end of a cycle and the end of a count chain and we cannot really be sure that the true count is 59, 60 or 61. But suppose a period meter is used, with the time base set to a 1 MHz frequency. The instrument can now resolve to an accuracy of 1 microsecond. A typical reading would be 16,668 microseconds per cycle. It is easy to tell the difference between say, 59.999 and 60.001 Hz. For very slow changes, say cycles per minute, the frequency meter becomes worthless, but the period meter is useful. On the other hand, the usual period meter is inaccurate for periods such as 1 microsecond. Both meter forms are needed.

It is evident from these figures that a useful counter requires addition of just two elements to clock chips. One is the time base, and the other an ON-OFF circuit. With a little switching added, one can have such elements as a counter, a frequency meter, an interval meter, and of course, a clock.

The characteristics of this counter will be surprisingly good. While specifications of common clock chips show that they will operate over a frequency range from 0 to 50 kHz, they are actually better. Most clock chips will operate at frequencies approaching 1 MHz. This covers the audio range, well into the supersonic range, and the lower ratio frequencies. This range can be extended into the upper radio frequencies by techniques described later. True, the clock accumulator indication needs a little interpretation to read; the instrument will count in units of 60 rather than in the more common ones of tens and hundreds. This is no great problem, and in many situations is no problem at all, since the interest is in changes rather than in absolute value and no mental transformation is needed. But even if the absolute value of frequency or time is needed, reading and converting is not enormous trouble. The low cost potential and ready availability seem much more important.

The large range that is covered by a clock means that a counter made from a clock chip can have great range. See Table 8-1 for a breakdown of the ranges available. For example, a twenty-four hour clock chip with fast set capability will give direct counter readings of zero to 86,400 events, by units, or zero to 4,320,000 events, by fifties. If the clock is used as a period meter, it can read from one second to 86,400 by one second units, or from one second to 1728 seconds by one-fiftieths of a second. As an interval meter, the counter can read periods of zero to 24 hours with one second resolution, or from zero to 28 minutes with one-fiftieth second resolution. Finally, as a frequency meter, it can read from zero to 86

**Table 8-1. Ranges Available with Clock Chips.**

| Nominal Quantity | Indicated Range | Maximum Count All Lower Digits |
|---|---|---|
| A. No Prescale (Fast Set), MM5316 Chip | | |
| Seconds | Does not operate | |
| Minutes | 0-59 | 0-59 |
| Hours | 0-23 | 0-1439 |
| B. No Prescale (Fast Set), MM5314 Chip | | |
| Seconds | 0-59 | 0-59 |
| Minutes | 0-59 | 0-3599 |
| Hours | 0-23 | 0-86399 |
| C. Prescale (50 Hz Setting) | | |
| Seconds | 0-59 | 0-2950 |
| Minutes | 0-59 | 0-179, 950 |
| Hours* | 0-23 | $0\text{-}4.31 \times 10^6$ |

*Maximum rate of count is limited by chip design.

kHz, with resolution to 1 Hz, or to nearly 1 MHz with a resolution of 50 Hz. And there are more complex chips which multiply the count range by 30, the clock-calendar chip.

With a little added circuitry, the counter can provide gates of one-tenth second, one second, one minute, one hour and so on, plus the reference oscillator frequency and its submultiples, as determined by the count chain used. Truly an impressive and versatile instrument.

## DESIGN CONSIDERATIONS

It seems that the two most important factors in chip selection for use as a counter are the desirability of reset to zero (00 hours, 00 minutes, 00 seconds) on demand, for initializing the counter, plus the ability to count cycles individually: this usually appears in a clock as "fast set." Other features, such as twenty-four hour display, flexibility of control and provision of alarm circuits are helpful.

Of the readily available chips, the MM5316 was used in this prototype. It has most of the required features and has very good flexibility, but does have three drawbacks. One is a relatively low frequency rating of 50 kHz as compared to 80 or so of some other chips. Secondly, it uses a four-digit nonmultiplexed display (however, all six of the normal clock digits are readily available). Finally, its control requirements are relatively complex. The chip is otherwise suitable for an audio-RF range frequency meter, and so was used. Of the other chips readily available, the MM5309 would seem to be the most suitable. However, it does not provide for auxiliary functions, such as gates.

The next design choice involves the time base. For best accuracy this should be a crystal, but a 60 Hz line base is far simpler, and can easily give a choice of the three bases, 1 second, 0.1 second, and 1/60 second. This seems quite adequate for audio range use, remembering that the instantaneous accuracy may be limited to about one part in $10^4$ or so, by the problem described by Fig. 2-4.

Additional design decisions are best shown by the complete block diagram, Fig. 8-5. Of the four basic elements, two are provided from the clock chip: the counter chain and the display. The additions are the gate and its input protective and amplifying circuits, and the time base. Of course, the switching employed by clock chips will need to be rearranged to suit the particular needs of the counter, and other switching will be needed.

In this simple circuit, straight counting can be obtained by providing a RUN switch which allows the counter-display to run

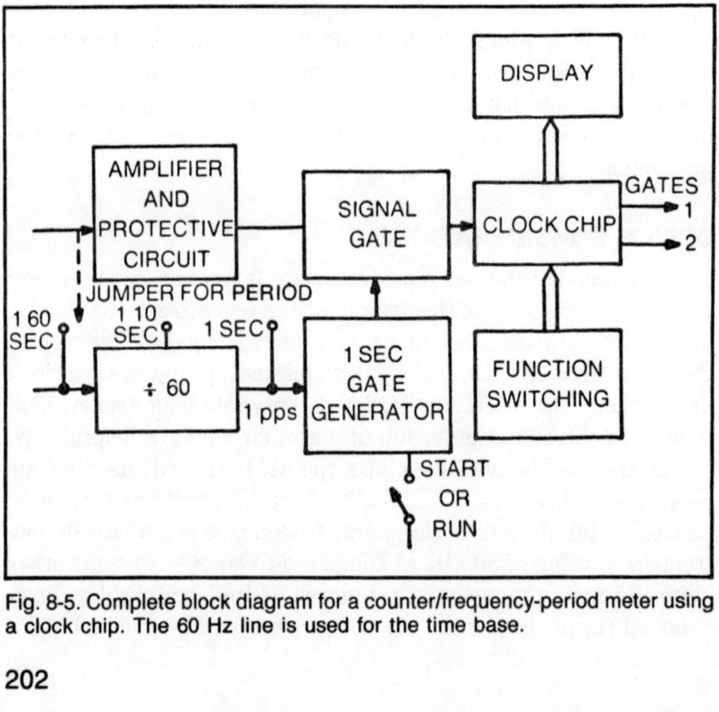


Fig. 8-5. Complete block diagram for a counter/frequency-period meter using a clock chip. The 60 Hz line is used for the time base.

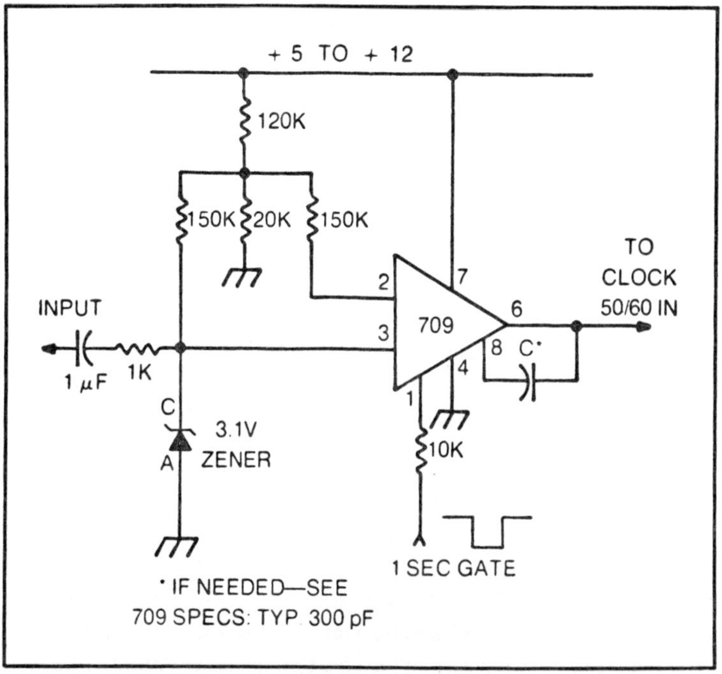


Fig. 8-6. Circuit of an input amplifier-limiter-gate using a 709 op-amp as a gated comparator. The circuit is usable to several hundred kHz. If higher frequency operation is needed, an LM361 or equivalent should be used as the active element.

continuously on demand. Period can be provided by using an external strap from the 60 Hz circuit to the input, using the RUN switch to start and stop the period: an external signal can also be used. Extra features can be provided by using the alarm and snooze features of the clock chip: these provide two gates which can be set to give a signal to a given count, or at a given time after start operation.

## THE GATE AND PROTECTION CIRCUIT

The simple gate circuit used is shown in Fig. 8-6. The heart of the gate is a linear integrated circuit, a 709 used as a comparator, and serving the combined functions of amplifier, limiter, and gate. Input protection for this IC is a Zener diode. Both inputs of the IC are biased to the same level, 1.8 volts nominal, about the midscale point of a TTL gate input. Applying an input of more than .05 volts to the 709 causes its output to saturate in the plus and minus direction, giving a swing of about 0.5 to 8.0 volts, ample for clock drive. Gating action is obtained by applying the one-second gate signal to the feedback input of the 709. When the gate signal is high, the 709 is biased to cutoff.

## GATE CONTROL CIRCUITS

The gate control circuit used in the prototype is shown in Fig. 8-7. In the following discussion it is assumed that gate control is generated from the 60 Hz line. It should be remembered, however, that the control voltage can be obtained from any source having several volts of amplitude. Of course, the scale factor changes with frequency, but this can be used to allow changes in resolution. Under normal conditions, by using frequency or period measurements as appropriate, the resolution of the counter is entirely ample.

To form the gate, the 60 Hz input signal is first squared up in two successive steps. The first squaring is done by a Zener diode used as a limiter. This also gives protection against accidental overvoltage. This partly squared wave is then fed to a Schmitt trigger, formed of two 7400 NAND gates with feedback. The output of this limiter is a good square-wave of excellent rise time.

This square-wave is then fed to a two-section counter. The first section is a 7492 divider, used to divide by six. This is followed by a 7490, used to divide by ten; the output of this is at the 1 Hz rate. This output is fed to a set-reset (S-R) flip-flop, formed from another pair of 7400 NAND gates. Set is initiated by a switch labeled START, which feeds the next occurring trailing edge of the 60 Hz square-wave to the S-R flip-flop, causing it to go to the set state. The low-going output of the flip-flop is fed to the 709 gate, as the gate control. It also feeds the reset of the counters; these immediately start counting the 60 Hz input. When a count of 60 has been reached, there is an output from the 7490. The trailing edge of this is differentiated and fed to the S-R flip-flop, causing it to reset. These reset the counters and turn off the gate, everything now being in preparation for a new cycle.

Note that the S-R multivibrator will retrigger on the next leading edge if the start switch should be held down for longer than one second. For proper count, the switch should be released as soon as count action is observed. Continuous RUN operation is secured by grounding one S-R gate. This causes the output to remain low, turning on the gate. When used for period, the indicated duration is the clock reading minus one second. Holding the RUN switch down for short periods allows a form of continuous running, which

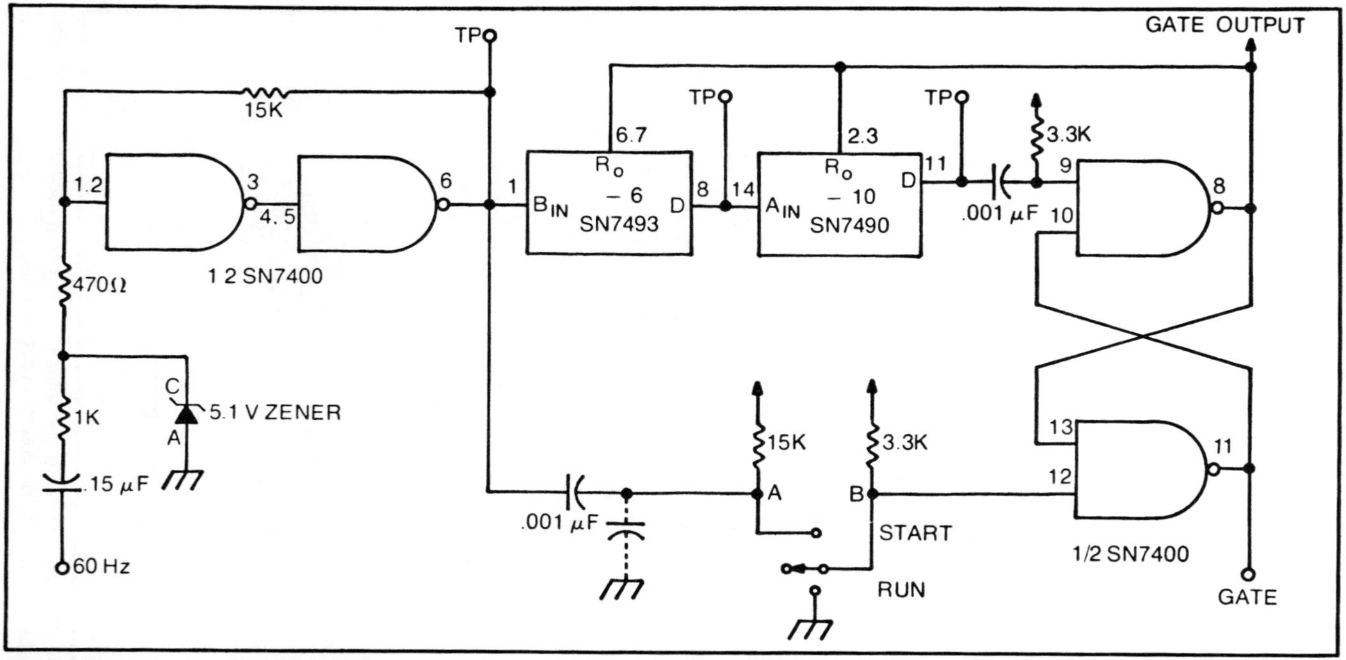


Fig. 8-7. Time base and gate signal generator for a clock counter. A one-second gate is formed when the start switch is closed. Alternately, the gate signal can be held on continuously by RUN. The time base points labeled TP can be used for period measurement.

corresponds to the "Period Average" of some expensive counters. Suppose the switch is held down for 9 + seconds, then released. The counter gate will be open for a total of ten seconds, so the clock chip will accumulate the unknown, but with ten times the reading of a normal one second period. Dividing the indicated value by ten gives the reading for one second.

## THE CHIP

The MM5516 clock chip is a large unit—40 pins. The labeling of these pins is shown in Fig. 8-8. The meaning of the labels is valid only if the pins are energized one at a time; if two are used, the effect changes. In addition, there are two priority lists involved. Some study is needed to fully understand the complexities of the priority. Table 8-2 gives the functions and priorities as listed by

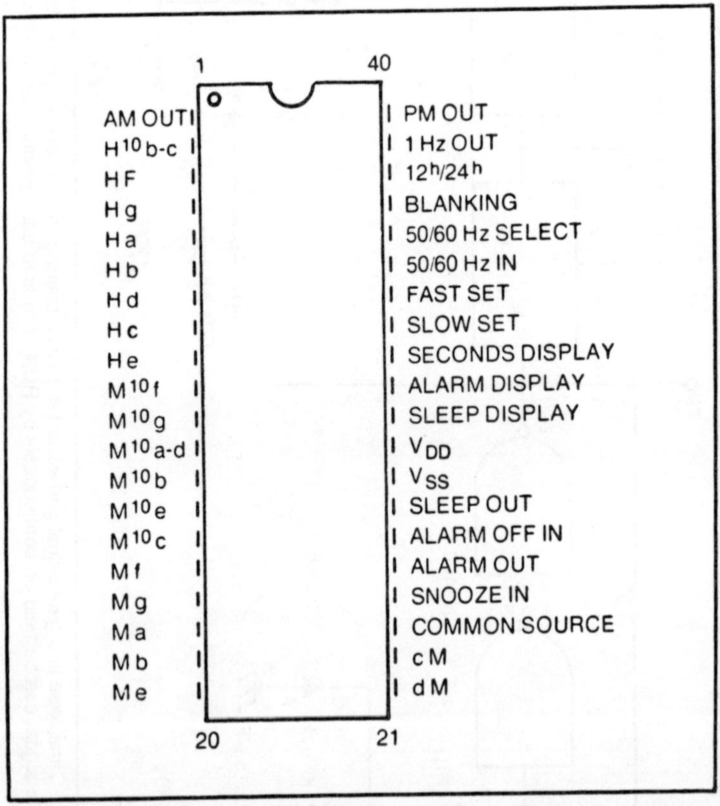


Fig. 8-8. Pin connections for the MM5516 clock chip used in the prototype. See Table 8-2 for data on control action.

**Table 8-2. MM5316 Display Modes.**

| *Selected Display Mode | Digit No 1 | Digit No 2 | Digit No 3 | Digit No 4 |
|---|---|---|---|---|
| Time Display | 10s of Hours and AM/PM | Hours | 10s of Minutes | Minutes |
| Seconds Display | Blanked | Minutes | 10s of Seconds | Seconds |
| Alarm Display | 10s of Hours and AM/PM | Hours | 10s of Minutes | Minutes |
| Sleep Display | Blanked | Blanked | 10s of Minutes | Minutes |

*If more than one display mode input is applied, the display priorities are in the order of Sleep (overrides all others), Alarm, Seconds, Time (no other modes selected).

MM5316 Setting Control Functions

| Selected Display Mode | Control Input | Control Function |
|---|---|---|
| *Time | Slow | Minutes Advance at 2 Hz Rate |
| | Fast | Minutes Advance at 60 Hz Rate |
| | Both | Minutes Advance at 60 Hz Rate |
| Alarm | Slow | Alarm Minutes Advance at 2 Hz Rate |
| | Fast | Alarm Minutes Advance at 60 Hz Rate |
| | Both | Alarm Resets to 12:00 AM (12-hour format) |
| | Both | Alarm Resets to 00:00 (24-hour format) |
| Seconds | Slow | Input to Entire Time Counter is Inhibited (Hold) |
| | Fast | Seconds and 10s of Seconds Reset to Zero Without a Carry to Minutes |
| | Both | Time Resets to 12:00 AM (12-hour format) |
| | Both | Time Resets to 00:00:00 (24-hour format) |
| Sleep | Slow | Subtracts Count at 2 Hz |
| | Fast | Subtracts Count at 60 Hz |
| | Both | Subtracts Count at 60 Hz |

*When setting time, sleep minutes will decrement at rate of time counter, until the sleep counter reaches 00 minutes (sleep counter will not recycle).

the manufacturers data sheet.

There are three provisions in the chip which are not covered by these tables. One is the inclusion of a power failure indication which causes one or more of the A.M./P.M. 1 Hz or 10s of hours segment (pins b/c) to flash when power is first reapplied. Activating any time-set function clears this flashing.

The alarm output goes high at the set time, and can remain on for 59 minutes. Energizing ALARM OFF stops this, and resets the alarm for 24 hours later. Alternately, activating SNOOZE inactivates ALARM for a period of 8 to 9 minutes. This action can be repeated, until the 59 minute period times out, or until ALARM is reset. This feature was not used in the prototype. Activating the SLEEP control sets a timer in operation, its range being up to 59 minutes. However, a second actuation will turn it off at any moment within this period. A third actuation starts a new period.

The various control inputs are high impedance circuits, normally at ground potential (with 2.5 megohm pull-down resistors). Taking them to the plus voltage causes them to be energized. The alarm and sleep outputs can supply only a small output current. If used, buffers are usually necessary.

## CONTROL CIRCUITS

To use the chip in the counter, the control circuits must be rearranged from normal clock use. Table 8-3 shows the counter features which were used in the prototype, and the particular chip circuit or circuits which must be energized to provide the function. The general performance of these functions should be clear, based on the discussion above; the specific performance and use is discussed later.

If all of the features listed are provided, several of the clock chip pins must be fed from several switches. Diode OR gates can be used to isolate these signal paths, so that the switching is nonambiguous. In doing this, some care is necessary, since these chip control circuits are very high impedance. Also, in laying out the switching, attention must be given to the fact that the clock chip has a priority system, which means that some functions will override others if simultaneous operation is attempted.

A switch design which considers these priorities is shown in Fig. 8-9. This follows directly from the table and assumes that all of the tabulated functions except the snooze feature are to be provided. In this figure, the four switches to the left are concerned with reading the counter, preparing it for operation and making

**Table 8-3. Relation—Counter Control to Clock Control.**

| Counter Function | Clock Function Used | | | | | |
|---|---|---|---|---|---|---|
| | Seconds | Set Slow | Set Fast | Display Alarm | Off Alarm | Display Sleep |
| Read Low Digits | x | | | | | |
| Read High Digits | | (None) | | | | |
| Clear Display | x | x | x | | | |
| Start | | Not on chip | | | | |
| Run | | Not on chip | | | | |
| Gate 1-Read | | | | x | | |
| -Set | | x | | x | | |
| Gate 2-Read | | | | | | x |
| -Start | | | | | | x |
| -Set | | | | | | x |
| Frequency-Period | | Not on chip | | | | |
| Prescale 50 1 | | (None) | | | | |
| Prescale 1 1 | | | x | | | |

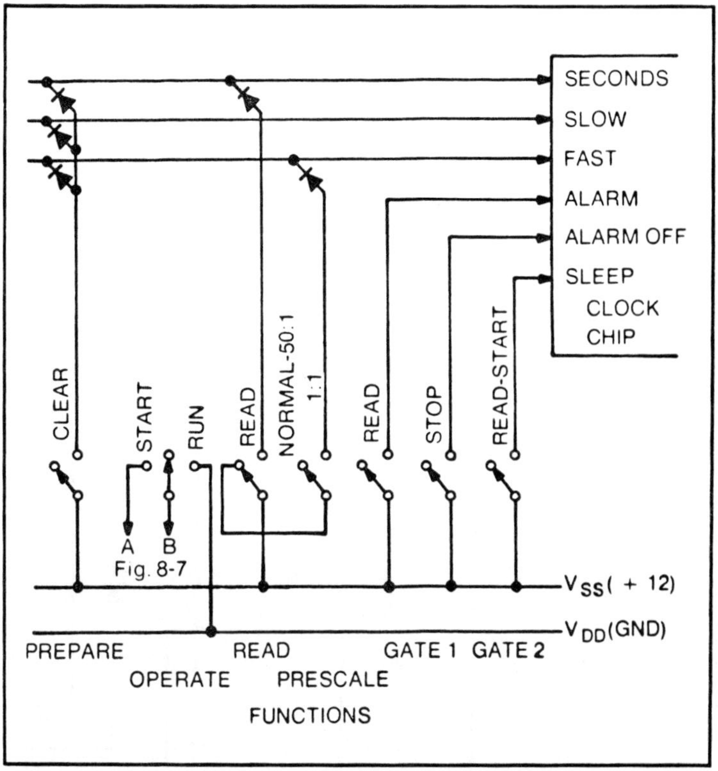


Fig. 8-9. Controls as used for the prototype. Diodes are used to isolate circuits. It may be necessary to add debounce R-C filters to the seconds, slow and fast input lines.

readings. The four switches to the right are concerned with the two additional gate functions, and allow for reading the setting of these gates and for changing the setting. These four switches can be omitted if desired. If this is done, the prescale input should be connected to the 12-volt line by the jumper indicated.

Because of their high input impedances, the control circuits are susceptible to noise. Figure 8-9 shows a method of combatting this, an RC filter in the fast set line. More may be needed. This filter also serves to "debounce" switch operation.

## DISPLAYS

The display design of the chip does not use multiplexing, so any of the common-cathode single-digit display designs is usable. There is also a feature which makes the chip especially useful with

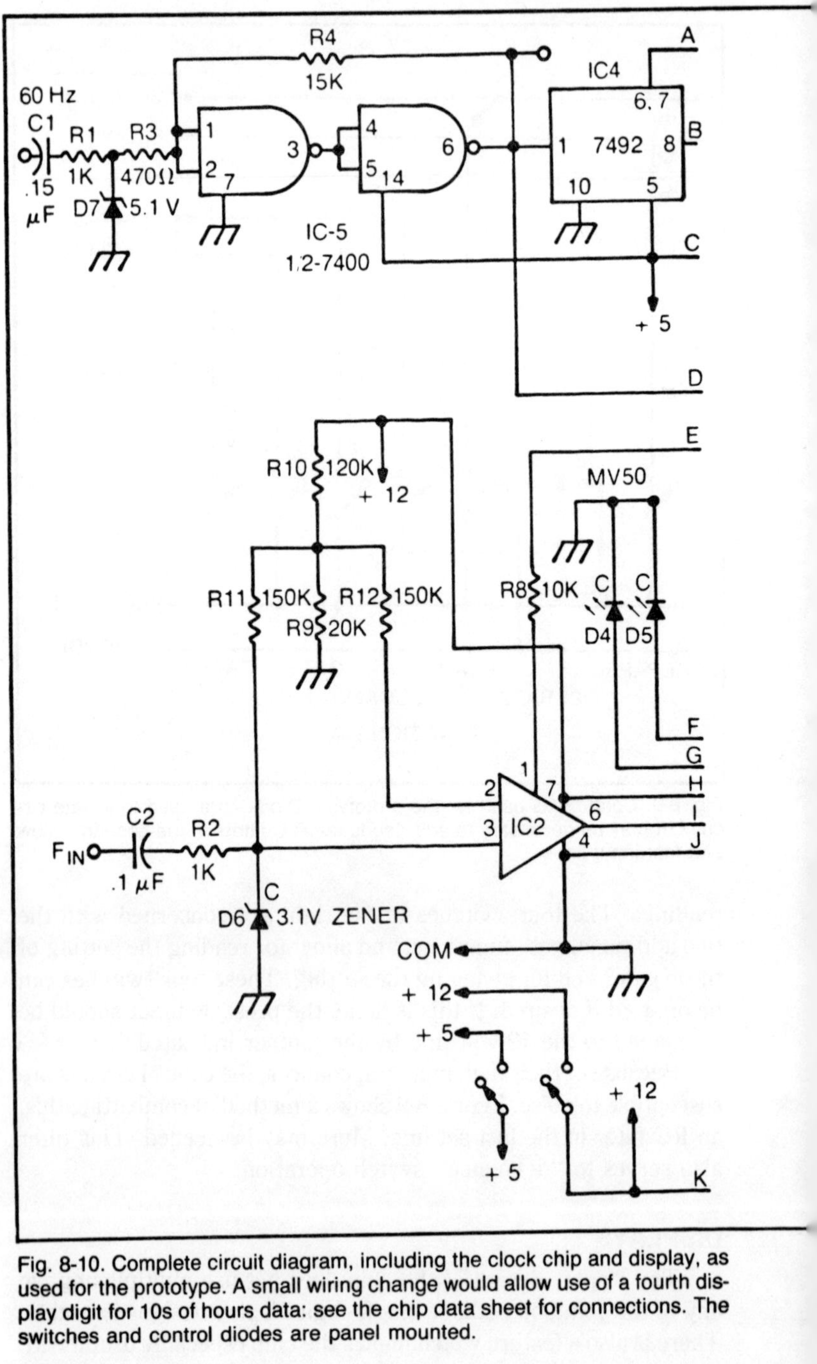


Fig. 8-10. Complete circuit diagram, including the clock chip and display, as used for the prototype. A small wiring change would allow use of a fourth display digit for 10s of hours data: see the chip data sheet for connections. The switches and control diodes are panel mounted.

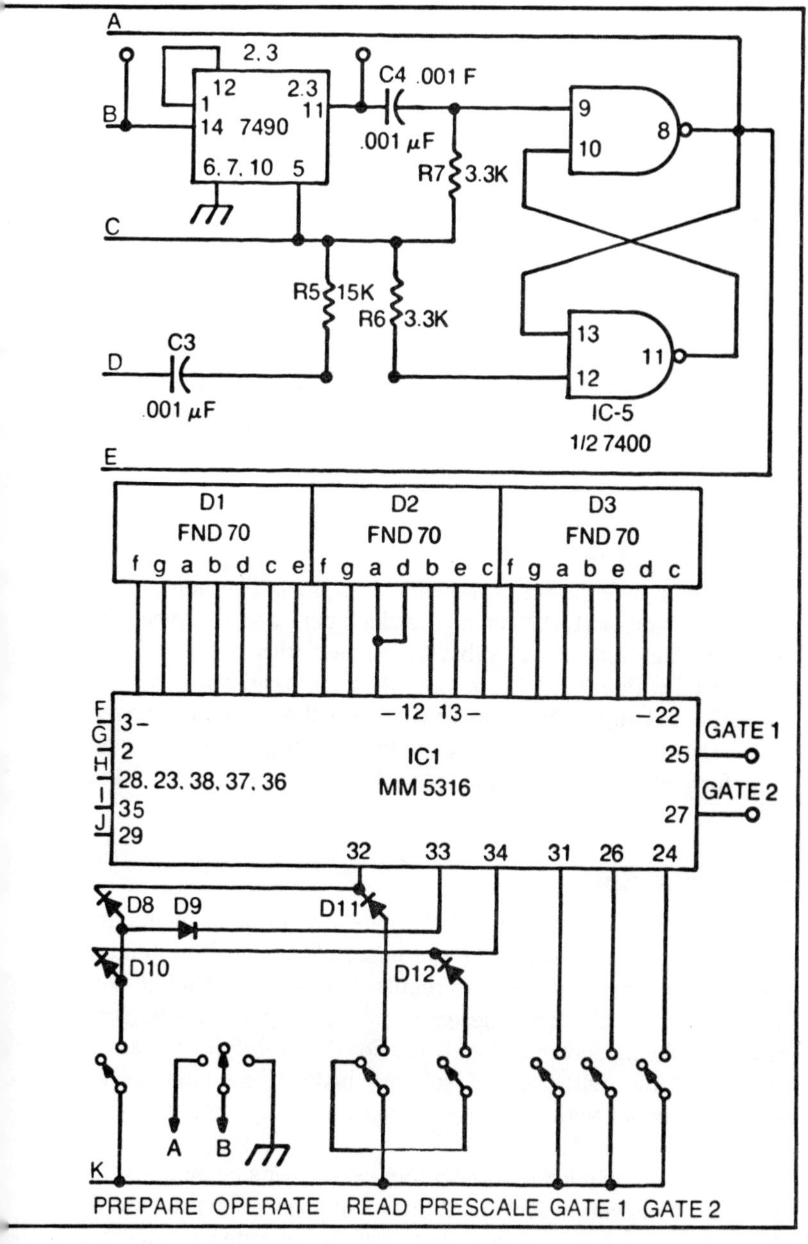

A
B
C
D
E
2.3
12
1
14 7490
2.3
11
6, 7, 10 5
C4 .001 F
.001 μF
R7 3.3K
9
10
8
R5 15K
R6 3.3K
C3
.001 μF
13
12
11
IC-5
1/2 7400
D1
FND 70
f g a b d c e
D2
FND 70
f g a d b e c
D3
FND 70
f g a b e d c
F
G
H
I
J
3–
2
28, 23, 38, 37, 36
35
29
– 12 13 –
– 22
IC1
MM 5316
GATE 1
25
GATE 2
27
32 33 34 31 26 24
D8 D9
D11
D10
D12
A B
K
PREPARE OPERATE READ PRESCALE GATE 1 GATE 2

LCDs; this is the common return, which simplifies feeding the LCD with ac, as needed for good life. Alternately, this return can be used for display intensity control.

For the prototype, the display was made of three FND-70 light emitting diode units. Two separate LEDs are wired to show the tens of hours (the coding is not the best however, 2 being lit for the first 10 hours, one for 10-20 hours, and none for the 20-24 hours reading).

The display normally shows minutes from 0-60 and hours from 0-24. Energizing the "seconds" circuit shuts off the two LEDs and the three display digits now read seconds from 0-60, and minutes from 0-9. This reading method is not a difficult process, but experimenters who do not need the gating functions of the MM5316 chip will probably want to change to the MM5309, with a six-digit display. This gives full range without switching.

## COMPLETE CIRCUIT

The complete circuit is shown in Fig. 8-10. The upper third is the time base part, the center the clock part, and the lower the controls. Each follows the subsystem schematics.

If used with another clock chip, the time base should be usable without change. The controls section will need to be changed to suit the chip. If the action is complex, it is suggested that a set of performance tables be made up, such as those in Tables 8-2 and 8-3.

## CONSTRUCTION

The chip, the displays and the gate generator elements are mounted on a single printed circuit board. This board is intended for a cabinet equal in size to two regular 3 × 4 × 5 inch miniboxes placed together. The board design is shown in Fig. 8-11, with parts placement being shown in Fig. 8-12. For the prototype, it was intended that the switching and isolating diodes be mounted on the instrument panel, with flexible leads going to the appropriate points of the printed circuit board.

It is recommended that the integrated circuits be mounted on the board by means of sockets. If this recommendation is followed, the sockets should be mounted first, then the discrete components. This should be followed by a check for good solder joints, with no solder bridges. Following this, the LED displays can be installed. After another solder check, the TTL ICs can be plugged in and

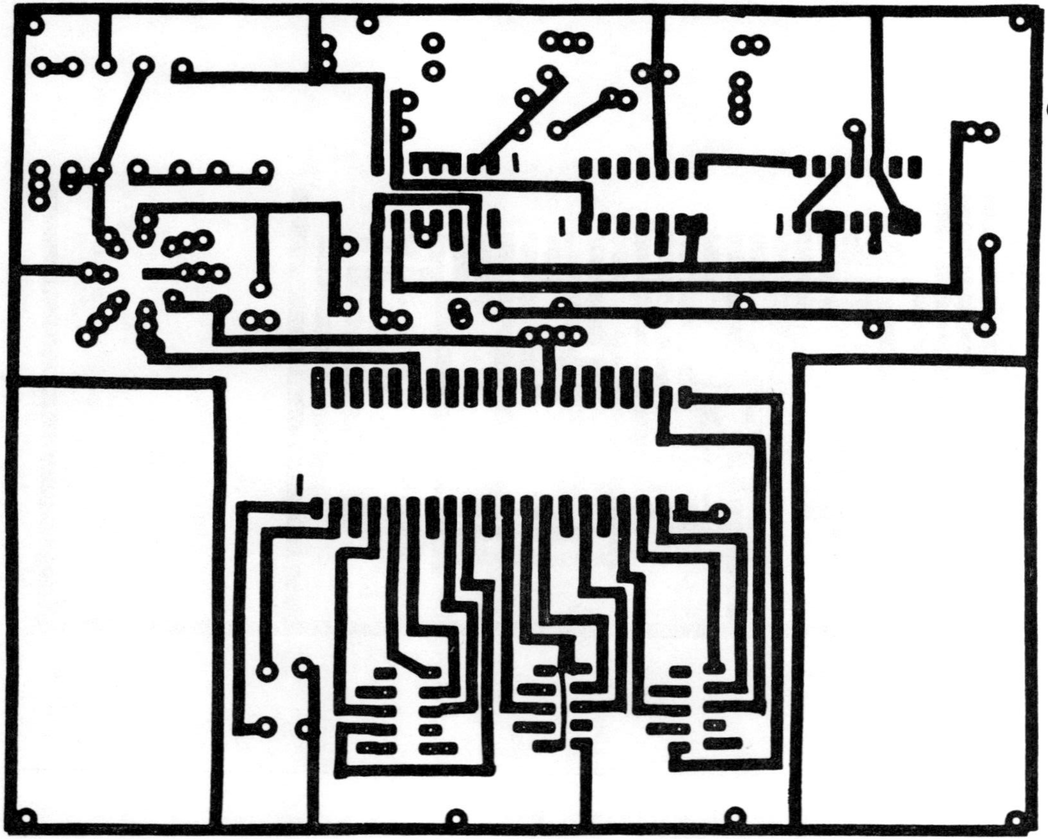


Fig. 8-11. Printed circuit board layout for the prototype. The blank areas are available for additional circuitry, or may be covered with resist. Original board prepared as for Fig. 4-7.

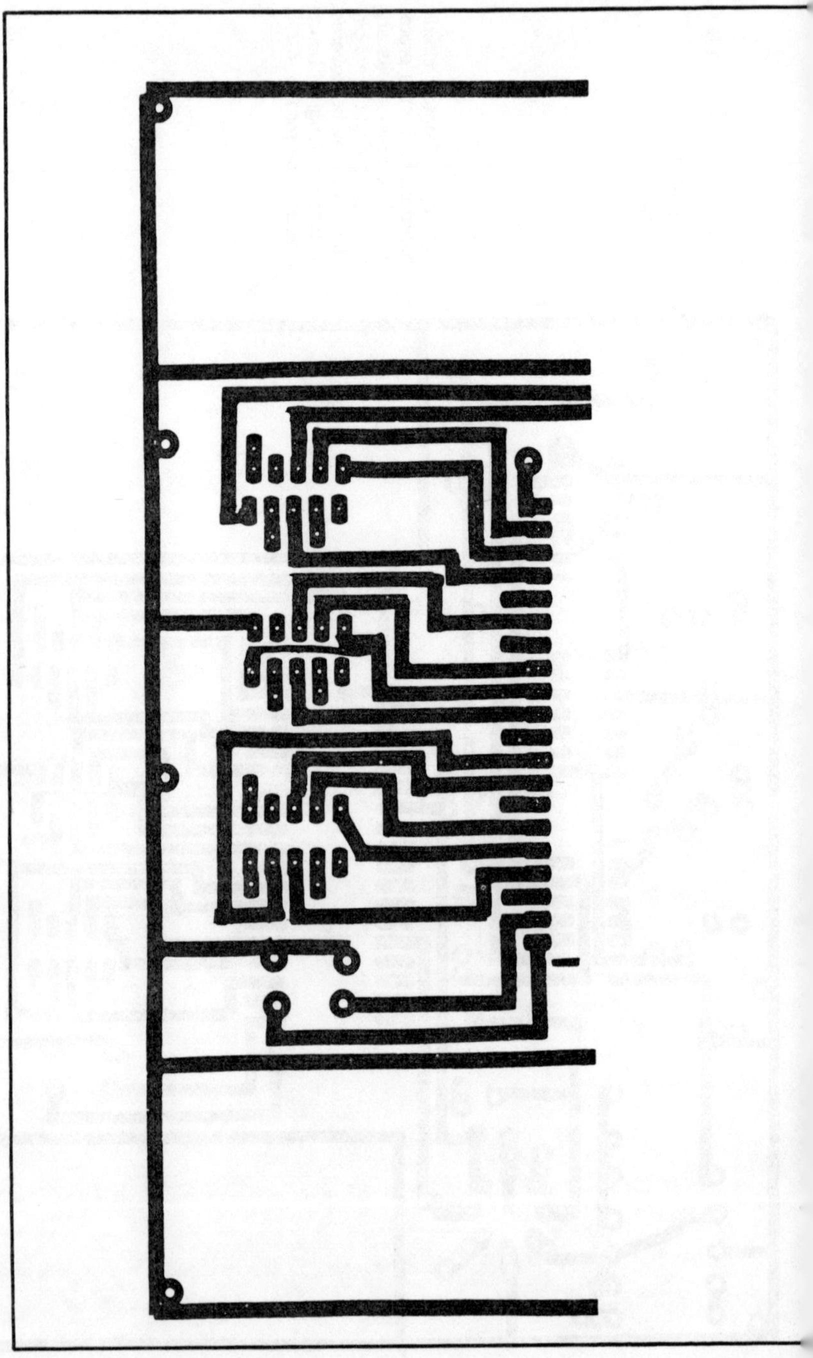

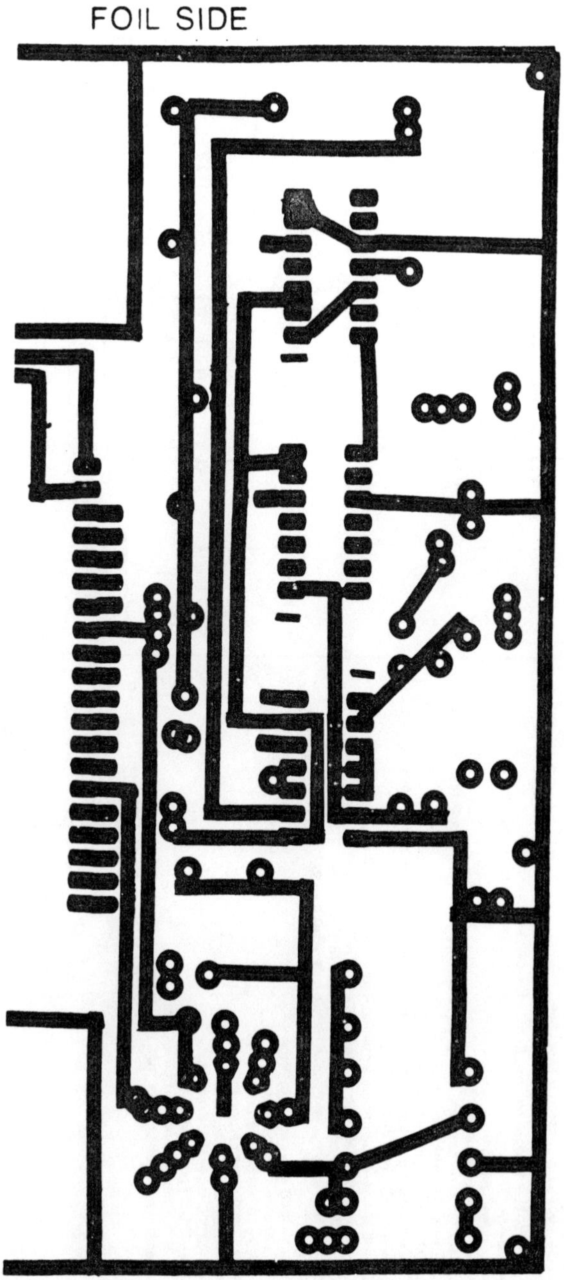


Fig. 8-11. Continued.

**PARTS LIST**

C1—.15 $\mu$F Mylar capacitor
C2—.1 $\mu$F Mylar capacitor
C3, C4—.001 $\mu$F disc ceramic capacitor
D1, D2, D3—FND70 7-segment single-digit LED readout (James Electronics, 1021-A Howard Avenue, San Carlos, CA. 94070) or equal
D4, D5—MV50 light emitting diode readout (James Electronics, 1021-A Howard Ave., San Carlos, CA. 94070) or equal
D6—3.1 volt zener diode
D7—5.1 volt zener diode, 1N5231 (ADVA Electronics, Box 4181K, Woodside, CA., 94062) or equal
D8, D9, D10, D11, D12—1N914 (Radio Shack 276-1144 or equal)
IC1—MM5316 digital clock chip (James Electronics, 1021-A Howard Ave., San Carlos, CA. 94070) or equal
IC2—709C linear operational amplifier (Radio Shack 276-017 or equal)
IC3—SN7490 decade counter (Radio Shack 276-1808 or equal)
IC4—SN7492 4-bit counter (Radio Shack 276-1819 or equal)
IC5—SN7400 quadruple 2-input NAND gate (Radio Shack 276-1801 or equal)
R1, R2—1000 ohm carbon resistor
R3—470 ohm carbon resistor
R4, R5—15,000 ohm carbon resistor
R6, R7—3300 ohm carbon resistor
R8—10,000 ohm carbon resistor
R9—20,000 ohm carbon resistor
R10—120,000 ohm carbon resistor
R11, R12—150,000 ohm carbon resistor
All resistors 1/4 watt or more, 10% tolerance

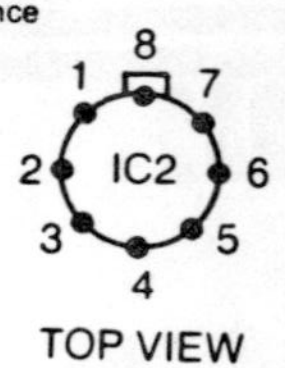


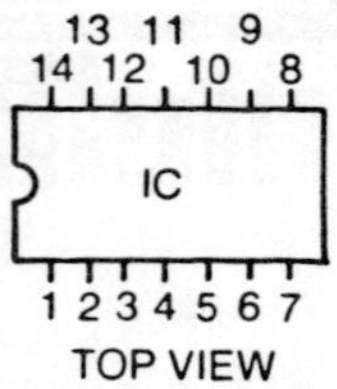


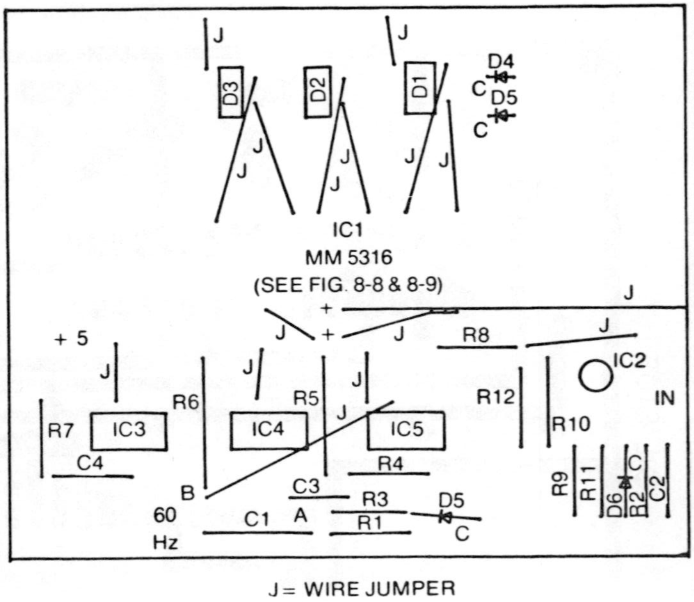


Fig. 8-12. Parts layout for the prototype board. See Figs. 8-8 and 8-9 for control connections.

power applied. The performance of the gate and gate-signal generator circuits can now be checked out. In particular, a check should be made of the duration of the gate—it should be exactly one second.

When the gate is operating, remove power, install the 709 and connect the 9-volt power. Gate action can be checked by jumpering 60 Hz to the input and observing the square-wave output from the gate. When all circuits are working properly, remove power, plug in the clock chip and apply power. Use the following section on operation as a guide to checking the various functions of the counter.

## CRITICAL CIRCUITS

The only circuit likely to give trouble is the comparator, which is quite sensitive to voltage levels. It may be necessary to change some of the resistor values to secure good limiting and gating action, especially if surplus ICs are used. It may even be necessary to change to a different IC, since surplus units are sometimes poor with respect to offset voltage.

Other than for this circuit, the problems should be the usual ones of digital and clock circuit construction, with difficulties most likely due to solder bridges or poor solder connections.

## SWITCH FUNCTIONS AND OPERATION OF THE INSTRUMENT

For the following discussions, refer to the switch diagram, Fig. 8-9, and assume that all switches are in the position shown on this figure. Assume, to begin, that a measuring cycle has just been completed.

Pressing the CLEAR switch clears the clock registers. However, occasionally one digit will remain set, this being due to the design of the chip circuits. If this happens, hold CLEAR and press RUN momentarily. After one second, release CLEAR. All counters will now read zero.

Assume that the measurement is to be made on the normal, or 50:1 prescale setting. Start the gate by pressing RUN momentarily. The display will blink and stop after one second.

To make the reading, first note the left digit. Add to this the number 10 if only the LED is lit, or add 20 if neither one is lit. The maximum is 24. These are nominally hours.

Next, note the reading of the two right digits. These are nominally minutes. Finally, press the READ SEC switch. The light emitting diodes will go out,and the left digit will change to the value of the right-most digit before the switch was pressed. Ignore this, and note the reading of the two right digits. These are nominally seconds.

The reading, in cycles per second (Hertz) is obtained from the sum:

| | | |
|---|---|---|
| Hours × 180,000 | = | ______ |
| Minutes × 3,000 | = | ______ |
| Seconds × 50 | = | ______ |
| Total | | ______ |

The entire process takes much longer to describe than to do.

If readings to the nearest cycle are desired, place the PRESCALE switch on 1:1. CLEAR and START perform the same operations, but there is a difference in reading, since the input is now connected directly to the minutes counter. The reading, in cycles per second (Hertz) is obtained from

| | | |
|---|---|---|
| Hours × 60 | = | ______ |
| Minutes × 1 | = | ______ |
| Sum | | ______ |

The maximum reading is 1439 Hertz.

When checking against the normal laboratory sources, repeated readings will commonly give different values. Digital counters are a good way to learn about oscillator stability. However, a good stable source should give readings to within a few hertz of each other. If there is much variation, check the 709 gate, and the gate generator output; it is likely that there is a problem with slow rise/fall of the gate, or with a parasitic oscillation.

Operation in the RUN position is the same as above, except that count is accumulated as long as RUN is on. The count may be reset to zero at any time by pressing CLEAR, or it may be ended by returning RUN to OFF. Reading in cycles or events is as above.

If timer operation is desired, jumper the input to the 60 Hz point, and proceed as for RUN, with the RUN acting as the start-stop of the period.

Clock operation is possible if a 50 Hz source is available, or the 50/60 pin can be brought to a switch. Proceed as for RUN. CLEAR sets the clock to zero hundred hours. Fast advance is obtained by throwing prescale to the 1:1 position. The clock can be stopped by returning the switch from RUN to OFF.

If gate 1 signal is desired, check its time setting by throwing READ GATE 1; if not correct, hold this switch on, and press CLEAR. The gate reading will go to zero. It can be advanced by holding READ on, and throwing PRESCALE to the 1:1 position. Stop by releasing PRESCALE when the desired "time" reading, or cycle count is obtained. Now the gate signal will go high whenever this indicated time count is reached. It will go low after the number of cycles corresponding to 55 minutes, or if GATE 1 STOP is activated momentarily.

Gate two is read in a similar fashion. This can be done at any time or cycle count, and automatically starts the gate. Pressing CLEAR in addition allows reset of Gate 2 duration. To stop the gate, press READ GATE 2 a second time.

## ADDITIONAL USES

In addition to these two major functions of counting and gating, this simple counter has some additional uses, which arise from the flexibility of the clock chip and the associated circuits. For example, it is often convenient to bring the output points of the divide by 60 chain in the gate control to the front panel. These outputs can be used to give periods of 1 second and 1/10th second. Also, these outputs can be used for prescalers. The TTL dividers can work to quite high speeds, up to about 20 MHz reliably, and the divide by 60 chain can be used above this range.

Additionally, the 1 Hz output of the clock chip (see Fig. 8-8) can be used to extend this counting range; with 60 Hz input, the output will be 1 Hz in the clock mode, and, of course, 1 Hz in the run mode with the prescale off. This output can also be used over an extended frequency range; however, due to the characteristics of the clock chips, the maximum frequency at the input to the divider chain is about 500 kHz.

It is also sometimes useful to bring the output of the gate to the front panel. This allows use of the input circuit for wave shaping, or for some complex waveform generation. With some attention to switching, a limited range of word generation for digital work is possible. Additionally, there are some other gating features in the particular clock chip used. See the instruction sheets of the chip for this. It is worthwhile reviewing carefully the chip characteristics; other chips may have features which are desirable.

## COUNTER VARIATIONS OF CLOCK CHIPS

In view of the great flexibility of clock chips, as can be seen from the characteristics of this counter, it is regrettable that counter versions of the chips are not available generally. It is likely that this lack will be remedied in the future. For example, MOSTEK has introduced a six-decade counter-decoder. Its major functions are apparent from Fig. 8-13. This chip includes two six-digit BCD registers, corresponding to the normal time and alarm. Both of these registers can be preset, and there is a comparator to give a gating signal. Both BCD and seven-segment multiplex outputs are available. Various gates and strobes are provided. Operating capability is quite good, the maximum frequency being 1 MHz, which is readable to 1 Hz. Some external circuitry is required for use of this chip, in the way of time base generation and gate control. For some applications, the counter itself can serve as a gate. Additionally, the usual display drivers are required. A major feature is the ability to reset the counters. This may be done simultaneously with a six-bank switch, or individually with a single switch, multiplexed to the scan signals.

## A PRESCALER FOR GENERAL COUNTER USE

The simple counters described above, and most inexpensive counters, have limitations in the upper frequency range. A method of extending this range which does not require much effort and which retains accuracy would be useful.

The usual procedure of installing a divide by 10 prescaler satisfies the requirement of simplicity of use, but does lose accuracy, since the uncertainty in the counter reading has been multiplied by ten. Cascading two prescalers does not increase the difficulty of use, but does increase the uncertainty by the prescale ratio.

There is a method of retaining the simplicity, while avoiding the loss of accuracy. This method depends on the mixing characteristics of a type D flip-flop. Suppose, as in Fig. 8-14, that an unknown input signal is fed to the D input of the flip-flop, and a known reference frequency fed to the clock input of the flip-flop. Suppose that the unknown frequency starts at a very low value, and then increases. For a time, the Q output of the flip-flop will follow the unknown frequency exactly. That is, for each complete cycle of

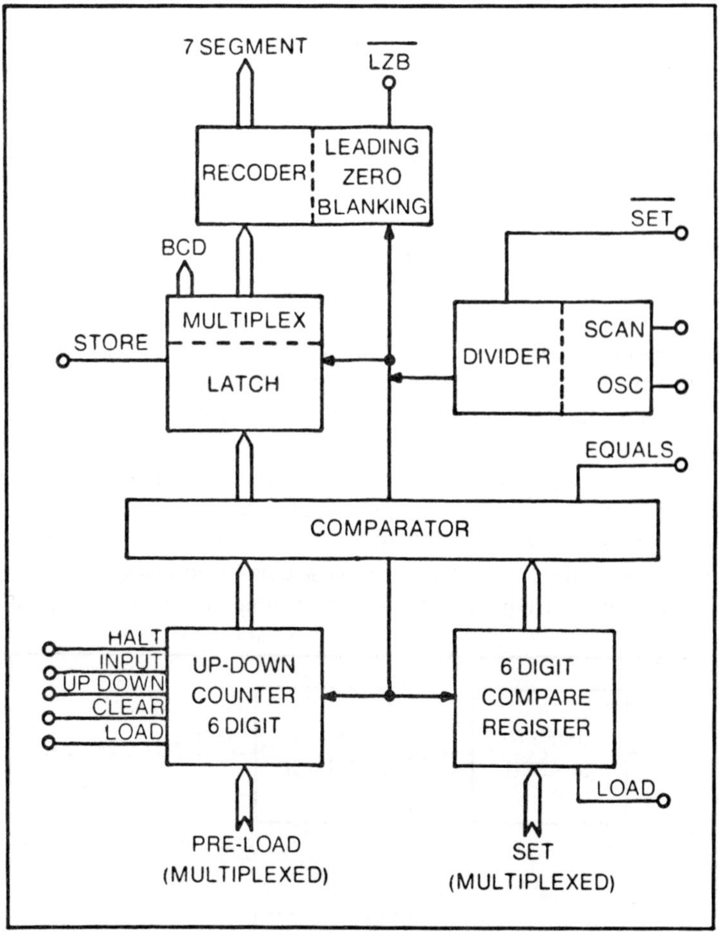


Fig. 8-13. Basic elements of a counter chip which uses clock chip technology, the MK50395. This six-digit counter-comparator will operate at 1 MHz.

the input, there will be a complete cycle of the output. However, when the input frequency reaches the reference frequency fed to the clock, this relation ceases. Now as the input frequency increases, the output frequency decreases. The decrease is exactly proportional, ten cycles increase in the input frequency causing a ten-cycle decrease in the output frequency. This new action continues until the input frequency is exactly twice the clock frequency, at which point the Q output of the flip-flop is zero. Now with a further increase in input frequency, the output frequency again starts to increase in exactly linear relation. Graphically the relation between input and output frequency is shown in Fig. 8-15. This figure also shows the mathematical relation governing the output frequency.

It seems clear that a very useful prescaler can be developed from this circuit. Suppose that the clock frequency is set to 2.0 MHz. Now the output frequency to the counter will never exceed 1 MHz. However, any frequency within the clocking range of the type D flip-flop can be measured. For example, a frequency of 18.976 MHz can be measured and will appear on a counter connected to the output as 976.00 kHz. The 18 must be determined by other means. In most cases this is not a problem, since it is usually known. If it is not known, simple circuits, such as indicating wave meters, can be used to determine the value.

For a counter of relatively limited range, such as the clock chip counter above, the clock source should be chosen with the maximum frequency range of the counter in mind. A frequency of 200 kHz would be suitable for the clock chip counter, although with some chips it may be possible to use 1 MHz. With normal six-digit counters, 2 MHz is probably the best clock frequency.

The upper frequency range is determined by the characteris-

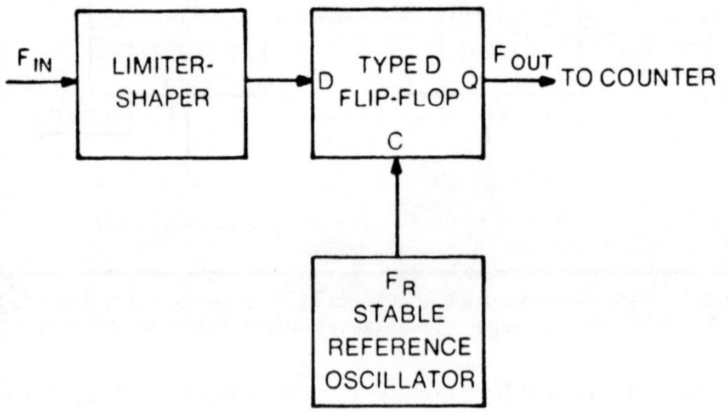


$$F_{OUT} \leq \frac{F_R}{2}$$

$$F_{OUT} = F_{IN} - \frac{mF_R}{2}$$

$$m = 0, 1, 2 \cdots$$

Fig. 8-14. Frequency translation by use of a type D flip-flop. The input-output relations are given by the two equations. Unlike the usual prescaler, this arrangement does not reduce counting resolution.

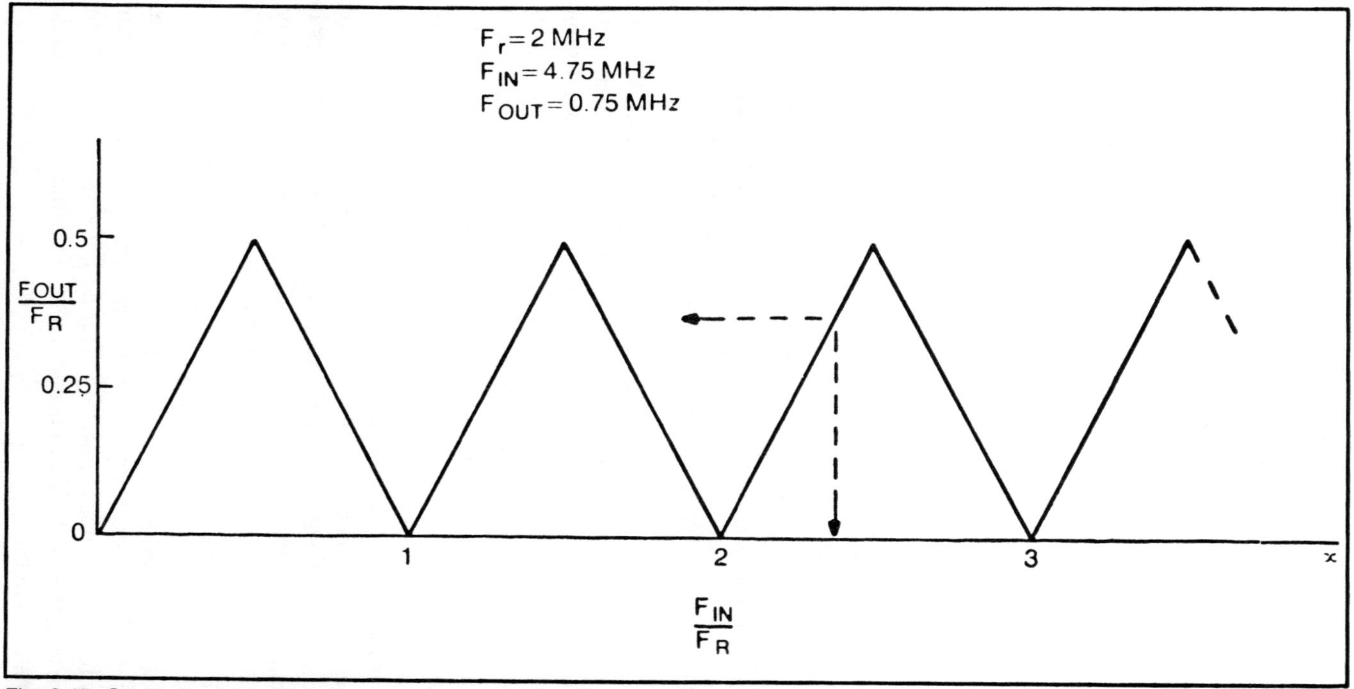


Fig. 8-15. Output frequency as a function of input frequency. The relation continues to the maximum operating frequency of the flip-flop.

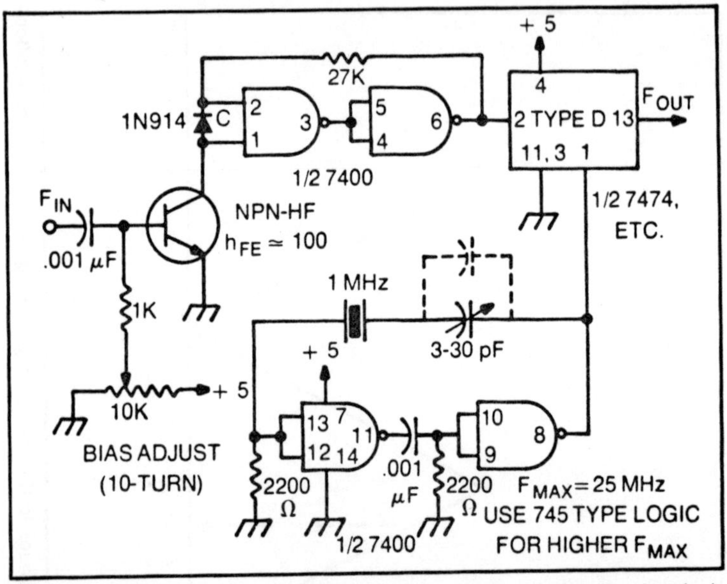


Fig. 8-16. Circuit for a practical frequency translator, for use with the clock-counter or other limited range counter. The input amplifier-Schmitt trigger can be adjusted to limit on signals as small as 50-100 mV. The accuracy is dependent on the accuracy of the 1 MHz reference frequency.

tics of the input shaper, and the type of flip-flop used. With MOS flip-flops, the upper frequency range will typically be 5 MHz; with TTL the upper range will typically be 20 to 30 MHz, although some Schottky types can reach 100 MHz. Frequencies of several hundred megahertz would be possible with emitter-coupled logic. There is no reason why two type D flip-flops could not be cascaded, for example, one giving output around 20 MHz with VHF inputs, the second giving outputs around 1 MHz for HF inputs.

One point of caution should be remembered if this circuit is used at low frequencies: the output is not a perfect square-wave, with each cycle of identical duration. Instead, the outputs vary in length, this being the variable which causes the particular behavior. It is instructive to examine the output waveforms for a variety of inputs (see the references).

A suitable circuit for this prescaler is shown in Fig. 8-16, for frequency ranges up to a few tens of megahertz. For higher frequencies this simple input circuit may need modification, primarily to increase input sensitivity. This design would be suitable for use with the clock chip frequency meter, or with almost any of the small counters.

# Chapter 9

# A Primary Standard of Frequency

This project covers the design and construction of a self-contained electronic clock which is also a primary standard of frequency. It is suitable for precise time measurements, for checking the accuracy of other clocks and watches, or for such functions as generating accurate time ticks. Equally, it is suitable for precise frequency measurements, for checking the accuracy of signal generators or similar sources, and for checking accuracy of working frequency standards.

## PRIMARY STANDARDS AND SECONDARY STANDARDS

To see why this unit is called a primary standard and why it is suitable for checking both frequency and time, let us look at the difference between primary and secondary standards, as applied to frequency. To start, look at secondary standards, the most common. Assume that we have just finished a kit communications receiver which contains a crystal calibrator, and wish to set the calibration frequency correctly. As shown in Fig. 9-1, most such calibrators feed the receiver RF stage, the signal passing through to the detector. To check the calibration, the receiver is tuned to a signal of known frequency, the highly accurate Standard Frequency Station WWV being the most common. This signal also passes through the RF stages to the detector. There the two signals mix or beat together, or against the locally generated beat os-

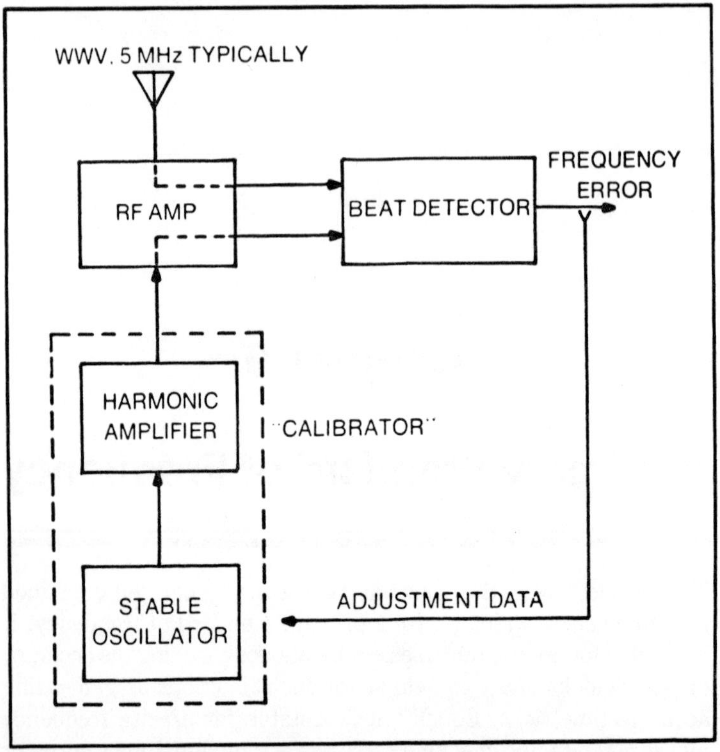


Fig. 9-1. Signal flow with a crystal calibrator being checked against WWV. The calibrator is a secondary standard, since it is second to WWV.

cillator; if there is a frequency difference between the calibrator and WWV, an audio signal comes out of the detector. If the two signals are very close in frequency, the detector output becomes very low frequency audio, and if they should be exactly the same frequency, the detector output becomes inaudible.

To adjust the calibrator, its frequency is varied, usually by a small capacitance in series with the crystal, until the audio beat note is zero. In doing this, we note that there is a small range in the condenser adjustment over which there is no perceptible beat. Considering this, we remember that communication receivers have poor audio response—frequencies below about 200 Hz are suppressed, as a way of reducing interference. We try to set the adjustment to the midpoint, approximating zero beat, and in fact, probably get to within 50 Hz of the exact frequency. The frequency difference between the calibrator and WWV is then no more than 50 Hertz out of 5 MHz, or about 1 in 100,000. This accuracy is typical of this class of standards used in this way.

While we could improve the accuracy of measurements by changing the audio system, or even by going to an audio amplifier with meter indications at the output of the detector, we cannot change one aspect of this type of standard. This is the fact that our calibration depends on another standard. We are dealing with a "secondary" standard. As long as we operate by comparing our locally generated signal to another, we are working on a secondary standard.

As an alternate, suppose that the output of a stable oscillator were fed to a frequency divider-clock combination, with division ratios such that the clock is supposed to keep correct time. Now, and as shown in Fig. 9-2, we have another way of checking, this time by checking clock time against WWV time. Suppose that this check is made only to the nearest second. If the check is made on two successive days, the difference will be that accumulated over

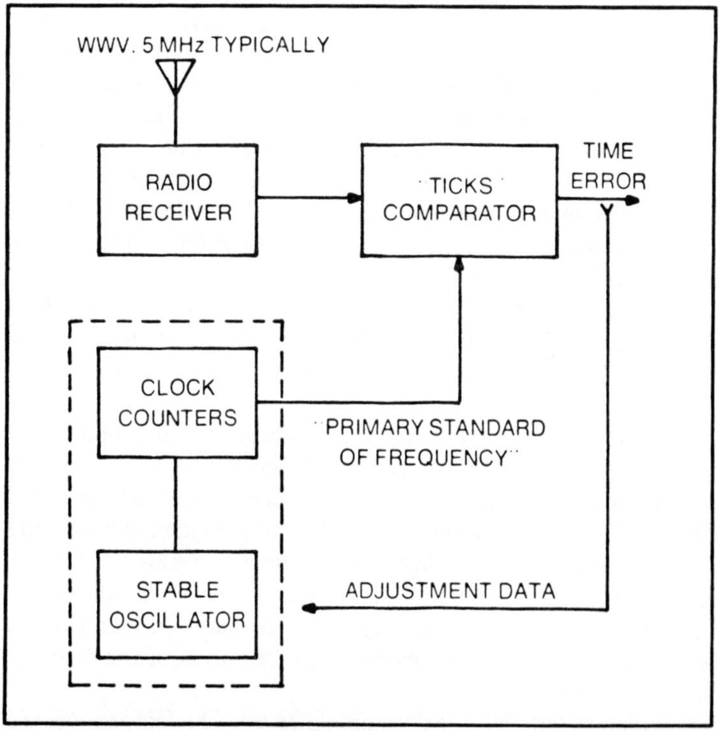


Fig. 9-2. Signal flow with an oscillator-clock being checked against WWV time ticks. Since frequency is defined in terms of time, a primary determination of frequency can be obtained. See text.

the number of seconds in a day, or the number of seconds of difference which accumulate in 86,400 seconds. We can measure the difference to within one part in 86,400, with our assumption of one second of time checking error.

Now suppose the clocks are checked at an interval of ten days time. Again we can detect the change accumulated in these ten days, or a change of one part in 864,000. We could extend the time still further, detecting smaller and smaller percentage differences.

Actually, it is no great problem to measure time difference to 1000th of a second. This means that two measurements a day apart are sufficient to detect differences of about one part in 10,000,000. One feature of this method of checking is that it is more sensitive to error. This is the result of the fact that we are dealing with the accumulation or integral of the error over a time period, rather than the error which exists at a specific instant.

There is another important result of this method of operating. Suppose we also construct an accurate sundial, say by using a fine stretched wire for the gnomen, with a large dial plate. With this we can measure time down to seconds. The sundial gives sun time, while we need mean time, which is what clocks are supposed to measure. The conversion from one to the other is done by the equation of time, as discussed in Chapter 1; this is known from thousands of years of observation, so we can get mean time without reference to anyone else's measurements. Since the basic definition of the second comes from the earth's rotation, this method of determination is in fact, first hand; it is "primary." We are getting our information "out of the horse's mouth," as it were. Furthermore, since we are transferring this primary measurement of time to our oscillator, we have both a primary measurement of time and a primary measurement of frequency.

These, then, are the two differences between primary and secondary time standards. One difference is that the secondary measurement of frequency is an instantaneous measurement and can detect only the differences existing at that instant, whereas the primary measurement, against time, can detect, accumulated differences. The more important difference, however, is that the secondary method depends on having a known source for comparison whereas the primary goes to the basic definition of the quantity being measured.

We must remember one thing, however. The fact that we are dealing with a primary standard does not guarantee that the primary standard is automatically better than a secondary standard.

Whether it is or is not depends on such factors as the stability of the oscillator, the existence of wrong counts in the divider chain, our accuracy in making the time comparison and so on. True, it is considerably easier to tell if these factors are affecting the timekeeping, and therefore are affecting the frequency. Assuming that you are working with a good stable oscillator, and being careful of all the other details, it is considerably easier to obtain the correction from oscillator frequency to true frequency. Always, the error will be better known, and changes due to drift or accident will be more evident. Because of this, a primary standard is a good device for any worker in radio or electronics.

## PRECISION AND RESOLUTION

Let us take a moment to review the factors of precision and resolution. They are very different, but it is easy to confuse the two and be misled into believing that a measurement is highly accurate when it is only made with high resolution.

Suppose we go to the lumber yard and ask for a strip of wood 36 inches long and use it in measurements. We could use this directly, lay it end to end to measure off distances and count the number of rod lengths. We could say, for example, that a length we measure is more than 64 times the strip length, but less than 65 times. We are working with a *resolution* of 1 yard.

We could also have asked for a one-inch length in addition, and used this to place marks on the strip, separating it into 36 equal spaces. Actually, we could also do this by geometric construction, using the principles learned in algebra. We can now make a measurement such as 64 yards, plus more than 3 and less than 4 inches.

For another step, we could divide each inch further, say into eight parts. However, we find that we are working against a limit. For one thing, the pencil we are using to make marks leaves a very broad line. Then too, the line position varies some, depending on how we hold the pencil. At this point, we are reaching a practical limit. It is in our ability to decide whether the distance we are measuring is at one mark or the next. We have reached our ability to resolve between two values. This is our resolution limit.

Now let us consider another matter. Is our "yardstick" really 36 inches long? Did the lumber yard really measure it, or just cut "by eye." If measured, did the saw go on the left or right of the mark, shortening or lengthening the stick by the width of the saw kerf? We don't know.

We try to resolve this question by borrowing a steel rule. Suppose we find that our yardstick measures one-eighth inch short. We can compensate for this by multiplying our measured values by the fraction that we are short and then add the total to our measurement.

What we have done is to use a working standard, the steel rule, to determine the accuracy of our measuring device, the yardstick. Note that this factor is completely independent of resolution, which hasn't changed by our work with the steel rule.

If we reflect a little, we see that we may still have problems with accuracy. Wood, for example, shrinks when dry. Was the stick green when we first used it, and did it dry out later? Or, if the wood was well cured, did we make the comparison with the steel rule after a period of high humidity, when the stick was longer? We don't know, unless we happened to keep records.

In addition to the difference between resolution and accuracy, this discussion points out the use of a working standard, in this case, the steel rule. There was another standard behind this, and so on back until there is a fundamental standard. This should not be confused with still another term, primary standard. For example, a primary standard of frequency must be based on time. The fundamental standard may be a primary standard. But not all primary standards are fundamental. In this chapter we will look at some working standards for time which can also serve as primary standards of frequency. Remember that all of this doesn't say anything about the quality of the standard. This will depend on many factors, chiefly on the care with which we work.

## CONCEPT OF THE PRIMARY STANDARD

From the above, the major elements of the primary standard are a stable oscillator, a clock system, and the necessary interface elements to link these together. Also, as indicated in the block diagram of Fig. 9-3, we need at least one more element, the power supply; several more elements are desirable, if we are to have an accurate standard. The power supply is a necessity, and since comparisons of time are needed over long periods, a standby power source is almost a necessity. Stable oscillators are not really easy to build; major contributors to frequency instability are variations of voltage and temperature, so regulators for these are needed if we are, in fact, to have an accurate standard. Finally, various controls, especially for the clock chips, are needed.

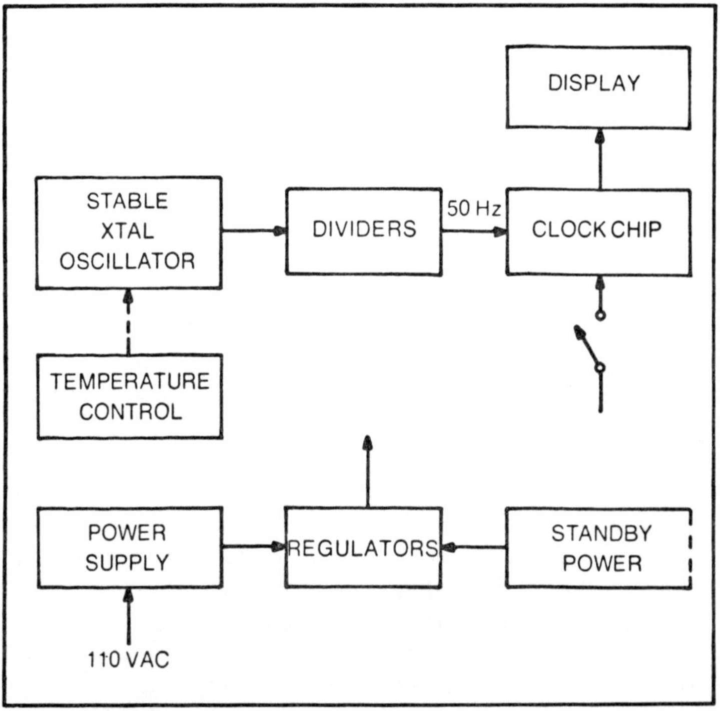


Fig. 9-3. Elements required for an accurate primary standard. Accuracy will be reduced if temperature and voltage regulation and standby power is omitted, but will still be better than for a simple calibrator.

The clock chip is essentially self-contained, and can be any of the types previously used, say the miniature display type of Fig. 4-21. To go with this we need the oscillator section, together with the dividers to reach the 50 Hz clock input. This can form one unit, equal in size to the clock board. Other units could be added for regulation, controls, etc.

## OSCILLATOR CIRCUITS

Of the easily built oscillators, a crystal oscillator using an AT-cut crystal is the most stable. The best range for these crystals is around 3 to 5 MHz. On the other hand, if we are to use a standard clock chip its inputs should be 50 or 60 Hz; for radio use, 50 Hz is a better choice, since it is much easier to obtain the integral frequencies which are most desirable for calibration. Evidently a large frequency division is necessary between the crystal and the clock input.

Suppose a 4 MHz crystal is used; this seems a good choice since the size of the crystal is reasonable, and good crystals are available at reasonable cost. Also the frequency is in the range for maximum stability. Choice of this frequency requires an 80,000-1 division. This could be done with a scale of 4 counter, then a cascade of scale of 10 counters, followed finally by a scale of 2 counter. This would have the advantage of furnishing a set of commonly used calibration frequencies. However, the package count would be rather large.

We can reduce the package count by using one of the multistage binary counters available in low power drain MOS integrated circuits. These can be converted to any division scale by using AND gate resets, as described in Chapter 2. Figure 9-4 shows use of such a counter, giving 8,000 to 1 step-down in one package and 10 to 1 in the second. The AND gates can be diodes, so the complexity is low and the power drain is no greater than for the two packages. If other oscillator frequencies must be used for any reason, a change in diode location will give the necessary change in division ratio. The table with Fig. 9-4 shows the scaling factors for some of the common crystal frequencies. These are by no means the only possible combinations, but it must be remembered that for radio use, frequencies which are integral MHz are preferred.

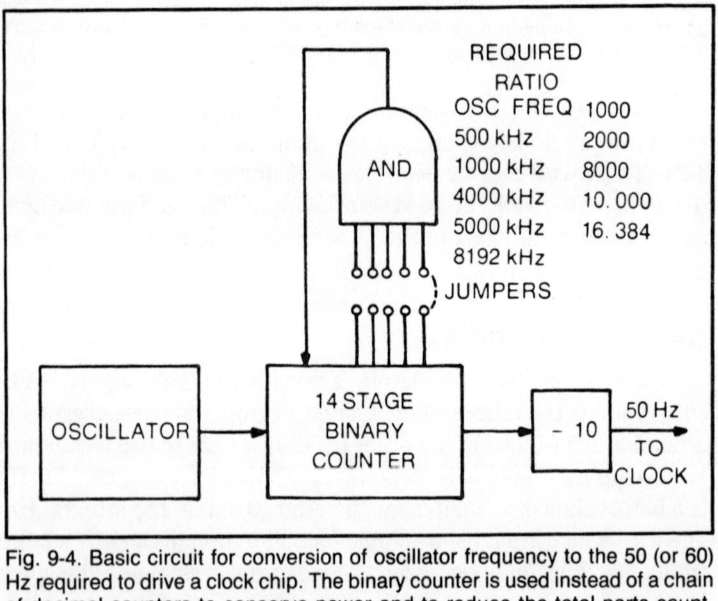


Fig. 9-4. Basic circuit for conversion of oscillator frequency to the 50 (or 60) Hz required to drive a clock chip. The binary counter is used instead of a chain of decimal counters to conserve power and to reduce the total parts count.

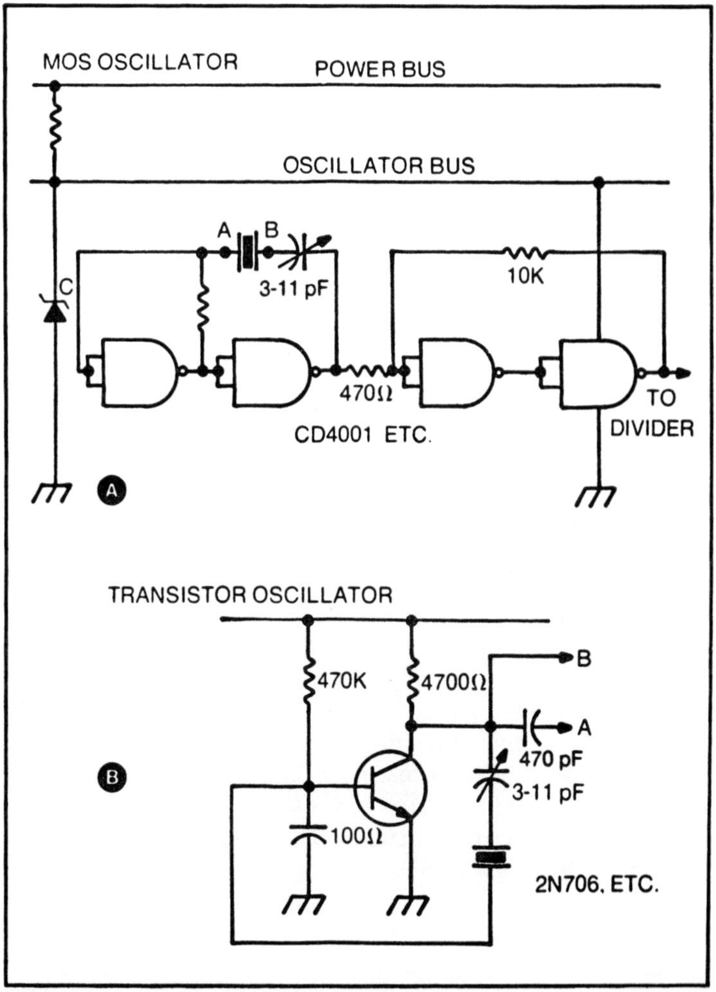


Fig. 9-5. Two simple crystal oscillators for a medium accuracy standard. The maximum frequency for the MOS type (A) will depend on the chip and operating voltage. The alternate transistor oscillator (B) has better high frequency characteristics.

Since low drain MOS dividers are used, we can use the same type of elements in the oscillator. A common digital IC is shown in Fig. 9-5(A). It is basically a multivibrator, followed by a Schmitt trigger. This provides a good square-wave output, and isolation from following stages. Stability is enhanced by providing a Zener regulator for the oscillator-limiter stages. The oscillator is not the most stable type which could be used, but with a good zero temperature coefficient crystal, good components, and care in construction, will be accurate enough for the intended purpose.

We could increase the stability by using an oscillator which operates at low level, using a buffer amplifier for isolation and better stability. This would also eliminate a limitation which sometimes arises: MOS digital circuits are usually appreciably poorer in performance at high frequencies, and the simple oscillator just described may need a low internal resistance or "active" crystal to give proper output at the high frequencies. The separate oscillator could use a high gain transistor or FET, and will not have this problem.

It is no great problem to provide for both. A suitable circuit for the separate oscillator is shown in Fig. 9-5(B). To use it, the pairs A-A′ and B-B′ are jumpered. Usually, the digital oscillator would be used with low frequency crystals, and the separate oscillator with higher frequency units.

With this design choice, first experimental work can be done with the simple MOS oscillator. Later, the separate oscillator can be used. As an ultimate, one of the super-stable types described in the references could be installed.

## DIVIDERS

The division ratio of the divider chain must be determined to suit the crystal frequency chosen; for the prototype this was 8,000 to 1 in one stage and 80,000 to 1 over all. The 8,000 to 1 requires a minimum of twelve binary stages. The closest commonly available MOS package is a 14 stage design, the RCA SD4020, but not all of the early stages are brought out to pins. To check if this package is usable, it is necessary to determine which stages are to be connected to the AND gates. This is done by successive subtraction, as shown in Table 9-1. The necessary division ratio is first entered and the next smaller binary number subtracted from this; the exponent of the binary number expressed as a power of two gives the last stage which must be connected. This process is repeated with the remainder, and repeated again until zero is reached. For the prototype, it was found that stage six plus stages seven through twelve must be connected to AND gates.

The RCA SD4020 COS-MOS counter has pin connections for these stages, and so is used. The detail connections for this counter and for the following SD4017 divide by 10 counter are shown in Fig. 9-6. The output of the 4017 goes directly to the clock chip.

**Table 9-1. Example of Divider Ratio Calculation.**

Oscillator Frequency — $4 \times 10^6$ Hz
Output Frequency (to ÷ 10) 500 Hz

| | | | |
|---|---|---|---|
| Required | Ratio | = | 8000 |
| | $2^{12}$ | = | 4096 |
| Subtract | | | 3904 |
| | $2^{11}$ | = | 2048 |
| | | | 1856 |
| | $2^{10}$ | = | 1024 |
| | | | 832 |
| | $2^{9}$ | = | 512 |
| | | | 320 |
| | $2^{8}$ | = | 256 |
| | | | 64 |
| | $2^{6}$ | = | 64 |
| | | | 00 |

Therefore, connect AND gate to
6th, 8th, 9th, 10th, 11th and 12th stage outputs.

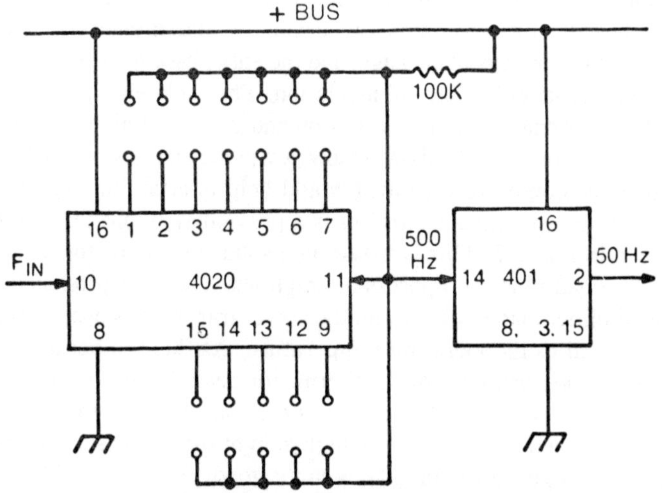


| FOR DIVISION RATIO OF 8000, CONNECT DIODES TO: | PIN | STAGE |
|---|---|---|
| | 1 | 12 |
| | 15 | 11 |
| | 14 | 10 |
| | 12 | 9 |
| | 13 | 8 |
| | 4 | 6 |

Fig. 9-6. Counter chain as used in the prototype, with crystals from 500 kHz to 4 MHz. The divide by 10 stage can be omitted if only low frequency oscillator crystals are to be used.

To allow for complete flexibility in crystal frequencies, pads should be provided for AND diodes for all available outputs of the 4020. The diodes to be connected are determined as described in Table 9-1.

## TEMPERATURE CONTROL

For best stability, temperature control is needed. This is important for the crystal, even though it is nominally zero coefficient, since the actual temperature coefficient may amount to a few parts per million of drift, either + or −, i.e., higher or lower. Temperature control is also important for the oscillator elements, in particular, those which change the circuit capacity "seen" by the crystal, or which could change the amplitude of oscillation. The method of control indicated here is to use a temperature sensitive controller on the crystal, with the combination of crystal and controller surrounded by insulating material and a metal shield. The printed circuit board which contains the oscillator and divider is placed adjacent to this crystal package, and the entire assembly surrounded by another layer of thermal insulation. This isolates the oscillator and divider chain from fast variation and gives partial temperature control for them, while allowing low power demand operation. (Two stages of temperature control would help in attaining stability.)

The temperature control uses a previously published circuit, shown in Fig. 9-7. This is based on a solid state detector, a thermistor bead plus a proportional amplifier. The heat produced by the amplifier load resistors and the power transistor is used to raise the crystal to its operating temperature. As this temperature increases, the resistance of the thermistor bead decreases, decreasing the current through the transistor, and thereby reducing the heat produced. Because of the high gain of the two cascaded transistors, temperature stability is quite good.

## THE CLOCK CHIP

Nearly all of the available clock chips have 50 Hz input capability. Some of the other features which are desirable are low power drain, simplicity of set, sufficient power output to drive a display directly, and in some cases, binary coded output to allow use with other equipment. The exact chip should be chosen to suit the requirements in mind; for the prototype, the design of Fig. 9-8 was used, actually a repackaging of the experimenter's clock of Chapter 4, laid out with this particular application in mind. The major

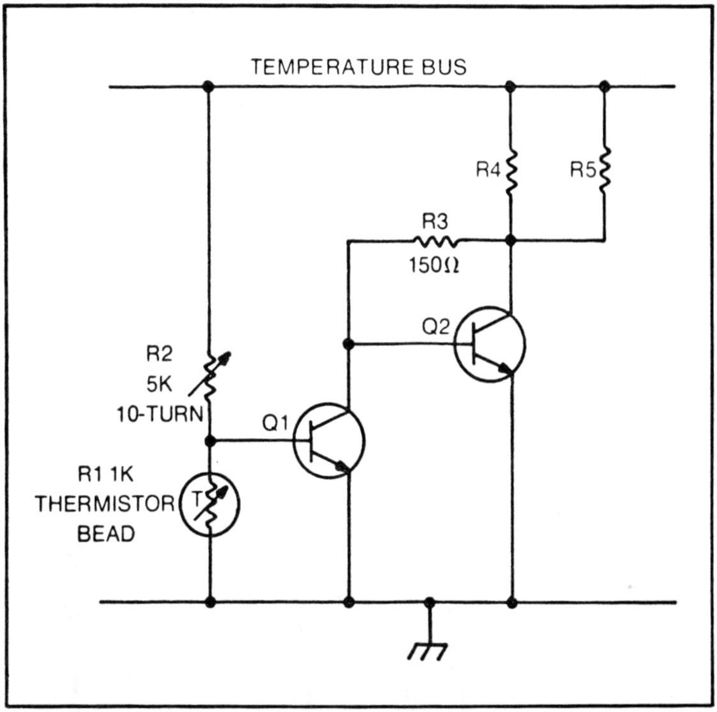


Fig. 9-7. Temperature regulator for the crystal. Regulation stability depends on the gain of the transistors. For further information see "Ham Radio Magazine," April 1976.

difference is that a small six-digit display is used, of the type used in small calculators. The segments of this can be driven directly. See Fig. 4-21. The exact arrangement of this chip is unimportant to the functioning of the primary standard, and others can be substituted.

## POWER SUPPLY

Power line operation of the primary standard is desirable, especially in view of the need for long term operation, to enhance accuracy and to make accurate checking easier. However, ac lines do fail at times, so some standby power provisions seem necessary. This standby capability can be provided by a 9-volt rechargeable battery. A circuit for this is shown in Fig. 9-9. The 110-to-12 volt step-down power transformer can be one of the type built into the power cord; this is good practice since it avoids having a potential heat source close to the clock.

Following the rectifier, note that there are five power paths:

—plus bus to clock and oscillator
—plus bus to display
—plus bus to battery for charge
—battery to clock chip and oscillator
—battery to display

Diodes are used to isolate the various circuits. The battery is left

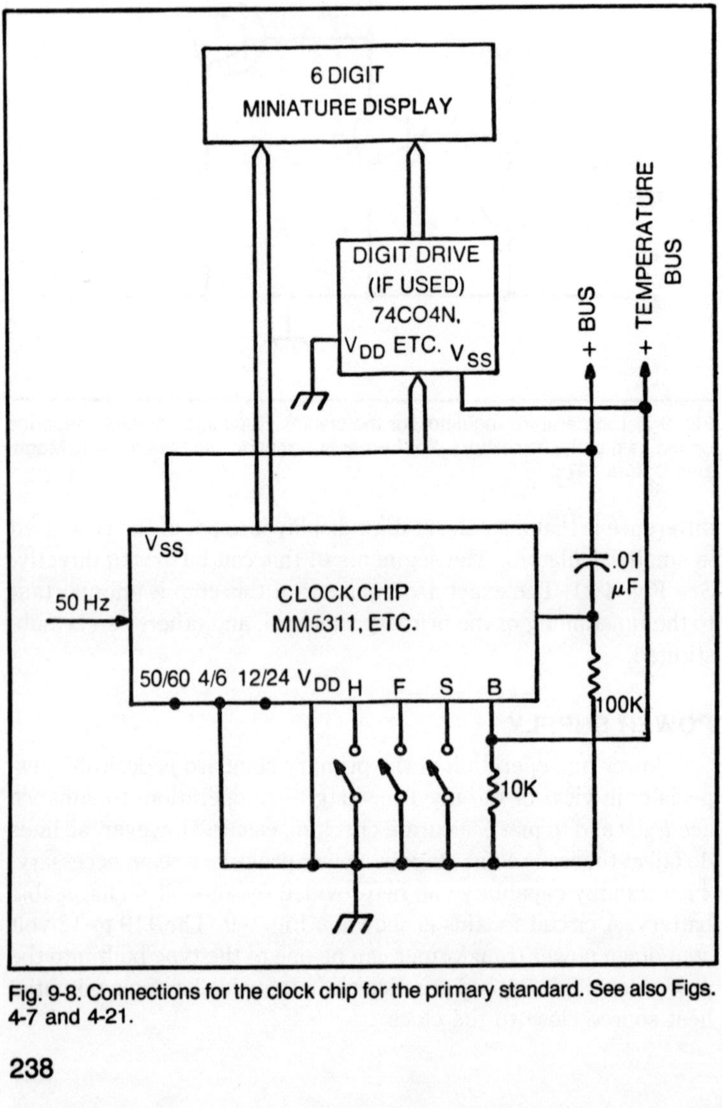


Fig. 9-8. Connections for the clock chip for the primary standard. See also Figs. 4-7 and 4-21.

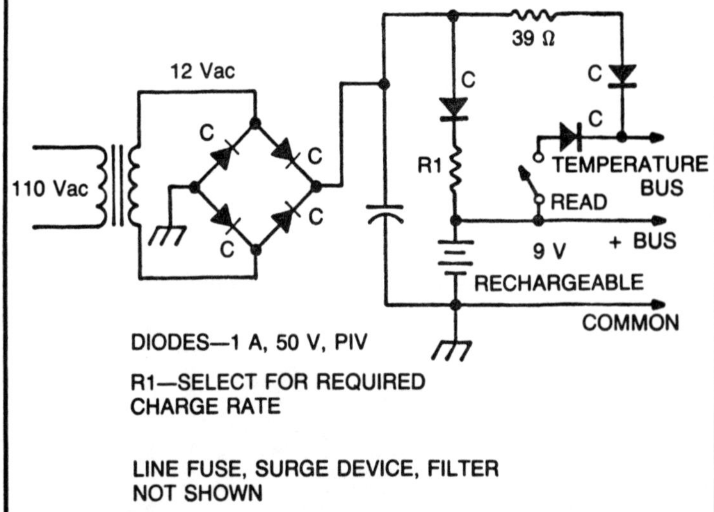


Fig. 9-9. Suitable power supply for the primary standard. It is suggested that the rechargeable batteries be obtained first, and the power supply built around these. Avoid thermal coupling to the crystal and oscillator when laying out the mounting.

across the line to the chips and oscillator, serving the dual purpose of voltage regulation and emergency supply. The switching diodes are arranged to supply power to the display circuit when the ac line is energized, but to remove this power if the line input fails, as a way of conserving energy. This action also serves as a power off signal and could be used to energize an alarm. A switch is provided to allow the displays to be operated from the battery.

## CIRCUIT BOARD LAYOUT

The layout of the two special circuit boards, the oscillator-divider board and the regulator board, are shown in Figs. 9-10 and 9-11; the layout of the clock board has previously been shown in Fig. 4-21. Parts placement for the oscillator-divider is shown in Fig. 9-12.

The circuits are straightforward; follow the circuit diagrams above. Either of the two oscillators can be used, with jumpers being used to make the connections. Alternately, a separate oscillator-buffer can be used; this would be done if the ultimate in stability were being attempted. For medium-accuracy operation however, the board oscillator should be satisfactory, with the crystal connected to the oscillator by its leads.

A terminal is provided to take the limiter output to an external circuit, if desired. Good practice would be to use a buffer amplifier for this circuit, to prevent possible loading changes from affecting the oscillator.

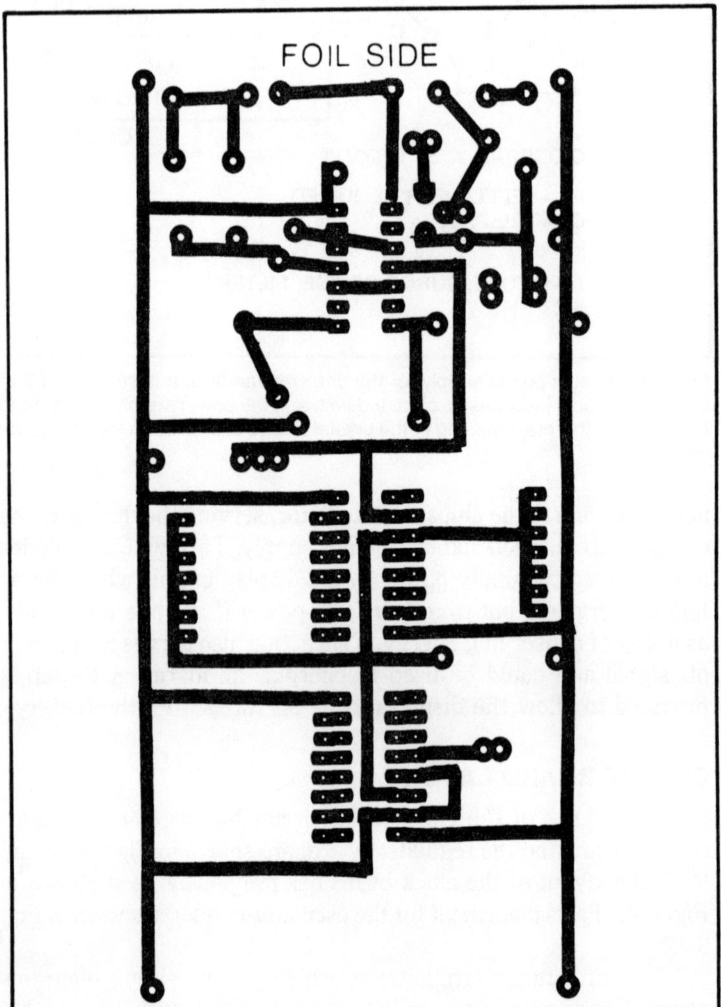


Fig. 9-10. Circuit board layout for the frequency standard. There is a small rearrangement of parts for the transistor oscillator as compared to the prototype. Original board prepared as for Fig. 4-8. Foil side; shown full size.

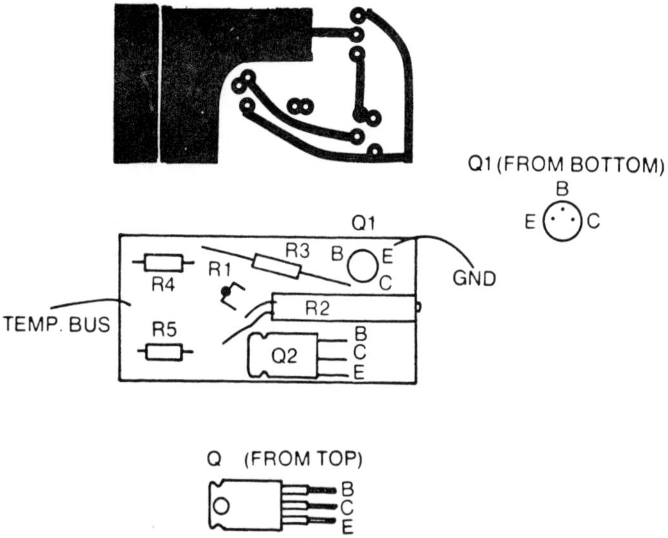


PARTS LIST

R1—thermistor, 1000 ohm + 25°C
R2—5000 ohm, 10-turn trim-pot (Bourns or Dalohm)
R3—150 ohm, 1 2 watt carbon resistor 20% tolerance
R4, R5—82 ohm carbon resistor 1 watt, 20% tolerance
Q1—high gain audio transistor silicon NPN (Radio Shack 276-2010 or equal)
Q2—power audio transistor silicon NPN (Radio Shack 276-2018 or equal)

**Fig. 9-11. Circuit board layout and parts placement for the regulator board. The thermistor bead is to be placed in good physical contact with the crystal case.**

All output terminals of the two dividers are available. Many of those on the 14-stage divider are needed for the reset AND circuit. However, these and the others can be used for output if needed. Again, good practice would be to use buffers for all external circuits.

The regular board is also straightforward. The space between the load resistors is for the crystal. The thermistor bead is lead supported, to be in good contact with the crystal case. Use silicon grease to ensure this. Also, large area pads are used to give heatsink action for the power transistor and the two load resistors. Good thermal contact between the power transistor and foil, and between the load resistors and foil, is needed and silicon grease or thermal conducting cement should be used. Likewise, the crystal case should be thermally bonded to the foil by silicon grease. In other respects this board layout is straightforward.

## CRITICAL CIRCUITS

The circuits shown are the result of some work directed at ensuring good accuracy while retaining simplicity. They are certainly not the ultimate, but should be adequate for first experience.

As far as can be determined from the prototype, the only element which is likely to show any problem is the digital oscillator, which may become "touchy" with some high frequency crystals.

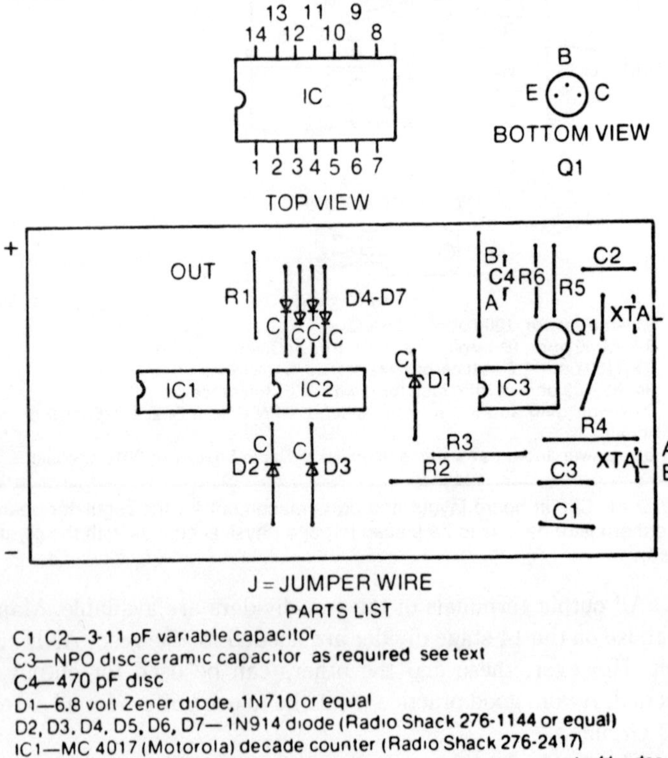


PARTS LIST

C1 C2—3-11 pF variable capacitor
C3—NPO disc ceramic capacitor as required see text
C4—470 pF disc
D1—6.8 volt Zener diode, 1N710 or equal
D2, D3, D4, D5, D6, D7—1N914 diode (Radio Shack 276-1144 or equal)
IC1—MC 4017 (Motorola) decade counter (Radio Shack 276-2417)
IC2—MC 4020 (Motorola) 14-stage ripple-carry binary counter/divider (Radio Shack 276-2420)
IC3—MC 4001 (Motorola) quadruple 2-input NOR gate (Radio Shack 276-2401)
R1—100.000 ohm carbon resistor
R2—10.000 ohm carbon resistor
R3—470 ohm carbon resistor
R4—1 megohm carbon resistor
R5—470.000 ohm carbon resistor
R6—4700 ohm carbon resistor
All resistors 1/4 watt or more. 10% tolerance

Fig. 9-12. Parts placement for the main board. The crystal leads must reach the regulator board, with a layer of insulation separating the two.

If the problem should arise, change to the other oscillator circuit.

However, there is a factor to be remembered. Workers in precision, and especially in precision radio frequency measurements, quickly learn that there are no noncritical circuits when precision is needed. Every value must stay put. Only the best grade of components, and the best type of mountings are suitable. It is extremely important that there be no free motion around any of the RF components—not just of the RF stages themselves, but of any elements within the influence field. Good regulation of temperature and voltage is necessary. The circuit must be free of parasitic oscillations and there must be isolation from load changes. All of these are necessary, and all may become critical.

Don't let these cautions be disturbing. Usable results will be available from this project immediately; the initial accuracy will be of the same order as that of a crystal calibrator even though no additional steps are taken. But over a period of time, by checking the error, adjusting it towards zero, and tracking down the sources of any drift or instability which may appear, a truly accurate system will develop. The limit accuracy is essentially that determined by the care taken in this continued check and adjust process. With good quality components, it should be possible to detect the difference between WWV's apparent frequency error due to propagation factors, and the true time as determined from the clock readings.

## CONSTRUCTION AND CHECKOUT

Construction of the basic circuit boards is straightforward but some precautions are needed in addition to the normal ones. For good stability everything on the RF board must be firmly fixed in place, so that it cannot move, introducing microphonics or frequency jumps. This same care is necessary when mounting the board into its case. High stability devices intended for severe environments, as in space equipment, are sometimes cased in foam plastic as a way of preventing any motion. This could be done as a final step, after the standard has originally been brought to stable operating conditions.

Be sure that the AND diodes are correctly connected, especially if the prototype division ratios are changed. An incorrect connection will give the wrong division ratio, and may be difficult to track down, because of the large ratio used. Also, even though the MOS chips used are gate protected, take precautions against static electricity.

The suggested first order of construction and test is the oscillator board. Checkout of this can be done in stages, starting with the oscillator. A useful trick is to use an old crystal in the same frequency range for the initial checks, changing to the precision crystal after everything is working. During checking, the oscillator frequency should be beat against WWV, and the series capacitor padding condenser adjusted so that the correct frequency is obtained at the middle of the variable capacitor range. As this is being done the divider can be checked by checking the duration of the one second output against WWV's time ticks. An oscilloscope synchronized to display the tone of WWV and the 500 Hz output of the counter chain is a big convenience here.

With this board working, the temperature regulator can be built up and checked by using a thermometer in place of the crystal. In open air the regulator should hold to an accuracy of about 1 ° for temperatures not too far above room air temperature. The power supply and a clock chip display board, if not already fabricated, would complete the basic assembly. With these all working as a breadboard, the overall packaging should be chosen, remembering that there should be about a one-inch layer of foam plastic around the crystal-regulator, another one-inch layer around the oscillator-crystal combination, then a metal shield (partly for electrical shielding and partly for temperature equalization). Another layer of insulation should surround this assembly. It is a good idea to have an outer shield. Leave a small access hole to adjust oscillator frequency, then closed by a block of insulation. Figure 9-13 shows a suggested layout of the overall package. The clock and power supply can be separate.

With assembly complete, final calibration can be started. Again, the first step is to make a preliminary calibration against WWV, by setting the oscillator frequency to give zero beat. Preferably this would be done with dc coupling to the detector, to allow setting to within one cycle in 10 or so seconds. With this done the clock can be set to WWV time. The best way to start checking and correcting for accuracy is to start logging the clock error, in seconds, at least once each day. If a steady drift in the error appears, the clock frequency should be adjusted to bring this drift to zero. Usually it is best not to change the clock setting, but instead to record the clock error as well as the change in error in time.

As accuracy increases, the error will amount to less than a second of change in a day. An oscilloscope can be used now, to permit counting the number of cycles of the WWV tone which occur be-

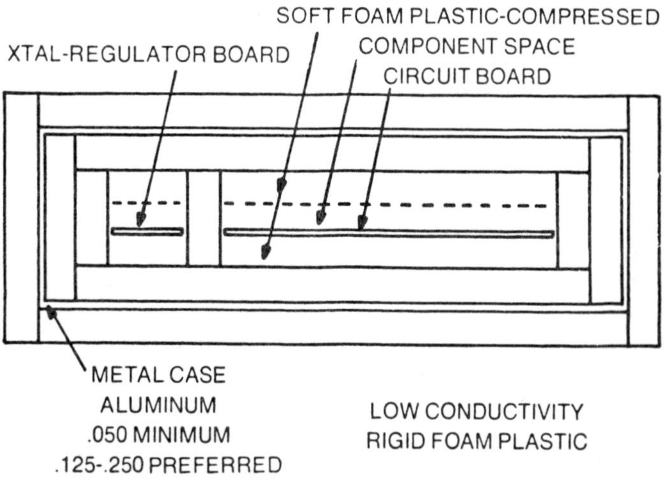


Fig. 9-13. Preferred case construction for the crystal and oscillator, shown in cross-section. The interior soft foam plastic acts as insulator and support, preventing microphonics. The outer plastic layer can be cemented to the metal shield, which also serves to equalize the temperature distribution. Caution—many cements dissolve plastic—casein or "white" household cement is safer.

tween the WWV tick, and the clock one second output. Later, with still further improvement, the change can be estimated to fractions of a cycle.

Actually, if time is the important factor, the clock stability is more important than the fact that the clock rate, the frequency, is not exactly correct. Good stability is shown by the fact that the change in time error, measured in seconds per day, is a constant. True time is always available by adding this error times the number of days to the clock reading.

For frequency, however, it is important that the frequencies be easy to use, which requires that they be the design value. This requires that the error per day be zero. Of course, when this is obtained, good timekeeping is also obtained; the clock can be reset, reading true time and keeping true rate. Thereafter the clock will keep nearly correct time. The word nearly is used, since crystals do age and change frequency. Parts do change, and even accidents happen. Moving a standard will probably make an appreciable change in its error rate. However, with reasonable precaution, the error can be brought to no more than a few seconds per month.

With great care, the error can be held to fractions of a second per month, which is far, far more accurate than the usual crystal calibrator.

## ALTERNATE CALIBRATION METHOD

Times change, sometimes bringing surprising changes. One of the current surprises is that nearly every home has in it a precision standard of frequency, slaved to an atomic standard, and typically available sixteen to twenty-four hours a day. This standard makes a usable alternate for calibration, and can in fact, be a worthwhile standard in its own right. This precision standard is the color demodulation reference signal of the home color television set.

The television signal contains a short reference burst of frequency, which is used to lock a local oscillator, by means of a phase-lock loop. The network program signals use an atomic frequency standard to generate the frequency which is gated to produce this burst. The signal is essentially as accurate as WWV when generated. When received at the home, however, it is probably more accurate than WWV as received. The reason for this is that the television signal is transmitted by coax and microwave relay to the local station and then by VHF to the home. The frequencies employed are nearly free from transmission disturbances, whereas the HF frequencies used by WWV for long range transmission are always subject to some disturbances, which become considerable at times. Under really disturbed conditions, the WWV frequencies can have instantaneous errors on the order of cycles. This, of course, is not true for the average error, which is the major reason for developing and using a primary standard of frequency.

The frequency of the television color burst reference signal is 3,579,545.454 Hz. This appears to be a strange value, but it is actually logical—it is 455/2 times the horizontal scan frequency. The reason for this choice is that it puts the color reference frequency in a part of the video spectrum where the interference between the black and white information and the color information is the least.

The obvious way of using this frequency for calibration purposes is to measure it with a frequency counter. If the counter reading is not the value above, the counter time base is in error, and needs to be readjusted. Most counters have some provision for using an external time base, so the frequency output of the primary standard can be used with the counter, thereby getting a direct reading of the calibration error. Using this technique, measurements to better than one part in $10^6$ can be made directly, and by counting for ten or one hundred seconds and averaging, the error can be checked to about one part in $10^8$. It may even be possible to bring the counter time base to this order of accuracy, although it could be expected that there will be some drift, since few counters use temperature regulated reference crystals.

While this method is usable, for most purposes it would be better if the frequency reference were a round number, say an integral number of MHz. Surprisingly, it is not really difficult to develop such frequencies from the color burst. As shown in Fig. 9-14, the very convenient frequency of 5 MHz can be obtained by first dividing the color reference frequency by 63, and then multiplying by 88. Check this if you don't believe it; but do it by hand, since calculators have a round off error and are likely to give values such as 4,999,944 instead of the true value.

Scale of 63 counters are not difficult to construct, using the principles discussed in Chapter 2. See Fig. 9-15. The required ratio can be obtained from three cascaded counters, which can conveniently be the 7490 decade counter. This unit is convenient because the internal gates can be connected to obtain the divide by 3 and divide by 7 ratios needed. A number of other counters can accomplish the same ratios with no external parts, and any counter with reset can use external gates.

Probably the simplest way to develop a times 88 frequency multiplier is to use a phase-lock loop such as the 560, as shown in Fig. 9-16. The free-running frequency of the internal oscillator in the PLL is set to 5 MHz, and the pull-in range is set to a convenient value such as 10 MHz—see the specification sheets for the PLL for the values needed to do this. The output of this PLL is at a rather inconvenient level, so amplification and level changing are needed.

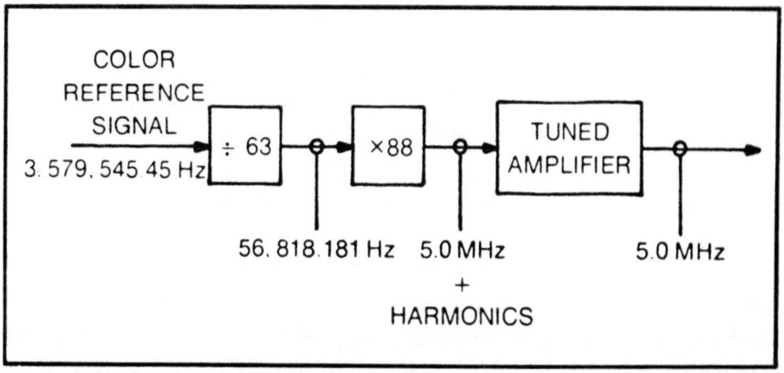


Fig. 9-14. Conversion of the color reference signal to a 5.0 MHz signal, suitable for use as a frequency standard.

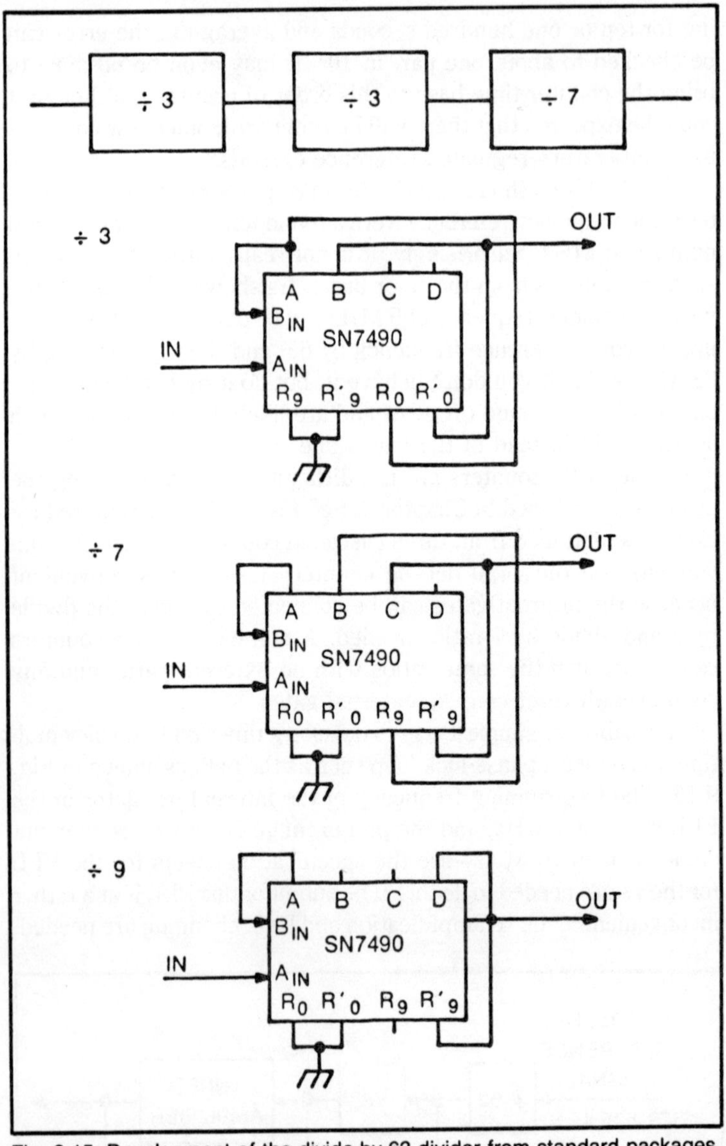


**Fig. 9-15. Development of the divide-by-63 divider from standard packages.**

Figure 9-16 shows the use of a 710 comparator to accomplish this level changing, and gives connections and component values for the PLL. The overall output is a good 5 MHz square-wave, of about 8 volts peak-to-peak, which is capable of driving a few thousand ohm load. The rise time is fairly good, and the output contains good

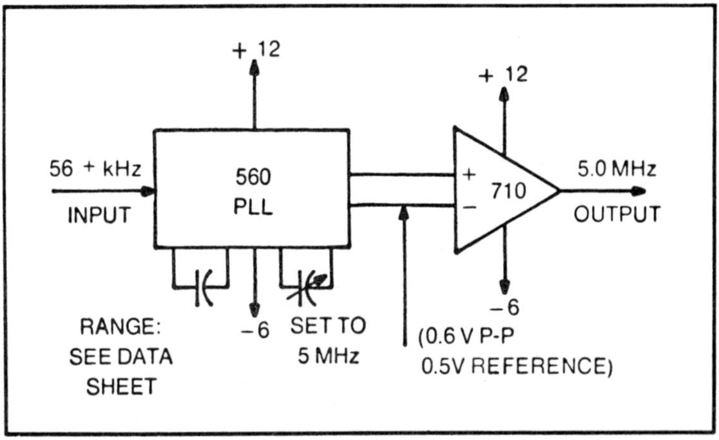


Fig. 9-16. Development of a times-88 multiplier by a PLL. See the PLL data sheets for connections and values.

harmonics. The output will drive a down counter if this is desired.

An alternate method of frequency multiplication is to use digital rate multipliers. These can be set to give an output pulse rate equal to N/10 times the input pulse rate. Two cascaded in accordance with the specification sheet can be used to give a multiplication ratio of 88/100, giving an output of 50 kHz when driven by the divide by 63 circuit. The connections for one family of rate multipliers is shown in Fig. 9-17. Of course, this can be followed by

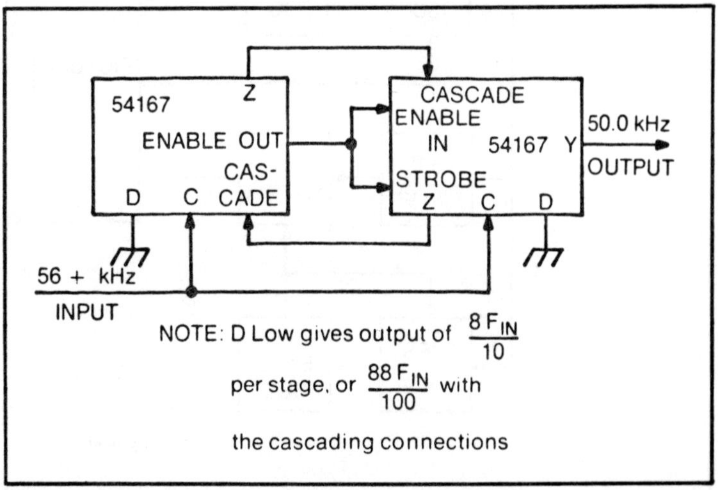


Fig. 9-17. Alternate multiplier using the 54167 digital rate multiplier. The 50 kHz output can be used directly or with tuned amplifiers.

an amplifier, or by count down stages as needed.

The location of the color reference signal in a typical color television set is shown in Fig. 9-18. If you should attempt to tap into this oscillator, remember two factors. First, the operation may be dangerous, since there are very high voltages around the set. Secondly, the reference oscillator circuit is quite critical, and changing the loading may require readjustment, which requires good equipment to do properly. If the set should happen to be a type which uses series filaments which are line operated, the tap should not be made. Also, if it is attempted, the set will not be legally salable unless the additional circuit is removed.

In many cases it is not necessary to make a direct connection. Most TV sets are very poorly shielded, and there is usually enough

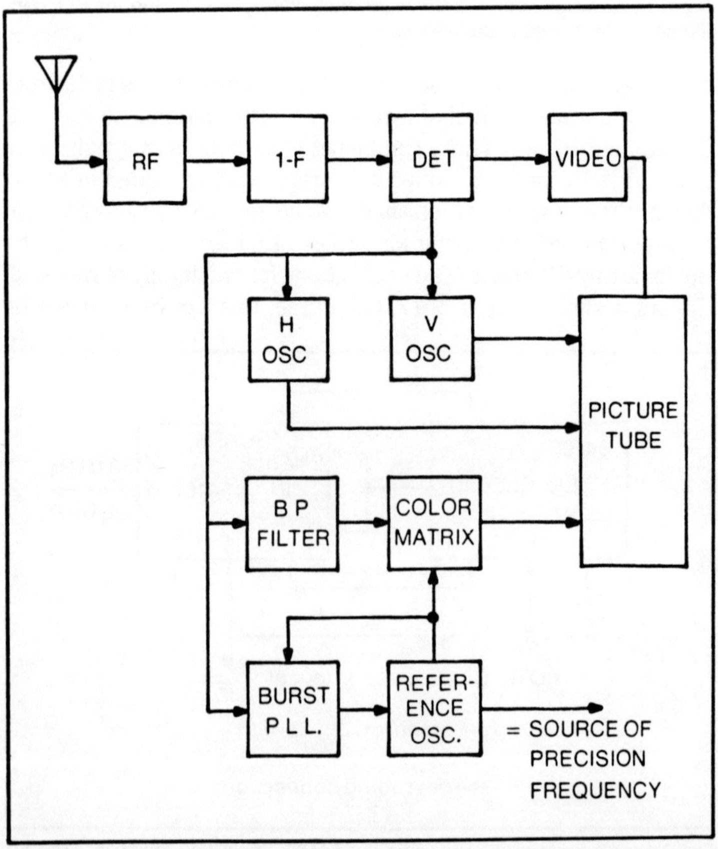


Fig. 9-18. Location of the color reference signal in a typical TV receiver. See text for cautions in making connections.

stray 3.58 MHz signal floating around that it can be picked up. A conventional communication receiver can be used as a probe, to find a "hot" location. A small tuned circuit and a following amplifier should give ample signal. Alternately, the tuned circuit can be inductively coupled to the 3.58 MHz signal, remembering that the coupling should be loose to avoid pulling the oscillator circuit unduly.

## A SIMPLE WORKING STANDARD

The primary standard of frequency just described does get to be a project of considerable size. Something simpler can be used if we assume that we are going to work with a multiple digit frequency counter. Now we don't have to have a precise value, such as 5.000 MHz. We can work with any convenient value, the only requirements being that this be known, and that it be stable.

A convenient value is something which will work with a 60 Hz clock. This takes care of all of the accumulator problems, the presentation, and makes comparison with a true standard (WWV time signals) easy.

To make this approach possible at low cost, the National Company makes a special form of their COSMOS 17-stage binary divider available. This is internally connected to create an output wave having 28,875 input cycles for the high stage, and 32,784 input cycles for the low state, for a total of 59,659 input cycles per output cycle. Multiplying this by 60 Hz gives 3,579,540 Hz as the required input frequency. This is almost exactly the frequency of the color burst signal of color television, specified to be 3,579,545 Hz. This allows use of a low cost color reference crystal, the intended step.

If exactly on frequency, the crystal would give a time error of just over one part per million. In principle this is easily compensated, since the suggested crystal circuit, Fig. 9-19, can "pull" the crystal frequency as much as one-hundred parts per million. Practically, however, the frequency setting accuracy will be nearer $\pm 2$ ppm, unless extremely stable and easily adjustable components are used.

The circuit of a working standard crystal driven clock is shown in Fig. 9-20. This provides battery back-up during power failure, and automatic return to ac power when it returns. (In this simple design there will be a small change in oscillator voltage when power fails.)

While simple, the performance of this circuit can be improved

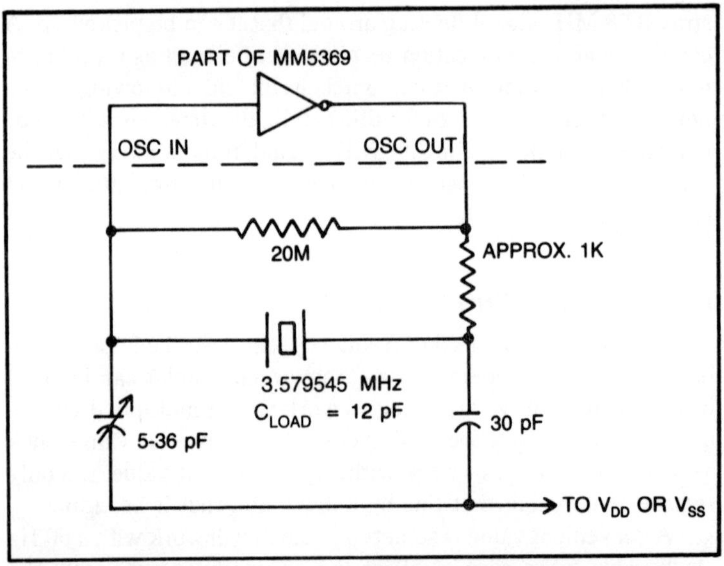


Fig. 9-19. Basic elements of the crystal oscillator circuit of the MM5369 chip. The amplifier is internal to the chip, the other components external. The 1K resistor may not be needed. The frequency can be "pulled" by adjustment of the variable capacitor to secure correct timekeeping.

by a factor of 10 or so by applying the principles of the precision standard. The crystal should be temperature controlled, with a separate back up battery for the heater circuit. A separate crystal oscillator, with bias and operating current adjusted to give minimum frequency change between the ac line and battery conditions should be used. And, of course, the crystal components should be selected for low drift.

Whether you use the simple version of this circuit, or a more complex temperature controlled version, you might want to look at time base kits which are available, one being shown in Fig. 9-21.

It should be noted that some quartz crystal clocks operate at 32 kHz. Stability of the order of five seconds a month is not uncommon, even in low cost commercial designs. These can be used for calibration, but good resolution will require measuring periods of 10 seconds or more.

## CONCLUDING REMARKS ON PRECISION MEASUREMENTS OF FREQUENCIES

Really precise frequency measurements can be both exasperating and rewarding. The exasperation arises from the importance

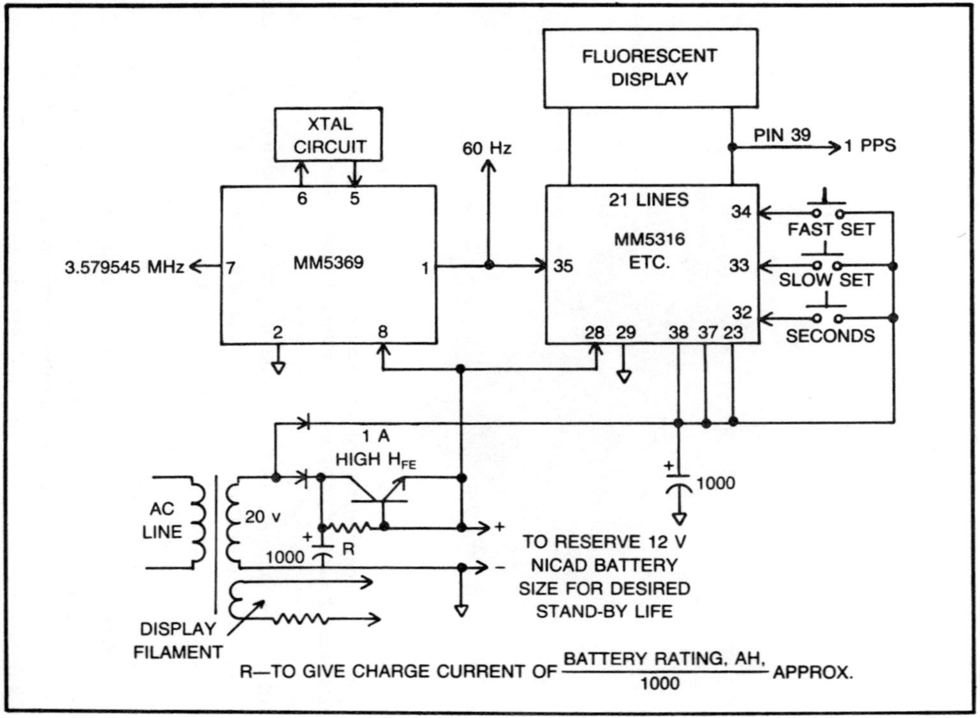


Fig. 9-20. Interconnection diagram of a simple working frequency standard using MM5369 and 5316 chips. For best accuracy, the crystal oscillator components should be high quality, and should be temperature controlled. With a battery of reasonable size, power loss of several days can be tolerated, although there may be a small rate change due to loss of thermal control, and to voltage changes.

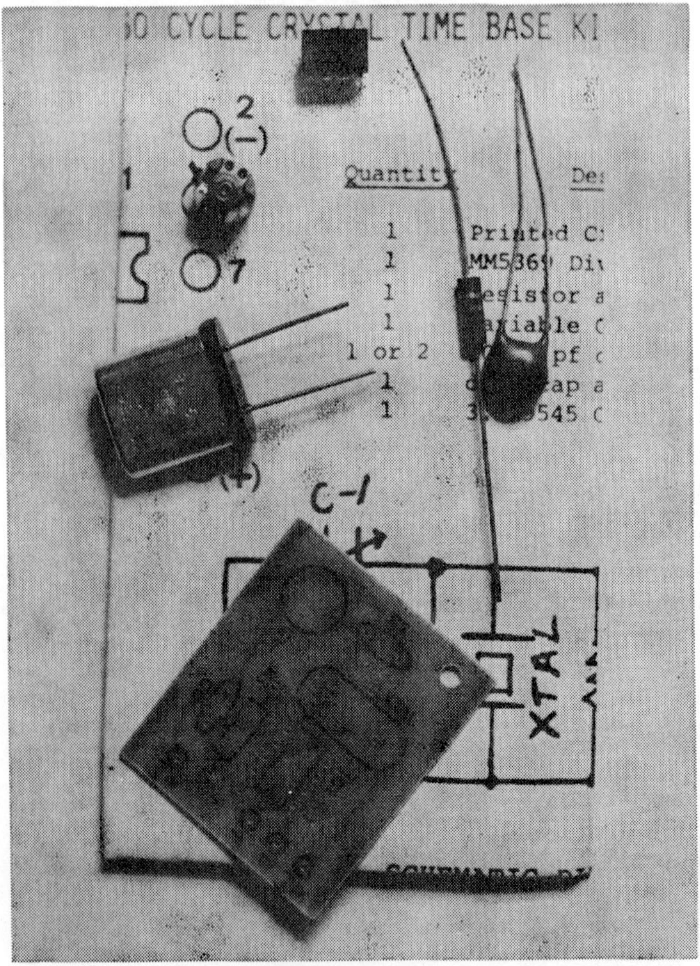


Fig. 9-21. A commercially available time base kit. Cost of these kits may be less than the cost of components purchased separately.

of detail—components which otherwise check as perfectly good can cause severe drift or jump in frequency—independent of regulation, time stability or freedom from shock. Finding one of these components can be a very troublesome problem and if there should be two in the circuit it is even worse, to use an understatement. The probability is very low with good quality components, and with care in construction, but the chance is by no means zero. And, of course, in any event considerable pains are necessary to adjust a standard to zero rate error and zero time error.

There are real rewards however. The first being the realization of improvement and of attainment of a difficult goal. It is quite a thrill to catch drift in the WWV signal as received, even though this is only due to propagation changes. (The worst drift situation occurs around sunset, and it is suggested that this be the one to look for first.)

# Chapter 10

# Clocks for the Handicapped

The standard digital clock has been developed for people with normal hearing, normal vision and normal power of manipulation. The designs available work fine for these people, although there are sometimes complaints about inconvenient controls. For many people who have handicaps the standard clock design ranges from poor to totally useless.

The rest of this chapter will be devoted to special clock designs which are intended to make life easier for the handicapped. None of the designs shown have been carried to practical operating form by me. Some have been by other experimenters but it won't be possible to give due credit because no notes were made at the time they were encountered. Others result from observation of problems encountered by the handicapped, and still others from discussions with handicapped persons, therapists, and teachers.

This entire field is almost untouched. It's a fertile field for experimenters and developers of ideas. Done carefully, there should even be a reasonable business potential for a good line of special products.

## CLOCKS FOR THE HARD-OF-HEARING

Some 20 years ago, a special electric alarm clock appeared on the market. It had an alarm feature which was quite useful for the hard-of-hearing. In addition to the regular alarm, there was an alternate circuit which flashed a lamp which was built into the clock.

The idea was that the flashing, rather bright light would awaken the sleeper. During waking periods, the light could serve as a normal bed light, for reading or general illumination.

After this special clock had been on the market for some time, the manufacturer was surprised to find that the idea had been patented a number of years before, specifically with reference to its use for the hearing impaired. The first inventor and the manufacturer finally worked out a licensing agreement.

While production of this unit continued for some time, I haven't seen a clock using it lately, nor do I know of the current state of the patent. These factors are no bar to development of a modern version for yourself, your family, or a friend. (But if you are thinking of a business, proceed carefully.)

Practically any modern digital alarm clock or clock chip already contains most of the elements needed for such a flashing light clock. Basically, the main requirement is an interface element capable of handling the voltage and power levels of the lamp or lamp circuit to be flashed. In some cases, a flasher will also be needed, but this is available in some clock chips. It can also be easily added to any external circuit. Alternately, it can even be made a part of the lamp. Figure 10-1 shows three basic possibilities.

As an example of a design, consider use of a National 5382 digital calendar-clock radio chip. The basic alarm circuits for this chip are shown in Fig. 10-2(A), and consist of a switch to turn the alarm circuit on, an indicator to show that it is activated, and the actual alarm output.

The capabilities of the alarm output are shown in Fig. 10-2(B).The indicator will drive an LED, sourcing as much as 15 mA. The alarm output itself is a 400-2000 Hz tone, gated at a 2 Hz rate, with a drive capability of 10 mA.

When used with a current amplifier, as shown in Fig. 10-2(A), there is ample power to drive a small speaker, or a headphone element used as a speaker, since a peak power of about 200 mW or more can be developed, depending on transistor gain. Considering that 60 mW is adequate power for radio listening with a good speaker, the alarm has adequate "wake-up power."

Instead of using the transistor circuit to drive a speaker, it can be used to drive a relay, as shown in Fig. 10-2(C). This can be a dc relay, with a coil rated at about one volt less than the supply voltage, and a coil current rating of 50-200 mA depending on the current gain of the transistor. The coil is shunted by a 0.01 mF capacitor, to reduce the tendency of the relay to buzz due to the

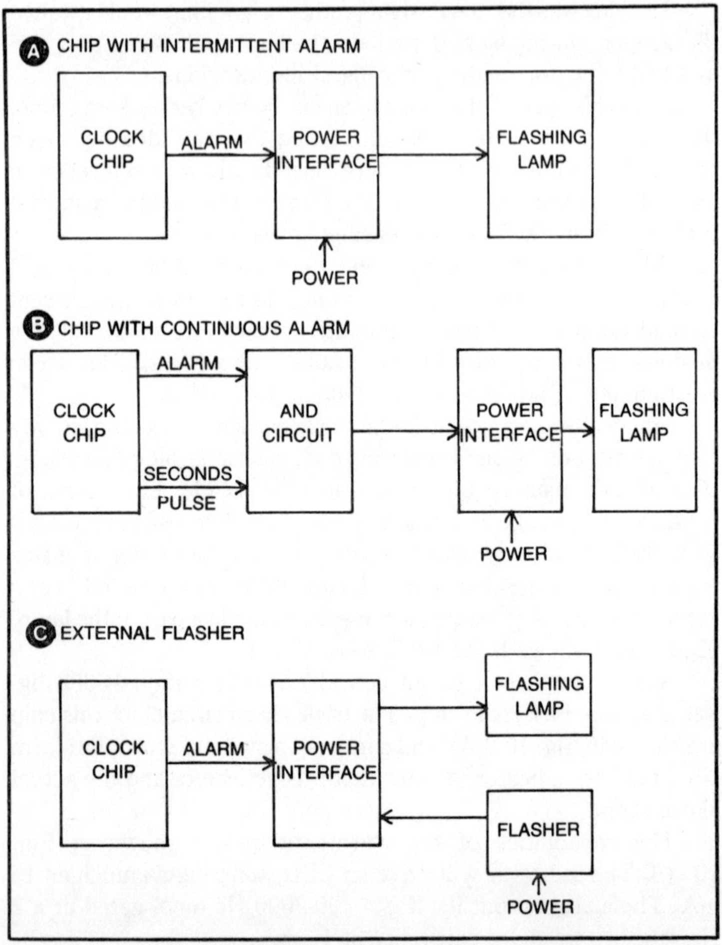


Fig. 10-1. Systems for obtaining a flashing light alarm. (A) Addition of a power driver to a chip with intermittent alarm output. (B) Use of seconds pulse to produce flashing. The AND circuit may be diodes. (C) Use of an external flasher.

400-2000 Hz signal component. The relay needs to be fairly fast, to follow the 2 Hz switching, which produces the lamp flashing.

The relay contacts must be rated for 120 Vac and for a current suitable for the lamp being used. (Remember that the inrush current of incandescent lamps is several times the steady state current.) A rating of 1 ampere is satisfactory for the 25 watt lamp shown.

The combination of the relay contacts and the lamp are in series across the 110 volt power line. The relay contacts are bridged

A BASIC ALARM CIRCUIT

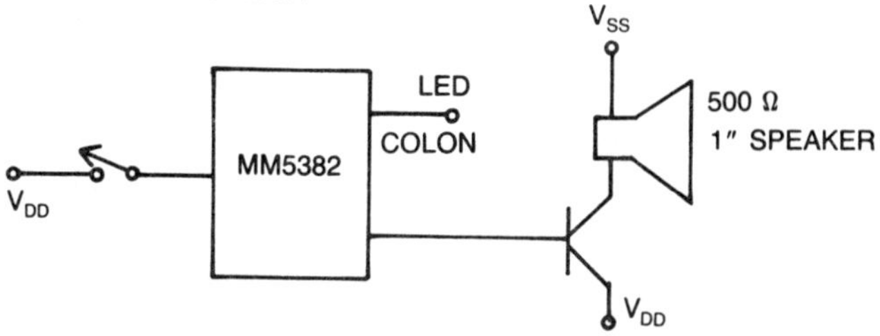


B OPERATING CONDITIONS

| FUNCTION | VOLTAGE | CURRENT, MIN |
|---|---|---|
| ALARM INDICATOR | $V_{SS}$ −2 | 15 mA |
| ALARM OUTPUT, HIGH | $V_{SS}$ −2 | 2 mA |
| ,LOW | $V_{SS}$ −15 | 10 μA |

ALARM TONE: 400-2000 Hz, MODULATED AT 2 Hz

C VISUAL CIRCUIT

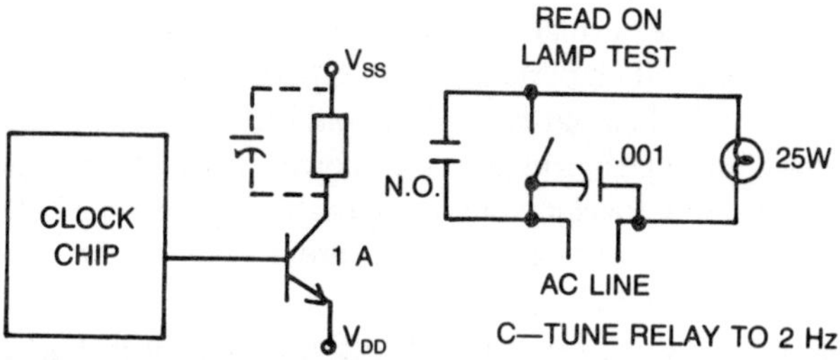


D ALTERNATE DRIVER

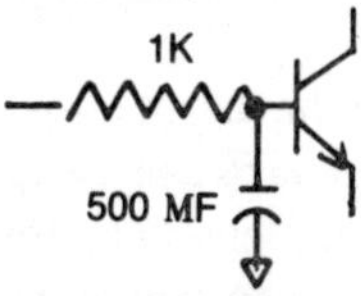


Fig. 10-2. Conversion of an aural alarm to visual, using the MM5382 chip as an example. The relay could be replaced by an optical isolator, but safety rules must be observed. The driver may be an audio or power tab transistor.

by a switch, which can be a momentary contact push-button type if the switch is to be used only for lamp testing, or it can be a toggle switch if the lamp is also to be used for reading. The small capacitor across the power line is to eliminate radio noise due to sparking at contact opening.

If you prefer, you could replace the relay by an optical isolator of suitable rating. This would mean using a bilateral isolator, or a lamp of special voltage for the normal isolator, which is also a rectifier if used on an ac circuit. The degree of isolation is extremely important, partly because the MOS clock chip cannot stand much in the way of surge voltages, but primarily because of safety. Good design practices must be used to protect the clock user against shock. For this reason, TRIAC type line control should not be used.

With the clock chip shown in the figure, the light will flicker at a rate of two flashes per second. If this is not acceptable, a large capacitor can be placed across the relay coil, holding the armature closed during the OFF period. Alternately, a series resistor-shunt capacitor can be used in the transistor base circuit to act as the holding element. Figure 10-2(D) shows these two elements. Only one need be used.

Other clock chips may not have the flash feature or a variable flash rate may be desired. A type 555 timer chip can be used in these cases. The basic circuit is shown in Fig. 10-3. The timer pulse rate is set by the normal resistor-capacitor arrangement, giving essentially equal off-on periods, which is best for incandescent lamps. The 555 control circuit is used to hold the timer chip in the inactive state until the alarm pulse is received at which time the timer operation starts. Flashing rate is adjustable over about a 10:1 range, so that it can be set for individual preference.

The 555 timer can handle about 200 mA of current, so many types of relays can be driven directly, as shown in Fig. 10-3. If necessary, a transistor driver circuit similar to the design used in Fig. 10-2 can be used. The timer output replaces the clock chip output. However, with reasonable choice of relays, the direct-drive would be ample.

There are other ways to use the alarm output besides flashing a light. A simple and effective one is to use the alarm signal to turn on a bed vibrator of the type often found in motel rooms to "soothe the tired traveler." These are not common in homes, but are easily constructed, or even improvised.

Basically, these bed vibrators are a motor with an unbalanced weight attached to the bed frame. The unbalanced condition

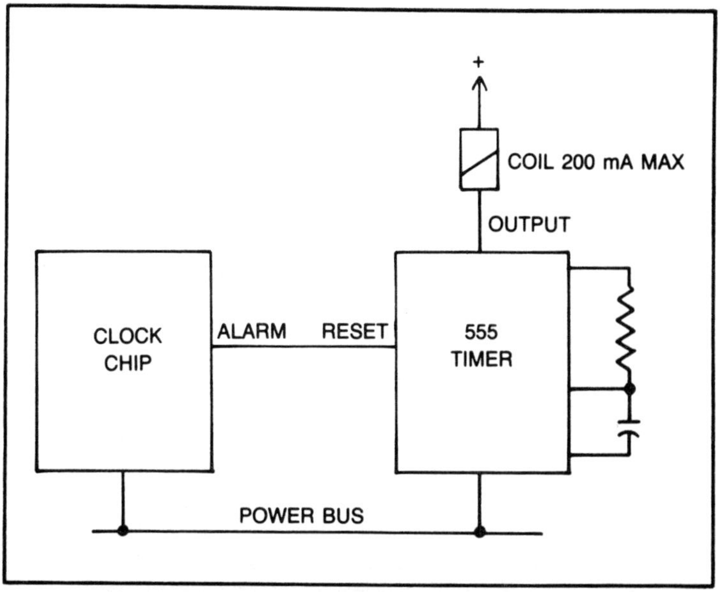


Fig. 10-3. Use of a 555 Timer to produce lamp flashing. Elements not shown may be handbook values, or taken from Fig. 10-2.

produces the vibration. It's exactly the same motion as produced by an orbital sander. The easy way to get the vibrator action is to fasten the sanding sole plate of a sander to the bed frame. Be sure that the motor housing is free to move and that there is no stress on the power cord.

It will probably be necessary to experiment to find the best location of the vibrator. For one thing, the sleeper may not be sensitive to motion in one direction, or may be very sensitive to any motion. Also, the sander orbital speed may be too high. In this case, use a small gear motor, of approximately 100 RPM. A brush type motor with a series resistor for speed control, (or even a lamp dimmer) could be tried, although the amplitude of vibration will decrease as speed decreases, unless the unbalanced weight is changed. This area is wide open for experimentation.

Normally, it seems, hearing impairment alone does not cause any great problems during waking hours. However, some persons with hearing problems might like a visual indication of time, corresponding to the chimes or signals found in some grandfather, mantel, or cuckoo clocks. This can be provided by flashing one or more lights in the house, shop or office area at appropriate intervals—hourly might be appropriate.

Signals at hourly and at ten minute intervals are easily obtained from any clock chip. Signals at other intervals are not so simple, but can be developed if sufficient external logic circuits are added. See the chapter on experimental clocks for some possibilities.

## CLOCKS FOR THE VISUALLY HANDICAPPED

The problems of the visually handicapped lie almost entirely in reading the actual time, and in setting the alarm to the desired time. The normal way of compensating for sight loss is to take advantage of the sense of touch which is developed to a very high degree in most visually impaired people.

There is a factor which may be helpful, but only if the problem is partial loss of sight. It may be found that satisfactory clock reading can be obtained by a suitable combination of size and color of clock digits. It seems that many people with visual problems can see the large numeral LED red figures with relative ease. This is not universal, however. Others, apparently a smaller number, see the fluorescent green much better. There may be focusing problems here, as well as actual changes in spectral sensitivity. In comparative tests, one person may report the green as clearly readable and the red as a diffuse blur, while another may find the red clearer. Still others may not see the green. I have had no experience with blue, but the same effect may be found. In any event, it would be worthwhile to experiment with different character colors.

If size alone is the important factor, there is a good choice in both commercial and home fabricated display units. Clock modules which require only a power supply are available with displays up to 0.7″ height. Still larger displays are available on special order. However, these are quite expensive, and home construction would seem preferable in most cases.

For home constructed displays, the large display technique of Chapter 6 of my Compulator Book can be used. Basically, this is composed of seven lighted strips arranged to form the two squares of a normal seven-segment display. The range of light sources is very wide. Possibilities are:

—Fourteen light bulbs, two per strip, with a reflector and diffuser to form the strip. A four-inch display could use miniature bulbs, a 12-inch display could use 7 1/2 watt, 120 volt bulbs, and so on. Without lenses, the visual intensity would be moderate.

—Seven fluorescent or lumiline bulbs can be arranged in the same shape. The smallest which can be built with commonly available bulbs would be a 12-inch character, using fluorescent bulbs. The intensity can be quite high.

—For intermediate sizes, the best current type is the multiple "bar" LED, intended to replace meter indications of volume level. A typical unit of two to two and a half inches length would give four to five inch characters, probably ample for most purposes. These types are made in red, green and mixed, but only the red is commonly stocked. Visual intensity is moderately high.

The major problem in these large displays is the drive circuitry. This is simplest for the LED bars, which only require current multiplication to function. This is easy to obtain by replacing the low power transistors often used by power Darlingtons. The basic circuit shown in Fig. 10-4. If the clock chip is designed to drive small LEDs directly, a power-tab transistor would be adequate.

The same principle can be used to drive small incandescent bulbs. A circuit which has been used is shown in Fig. 10-5(A), using standard IC type drivers. This is suitable for currents of about 25-50 mA. For larger bulbs, a power transistor can be used to obtain currents up to several amperes, as shown in Fig. 10-5(B).

In these incandescent lamp circuits, the light intensity is determined by the average current through the bulb. The peak current during the scan interval will be much larger, typically by a factor of five to ten. This is determined by the duty cycle established by the clock chip. Light intensity increases with current, but

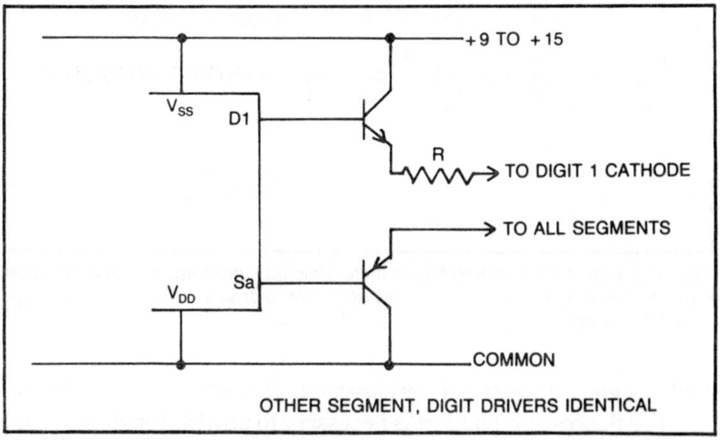


Fig. 10-4. Use of current multipliers to increase chip drive capability, to permit use of large, bright LED displays. Suitable for displays in the range 1.0 to 3.0 inches height.

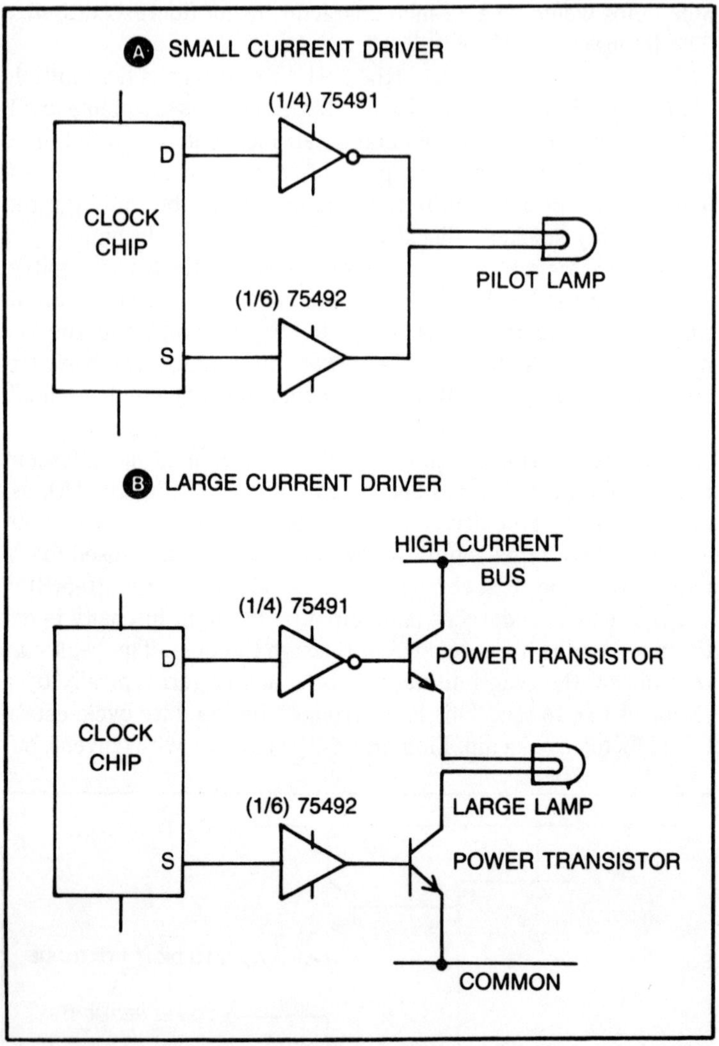


Fig. 10-5. Use of IC drivers to create incandescent lamp displays. Several bulbs can be used in series in each display segment to create displays in the range of 3.0 to 24 inch height.

bulb life goes down rapidly as current increases beyond rated value.

For fluorescent bulbs, it is necessary to provide for starting and for keep-alive during the off period of the scan cycle. A circuit which has been used is shown in Fig. 10-6. Note that these circuits are sources of radio noise, and a filter on the power line may be needed.

These oversize and giant displays are a fertile field for the experimenter. Not too much is known about the needs of the visually handicapped in this respect, so both electronic and physio-psychological experiments are in order.

In this connection, it would seem that a multiple use display should be considered. This would make such devices as calculators, TV channel indicators, bathroom scales and other items more available to the visually handicapped. There should also be applications in schools, for teaching, and for information in public areas such as airports.

## CLOCKS FOR TOTAL VISION LOSS

If there is total vision loss, the standard substitute is the sense of touch. Related to this is another, less commonly used—sensitivity to heat. I have found no references to this in connection with modern devices, so experiments will be needed if heat sensitivity is to be used.

### Design for Thermal Display

The driver circuits for the large displays can be used to form a simple thermal display, by replacing the lamps by small resis-

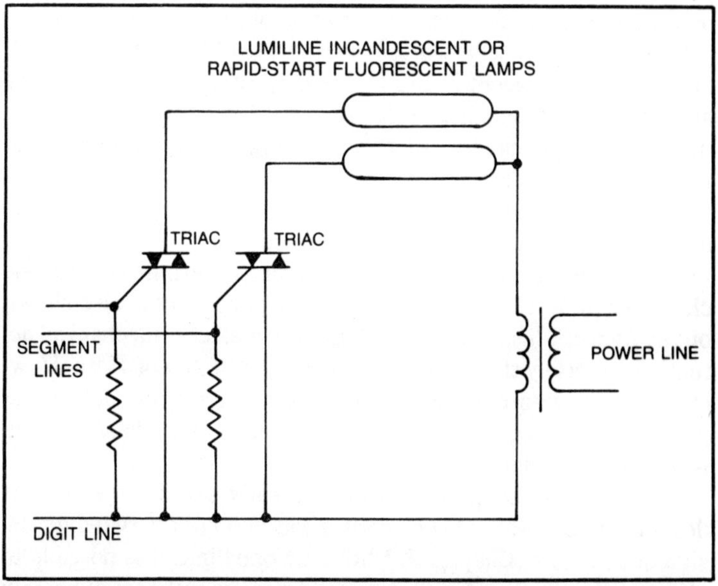


Fig. 10-6. Basic elements of drivers for lumiline or rapid-start fluorescent tube lamps. Displays from 12 inches to 16 to 24 feet height can be developed.

tors of quarter-watt size. Resistor current should be adjusted to keep the temperature rise to a few degrees above ambient, just enough to be easily detectable.

Special care to prevent contact with circuit leads will be needed. A method of doing this is to make cutouts in a thin piece of phenolic or glass-matt board, just large enough for the insulated resistor body to project slightly above the surface. The resistor leads would be soldered to the back-side foil. Another method would be to form a small copper strap around an insulated resistor, allowing the ends of the strap to project slightly through the board. With this technique the resistors could be larger than the actual character formed by the copper strap.

With standard clock chips and normal driver circuits, the only readout arrangement which seems possible is seven-segment ones. It might be better to work in Braille as described below. Again, experiments are needed.

## Design for Braille Display

The world standard for touch coding is the Braille character, formed by a six-position character of two vertical columns of three positions. The original Braille used raised dots for the "ON" state on the paper page. Smooth paper corresponds to the "OFF" state.

There is an excellent article on the use of Braille in the modern electronic concept in the December, 1981 issue of *QST*, the amateur radio magazine. Professor G.W. Horn, of Bologna, Italy describes developments in Italy, including a design by Professor Horn. Source data is also given. This article is recommended if any work is to be done on touch design. The following is essentially a summary of the major points of this article.

For numerical data, only the top four pins are needed for the character table of Fig. 10-7. The standard spacing is also shown on the figure. This is rather small for direct electro-mechanical actuation, precisely the problem addressed by Professor Horn. However, it isn't greatly different from the seven or nine dot matrix printer widely used for computer output, and perhaps these mechanisms could be adapted to Braille use.

The mechanical problem can be greatly eased if the clock is designed to present only one digit, a variation of the two-digit design described in Chapter 7. With only one digit, it is possible to place mechanisms on four sides of the display, rather than on only two sides if four or six-digit display is used. The change from tens

| DECIMAL | BRAILLE D1 | D2 | D4 | D5 |
|---|---|---|---|---|
| 0 | φ | 1 | 1 | 1 |
| 1 | 1 | φ | φ | φ |
| 2 | 1 | 1 | φ | φ |
| 3 | 1 | φ | 1 | φ |
| 4 | 1 | φ | 1 | 1 |
| 5 | 1 | φ | φ | 1 |
| 6 | 1 | 1 | 1 | φ |
| 7 | 1 | 1 | 1 | 1 |
| 8 | 1 | 1 | φ | 1 |
| 9 | φ | 1 | 1 | φ |

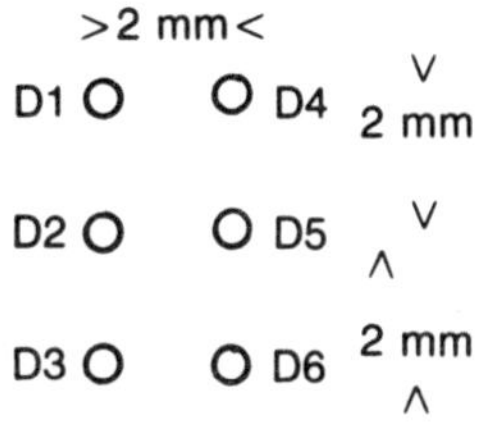


Fig. 10-7. Arrangement, dimensions and reading code for numerical values in the Braille code. Note that the coding is arbitrary, rather than in any binary form. The full alphabet may be found in encyclopedias, or in specialized literature.

of hours to hours to minutes could be done by a switching circuit, as in Chapter 7, but a design with three push switches would seem to be more convenient, as well as simpler. For good design, the top of the switches should be labeled in Braille.

A design which is within an experimenter's reach is shown in Fig. 10-8. Four small relays are modified to provide the pin actuation by removing the contacts, and adding a metal strip to the armature. The right end of this strip carries a round-end pin which projects through a cover. To allow close spacing, the arms project from the left side of the other two relays. Details of support and wiring are not shown.

For this design, the driver circuit must keep the relay energized for the null position of a pin. This is no great problem, and can be done by the same circuits which convert seven-segment to Braille coding. Two approaches to this are shown in Fig. 10-9.

It might be worthwhile to experiment with time in hex format rather than decimal. This allows a single six position block to present hours from one to twelve, and by a single switch motion to present minutes from 0 to 55 in five minute steps. It would, however, require additional memorizing.

## Decorator Digital Clocks for the Blind

A completely different approach is to retain the standard clock face, in the "Decorator Digital" form of Chapter 5. For the indicators, the LEDs of the original design are replaced by resistors for thermal readout, or by modified relays for touch reading. Single transistor drivers would probably be necessary for the resistors, but the TTL circuits can sink sufficient current for many miniature relays. There is also the possibility of using incandescent lamps

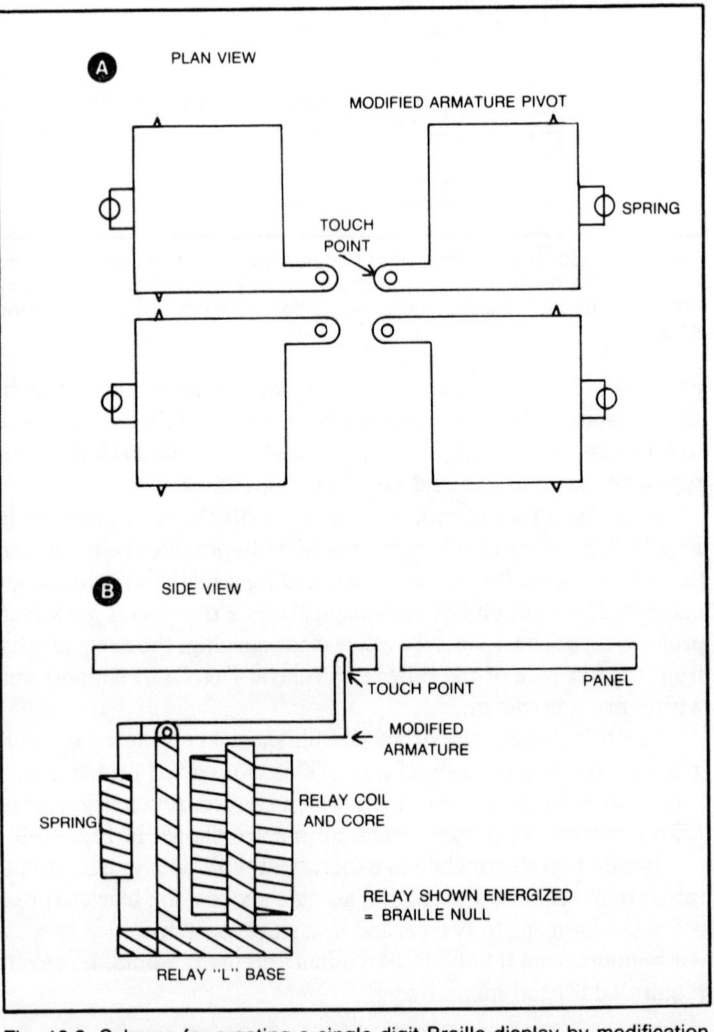


Fig. 10-8. Scheme for creating a single digit Braille display by modification of small relays. The mirror image arrangement allows close spacing of the touch points. With some rearrangement, six relays could give the full alphabet of 64 characters.

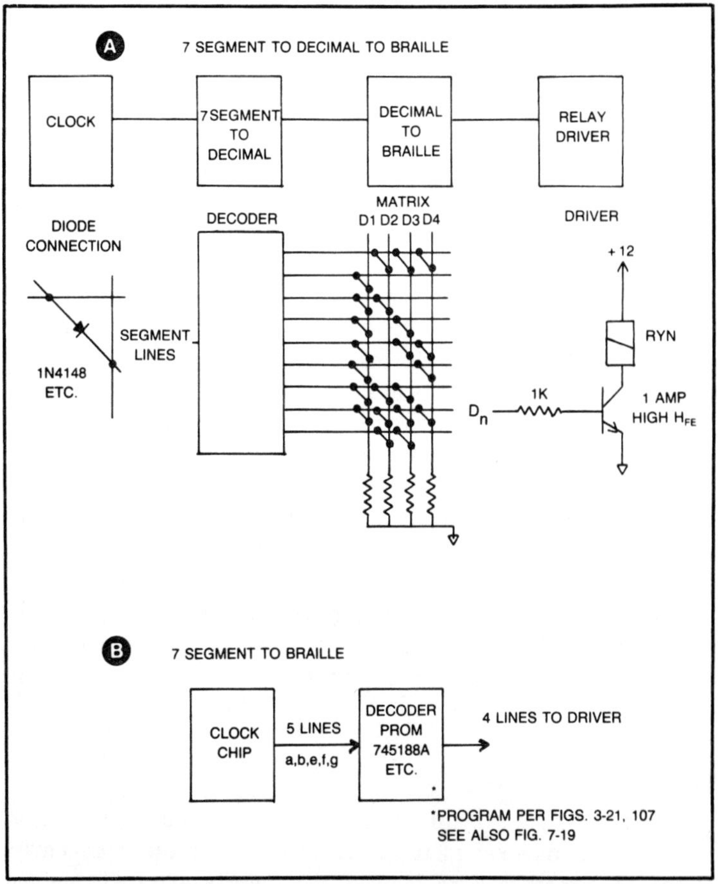


Fig. 10-9. Possible methods of developing Braille code from seven-segment clock output. (A) Use of a seven-segment to decimal decoder followed by a diode matrix. (B) Use of a PROM for direct decoding. Variations can be used for chips with BCD output.

instead of LEDs to produce a notable temperature rise. This would give a general use clock.

A possible design simplification is to use only a single decoder and a single circle of lamps, switching the four input decoder leads from the hour to the five-minute counter by a push switch. The minutes presentation could be eliminated also.

The physical design of such a clock will be influenced greatly by the presentation method. It would seem best to keep the diameter of the indicator circle fairly small, perhaps no more than two inches. This is easy with thermal readout, or no great problem with a single circle indicator. Careful design would be needed for a protruding pin design.

## STANDARD CLOCK ADAPTERS

A number of years ago a radio amateur device was developed, which makes it possible to read the position of a meter needle with fair accuracy. This is no different than reading the hand positions of a clock, providing the hands are appreciably different in length.

Basically, these instruments consist of a light source, a reflected light intensity meter, and a tone generator whose pitch changes with light intensity. When the instrument is swept across a meter face, there is a definite change in pitch when the black needle reflects less light than the white meter face. A good design can even detect the coarser calibration markings. By interpolating by feel, accuracy to one part in twenty or more is obtained.

Clocks for this use should have rather large black hands, and a circular rim for easy guidance. Braille markings on the rim would simplify reading. It would seem that a transparent inner circle would simplify hours reading. If no great precision in reading is needed, the minute hand can be removed. On a clock of any size, it is no trouble to interpolate to the nearest fifteen minutes if the outer rim is suitably marked.

## TALKING CLOCKS

Now that speech synthesizer chips are becoming common, it is natural to think about talking clocks. However, some caution may be needed in these designs. It seems that most people are not really interested in time itself, but rather in whether it is the time for a certain event, or whether there is time to complete some action before another action is due to start. "It is time for my favorite radio or TV program?" is a typical example.

Because of these habits, a talking clock may actually be resented. But it would seem that this resentment would disappear if several operational modes were established. In one, time could be announced at regular intervals, say, hourly. Alternately, time could be announced at some interval over a period, say each five minutes for the next 15 minutes. For still another alternate, only a specified time would be announced. Finally, this last could be repeated, to form an alarm mode.

Texas Instruments, National Semiconductor, and General Instruments make a number of speech elements, such as the actual processors, special ROMs to store the control, and various interface chips. GI's voice synthesis module combines a speech synthesizer, a microprocessor, ROM and a filter-amplifier on a single board. And don't forget the possibility of using a TI "SPEAK and SPELL" as the basic unit, perhaps with added memory for strictly clock use. The bibliography includes some references on the subject.

Building such a talking clock using available components would involve a number of format changes, as sketched in Fig. 10-10. For this reason it would become a major project, well beyond the scope of this book. It does promise to be interesting as well as challenging.

There are alternatives to talking with words which should be considered. Many people with vision problems have extra good hearing, and some have superb pitch detection. This suggests announcing time by a series of tones. Four, for example, would allow

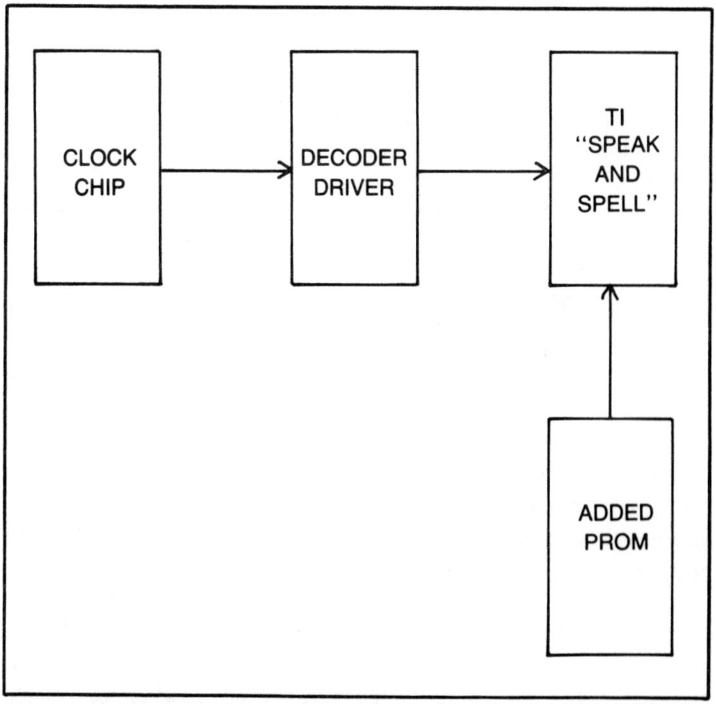


Fig. 10-10. Basic elements of a talking clock based on the TI "Speak and Spell" package. The added PROM will not be necessary if operation is restricted to simple number pronouncement. The decoder must activate the selection circuits in the correct order. TI, and others, have literature on the principles of voice generation available.

a separate tone for each quarter hour, and could be used in a BCD format to give the hour, perhaps by tones in sequence, or perhaps as a chord. Experimentation to find the most acceptable method is indicated.

An additional possibility is to announce the time in the form of Morse code. The numericals of this are arranged in logical order, and are no problem to learn (speed, of course, requires practice). A sequence of four characters would give time to the nearest minute.

All of these possibilities become rather complex in the way of hardware. A method of avoiding this hardware complexity, and to get other useful features, is to do most of the work by computer software for timekeeping and output generation. Quite good time accuracy is possible, since most computers use crystal control of their internal "clock."

The computer does not have to be very large—the small Timex Sinclair 1000, or the Commodore VIC 20 have ample memory, and also have provisions for audible serial output of sufficient level to drive a small speaker. Dedicated designs can be even simpler. The National 8073, with its built-in BASIC, would allow a two-chip design, one of which would be a programmable memory. (However, good tone quality might require addition of a quad OP-AMP chip.) These design approaches are another field for experimenters.

## CLOCKS FOR THE PHYSICALLY HANDICAPPED

For persons who have physical handicaps, the problems are not of seeing the time or hearing the alarm, but rather of manipulating the controls which set the time or alarm, and turn the alarm on or off. Commercially built clocks are all poor in this respect. Setting is often a two-hand operation manipulating small buttons on the back of the clock. Better designs have the buttons more accessible, but they are still likely to be small, and to require two-hand operation.

There is quite a range of physical problems. It may be very painful to exert any pressure. Fingers may be stiff or unable to move. The necessity for single-hand operation is not uncommon, and extreme problems may impose the necessity of using feet for manipulation, or even a rod held in the teeth. There may be problems of muscle control. A wheelchair may also be a limiting factor in accessibility.

Because of this range of problems, it is best to take the approach of adapting the machine to the person. Practically, this is best done by taking a suitable clock chip, or a commercial clock, and adding a set of external controls tailored for the particular physical problem.

To do this, it is probably best to separate the clock with its display, and the controlling switches. Connection between the two sections could be by a cable. There is no reason for the size of the controls to be governed by the clock case. They can be as large as necessary.

One approach to controls is to use the long-handled paddle switches found in dressed-up automobile sound and radio systems. Widely spaced, they are relatively easy to operate. Still easier to operate is a piano-key type, formed by a sensitive switch under a hinged flapper or key. A four inch square is easily operated by fist motion alone, if it is suitably separated from adjacent keys.

It would seem that down depression will be good for most problems, but a push motion might be easier for some. Some trials will show the best technique. They will also show if two-hand operation is possible, or if "one control at a time" design is needed. This will have an effect on choice of clock chips used, or, alternately, on the number of independent controls which must be provided. Some chips require two lines to be activated for some functions. It will be necessary to use diode OR circuits, or the multiple pole mechanical switch equivalents, to give the correct signals.

The techniques you work out are not limited to clocks. Manipulative ease improvement is needed for almost all standard mechanical, electrical, and electronic devices used by the handicapped.

# Chapter 11

# Clocks for Microcomputers

As the home or small office microcomputer markets and capabilities expand, more and more techniques are borrowed from those developed for the centralized computers, the giant mainframes. One technique that is just becoming common is the provision of a clock as an integral part of the computer installation.

It seems that these clocks were originally installed for two purposes. A big jump in use occurred when time-sharing was introduced. To permit charges to be assessed in proportion to use, it was necessary to keep track of the time allocated to each user. This was typically in blocks of a few seconds, so good accuracy and resolution were necessary. Even the very early time-shared systems included in the closing printout of a session both the ending time and the total computing time used during the session. Incidentally, it may be noted that it was often disconcerting to find that a program which had required several days of head scratching to create required only a few seconds of computer time to run.

The second method of clock use was for process control, using the clock to initiate some process which should occur at a particular time, or at definite intervals. At first, most of these applications involved dedicated, single purpose computers, so it was common to let the processor calculate time, usually by simple accumulation of clock pulses. Later, it was found more economical to use the computer for a number of purposes which required removal of the time function from the processor. The clock was used in two ways. First, it read the time as demanded by the computer control. Second, it generated an interrupt signal to the processor when time for a certain event occurred. This caused the processor to cease the work it was doing, and begin the process dictated by the interrupt. A whole family of these interrupts might be arranged in a priority scheme, so that the most urgent actions occurred first.

Later, a third use of clocks developed. As electrical power became more expensive, it became desirable to use clock function to control the computer and turn off sections not used at times. Or the clock placed these in a "power down" mode to save energy. The clock control could be applied to the processor or to a section of the computer installation, as a "last man out—turn off the lights" situation.

It is not likely that the home computer experimenter will need a clock to keep track of multiple users. However, it really is desirable to place at least the date on any permanent copy or printout. It saves much trouble later, for example, by making it possible to tell which was the latest listing of an important program. The dating could even be legally important, say in proving that an event occurred at the time claimed. It is also common to hear a wish that some member of the family could be reminded of the time, when one of those all evening—all night computer work sessions occurs.

The place where clocks and the home computer really get together is in the field of home control. It started in the form of the energy efficient house, although the simplest form of this can be done with sensors rather than time. As the computer assumes more functions such as appointment keeping, alarm clock, and home security, clock reference and clocked operations are a necessity. For example, a noise in the garage at 3 A.M. requires a different response from the same noise occurring five minutes after the school bus stops. It is these possibilities of time recording and event control that the clock-home computer combination should address.

## CONVENTIONAL CLOCK CHIPS AND THE COMPUTER

The clock chips intended for clock use only have been optimized for reasonable clock performance and low cost. These chips can be used with computers and there are one or two operations where they perform very well at very low cost. However, they become very unattractive for general use because of the complexity of the interfacing circuits necessary to adapt the specialized clock signals to the form used by computers. Following are some possibilities.

It must be kept in mind, however, that there are better techniques for the more complex ideas.

The simplest way to use a clock chip is to use the alarm function. This is a single lead operation, as shown in Fig. 11-1. The chip is used to generate an interrupt which goes directly to the processor. It could also go to a priority encoder, allowing combination with many other interrupts.

This simple technique is fine for a single daily event, but becomes unattractive for even twice a day actions. It would be necessary to reset either the clock or the alarm for each added event. This can be done manually which is a time consuming nuisance, or by adding some fairly complex elements under computer control.

A second, fairly simple use of a clock chip is to generate a signal at intervals by stopping the scan as shown by Fig. 11-2. This takes very little added equipment. But, as described in Fig. 3-22, it is not possible to develop the full range of intervals with simple components. The easiest by far is one pulse per 10 time intervals, that is, one pulse per 10 seconds, 10 minutes or 10 hours. Some additional intervals are available, but there is no regular pattern.

There are two other possibilities, which allow other patterns. One is to feed the seven-segment information to a converter, for example, to a 74C915 seven-segment to BCD decoder. The output of this can be used directly, or fed to another encoder. A wide range of intervals is now possible. In fact, by using programmable divid-

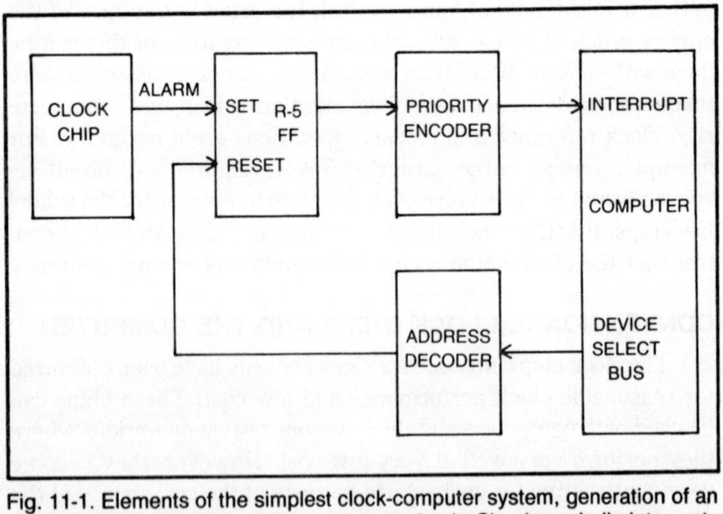


Fig. 11-1. Elements of the simplest clock-computer system, generation of an interrupt from the alarm, sleep, or snooze circuit. Simple periodic interrupts can also be generated with some chips, those having special outputs.

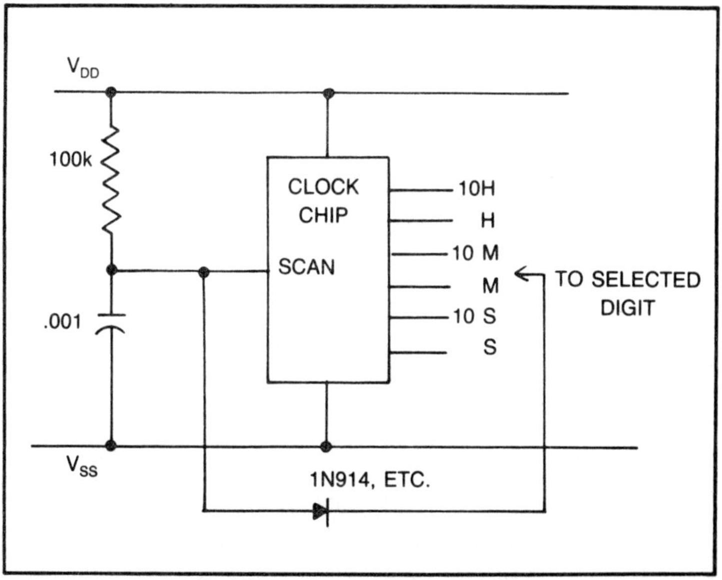


Fig. 11-2. Method of stopping the scan of any clock chip on a desired digit output. This permits use of that digit output to generate periodic interrupts. For example, with the minute digit selected, segment line six will produce an interrupt once each 10 minutes.

ers in addition, intervals can be set by remote control.

However, the easiest way of obtaining intervals of most values is to take advantage of the speed capability of the clock chip. Most are really three speed devices, the additional speeds being slow set and fast set. Typically these will give a 60:1 and a 3600:1 speed increase. There may be 50 Hz capability in the clock, giving three other ratios.

For other intervals, the easy approach is to divorce the time counting from the 60 Hz power, feeding the chip from an oscillator of suitable frequency. If no great accuracy is needed, a simple 555 oscillator is adequate. This will give intervals accurate to about one percent.

If greater accuracy is needed, the oscillator can be crystal controlled. A divider chain will be required with high frequency crystals, but these will not be necessary if the crystal frequency is kept below some critical frequency, which depends on the chip fabrication technique, but which is probably above 500 kHz. This is high enough to allow use of low cost crystals, and simple oscillators.

Another way to use the clock chip with the computer is to obtain time. This isn't bad if only a single digit can be read, two possibilities being shown in Fig. 11-3. In the first, the seven-segment data is fed to the data bus through an addressable gate. When required, the data is placed on the bus, read by the computer, then converted to the internal data form, BCD, ASCII, or other, by a

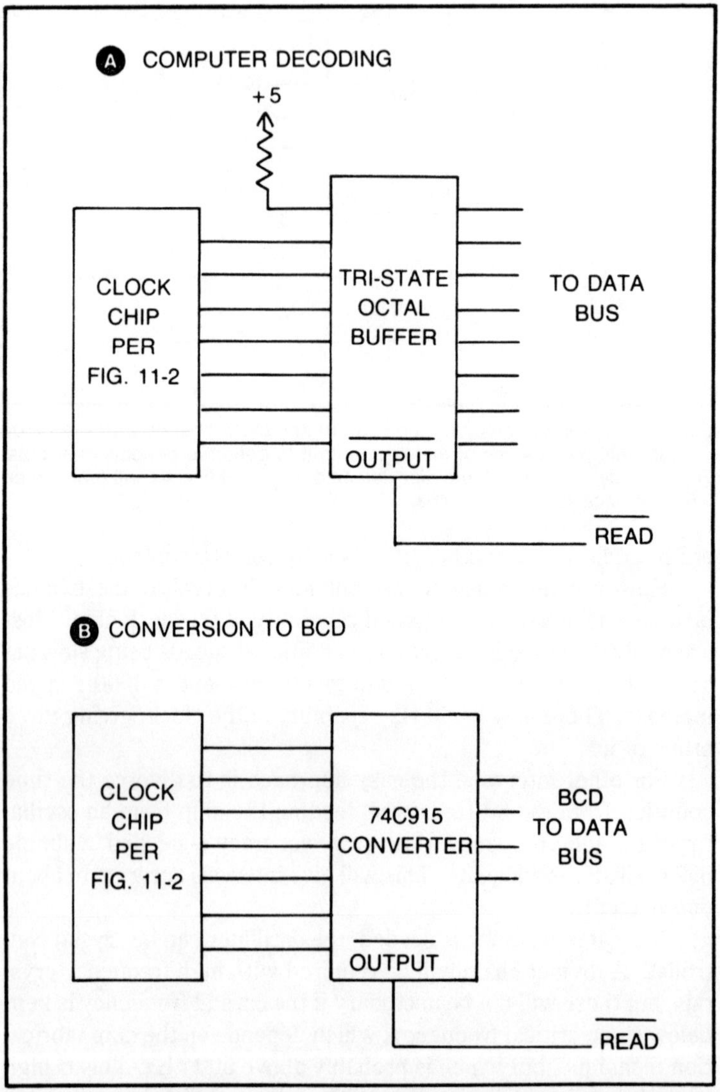


Fig. 11-3. Method of feeding time information to a computer. As shown, this works only for clocks with stepped scan. (A) Uses buffering only, seven-segment to BCD conversion being done by the computer. (B) Uses a seven-segment to BCD converter-buffer.

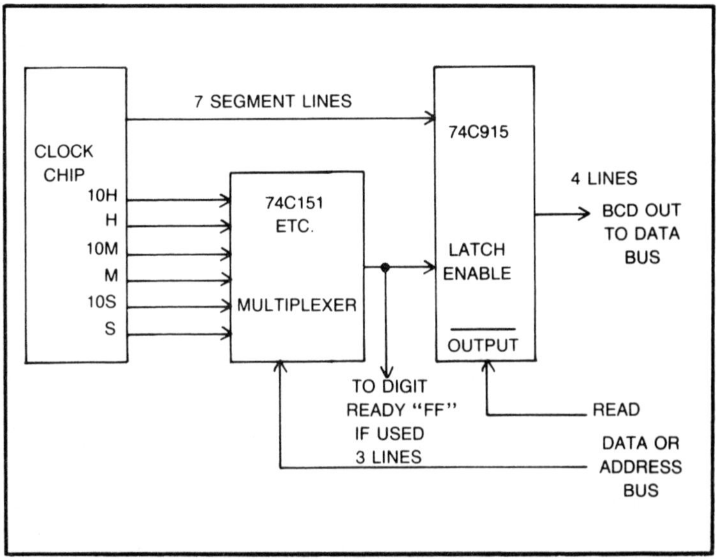


Fig. 11-4. Method of adding digit selection to the clock chip-computer interface. The multiplexer selects the desired digit. When it occurs, the data is latched into the '915 converter, then read out by the computer. Because of the time delay which must be allowed for digit selection, and complexity, special clock chips are better.

look-up table. The alternate is to use a seven-segment to BCD converter at the clock chip output. Either form is reasonably simple and operation would be fast and reliable.

When two or more digits are needed, complexity goes up. As shown in Fig. 11-4, a digit selection circuit must be added. Further, either a "digit ready" signal must be developed, or a delay between digit request and digit reading must be built into the program.

This figure also shows another possibility. The time setting function of the clock chip can be placed under computer control. The computer selects the digit to be observed during setting, starting with the highest, and the external circuit advances the clock until this digit is set, then stops. The computer then selects the next lowest digit, and the cycle repeats until all digits are set.

Of course, this method only works if the computer has access to a clock, so that it "knows" the correct time. This is alright for the alarm circuit, since, presumably, the computer will have built in program data on interrupt time. For the clock function, either manual setting or still further complication is needed.

These points add up to a very clear finding. Except for a few specialized and simple situations, the standard clock chip does not go well with computer operation. This is the reason a number of "computer-oriented" clock chips are now available.

## SPECIALIZED COMPUTER-CLOCK CHIPS

From the preceding discussion and drawings, we can set down a list of features a specialized computer-clock chip should have. The more important ones seem to be:

—Output in computer format.
—Date and time available as selected.
—Simple digit selection.
—Simple setting provisions.
—Compatible with computer speeds.

Some additional features which could be useful are:

—Greater than one second resolution.
—Settable interrupt.
—One or more settable internal timers.

Finally, if several designs were available, it would be nice to have a choice of other features, including 50/60 Hz or crystal time base, clock as well as computer presentation, battery operation and so on.

Manufacturers seem to have done a good job of providing these features. Table 11-1 shows the major characteristics of chips for which data is available. It is evident from this that there are areas of strong resemblance among the various designs, and also some considerable differences. Basically the chips fall into three groups, simple, very complex with many features, and an intermediate group essentially having the basic functions, plus a selection of "special features." From the view of the user, this range of choice is good, since he can have a choice of designs for his particular problem.

Let us look at two chips in some detail, one from the simple and one from the complex group. In doing this, we will try to cover enough data to be useful in design, although the normal practice of referring to the manufacturer's data sheets and applications notes should be followed.

## THE NATIONAL MM58174A CHIP

The National MM58174A is termed a microprocessor compat-

**Table 11-1. Major Characteristics of Specialized Clock Chips Designed for Simple Interface to Computers, or for Similar Applications. There Is Great Similarity in Basic Time Functions, but Considerable Difference in Added Features, Which Means That a Particular Chip May Be Superior for a Specific Application.**

| Type | CDP1879 | HI-8000 | MC14681P | MM58167 | MM58174 | MSM5832 | MPD1990AC | WD2412 |
|---|---|---|---|---|---|---|---|---|
| Maker | RCA | Holt | Motorola | National | National | OK1 | NEC | Western Digital |
| Bus Bits Data | 8 | 8 | 8 | 8 | 4 | 4 | 1 | 8 |
| Resolution Data | 488 μs | 15 μs | 31 μs | 100 μs | 1000 μs | 278 μs | 1 sec | 100 μs |
| Maximum | 1 year | 128 yrs | 100 yrs | 4 years | 4 yrs | 99 yrs | 1 yr | 99 yrs |
| Access Time | 450 μs | 250 μs | 1000 μs | 1050 μs | 850 μs | 150 μs | 40 μs | 250 μs |
| No. Interrupts | 1 | 1 | 3 | 2 | 1 | 4 | 1 | 2 |
| Features | A,S | P,Y,L,A,P | D,Y,L,A,P | 24,D,A | 24,D,Y,L | D,Y,P | 24,D,S | D,Y,L,A,C |

All units provide seconds, minutes, hours, date and month.

FEATURES LEGEND
A—Alarm
C—Clock Presentation
D—Day of Week
L—Leap Year Correction
P—Periodic Interrupts
S—Serial Bus
W—Computer Sleep/Wake
Y—Year
24—$24^h$ Time Only

ible real time clock. Design is quite straightforward, basically consisting of a number of counters arranged in groups. Some run free, others can be preset. An on-board clock circuit is provided. Control circuitry is minimal, but adequate for the design.

The basic block diagram of the chip is shown in Fig. 11-5. There are twelve groups of counters, covering the time span from tenths of seconds to tens of months. All of these can be read by program control from the computer, but only nine can be set—those for

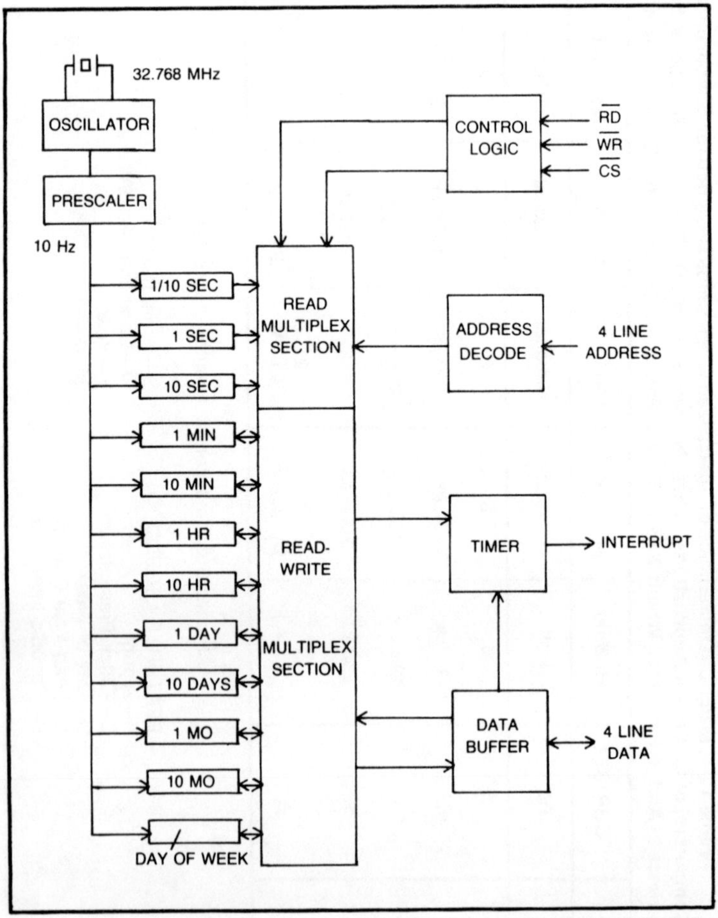


Fig. 11-5. Internal elements of a relatively simple specialized chip for clock-computer application. Each time unit has its own counter chain, all working from a 10 Hz line. This allows each counter to be set independently, but this capability is used only for the minutes and up time periods. As a result, while intervals can be resolved to 0.1 second, time reading can be accurate only to the nearest minute.

minutes or longer. The days register count will be reset after 28, 29, 30 or 31 days, depending on the setting of the months register, and on a special four year (leap-year) register. Leap centuries are not taken into account.

In addition, there is an interrupt circuit which can be set to generate a single interrupt after an interval, or repeated interrupts at the same interval, Only one of three interrupt periods can be chosen, 0.5, 5.0 or 60 seconds.

There is one special feature, intended for test. This feeds the crystal frequency to all counter groups in place of the normal ten pulse/second feed. This 3276.8:1 speed up is under computer control.

The connections to the computer are very simple, as shown in Fig. 11-6. Four data lines carry BCD data to read or set a counter, as selected by four address lines, and a read line or write line. Separately, the chip must have been selected ($\overline{\text{cs}}$) by selection logic. For the Z-80 type, for example, this can be generated from the I-O cycle, or it can be memory mapped for any computer. The interrupt line is separate and may go directly to the computer or to an interrupt bus.

The crystal circuit is very simple. The full equivalent circuit is shown in Fig. 11-7, along with the equivalent circuit of an NT or Tuning Fork, or XY or Bar crystal cut. (You might like to work out the value of equivalent inductance LI which is resonant at 32.768 kHz with the 0.003 pF equivalent series capacity.) Clock accuracy will depend on the exact frequency and stability of the crystal, of course, so it might be well to place the crystal and associated components in a temperature controlled enclosure (oven). Alternately, the clock can be reset under computer control at intervals, or, if available, a precision signal can be fed to the chip to replace the crystal.

Internally, the crystal output is first divided by the ratio of 15/16, to obtain 30,720 Hz. This is 60 Hz times $2^9$, the 60 Hz being obtained by a nine-stage counter. This is divided by six to obtain the 10 Hz signal which goes to all of the time-keeping counter groups.

Figure 11-6 shows the simple battery circuit suggested by National to keep the clock in operation during computer shut down or for power loss. Since the standby drain is only 10 mA maximum at 2.2 volts, the battery can be very small.

Table 11-2 gives the read-write information for chip control. Note that three of the possible sixteen states of the address lines

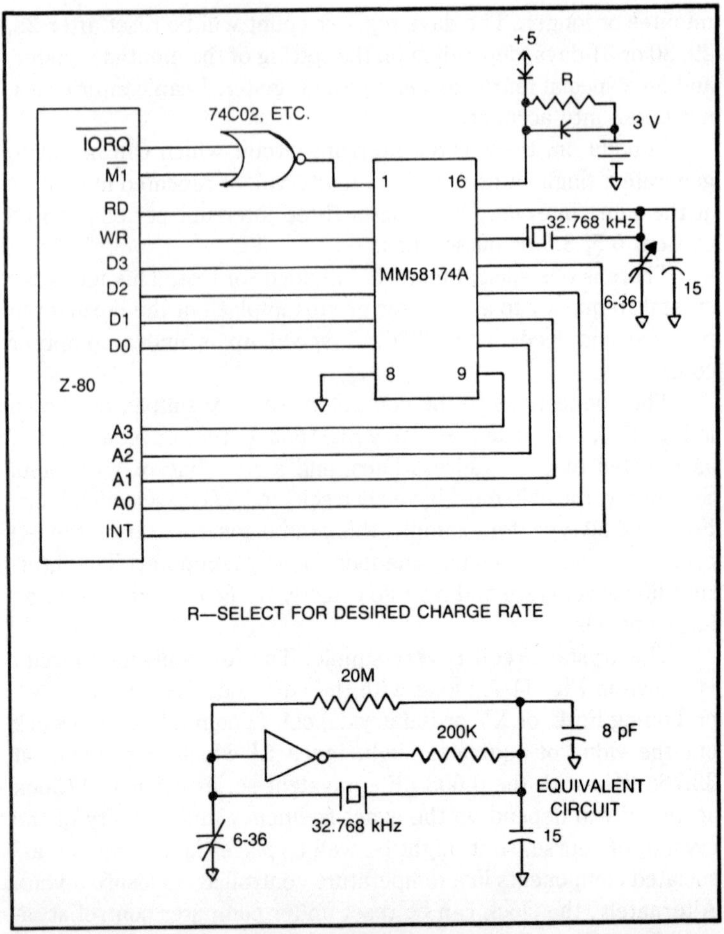


Fig. 11-6. Interconnections for the MM58174 clock chip to a microprocessor, the Z-80 being used as an example. Compare to Fig. 11-4 for relative complexity. Since this chip has low voltage operation capability, standby power is easy to supply.

are write-only, to set internal registers. As noted before, the three seconds registers are read only, and the rest are read-write, to give full clock set capability from minutes to tens of months. Other sections of the figure give the interrupt selection, read status and year status.

Unlike clock chips, the read/write speed is compatible with most microprocessors. This is shown by the diagram of Fig. 11-7. The values shown are specification extremes. Note that some signals, those with 0 as a limit, are specified to not occur before another signal. There is one limit condition since the internal registers are reset only at 0.1 second intervals. When reset is required, too fast a read after a reset may give invalid data. Pin connections are given in Fig. 11-8. The 16 pin package is smaller than commonly found in this class of chip.

## THE WESTERN DIGITAL WD2412 CHIP

The Western Digital WD2412 Time of Day clock chip is at or near the upper limit of capability and complexity. This is reflected,

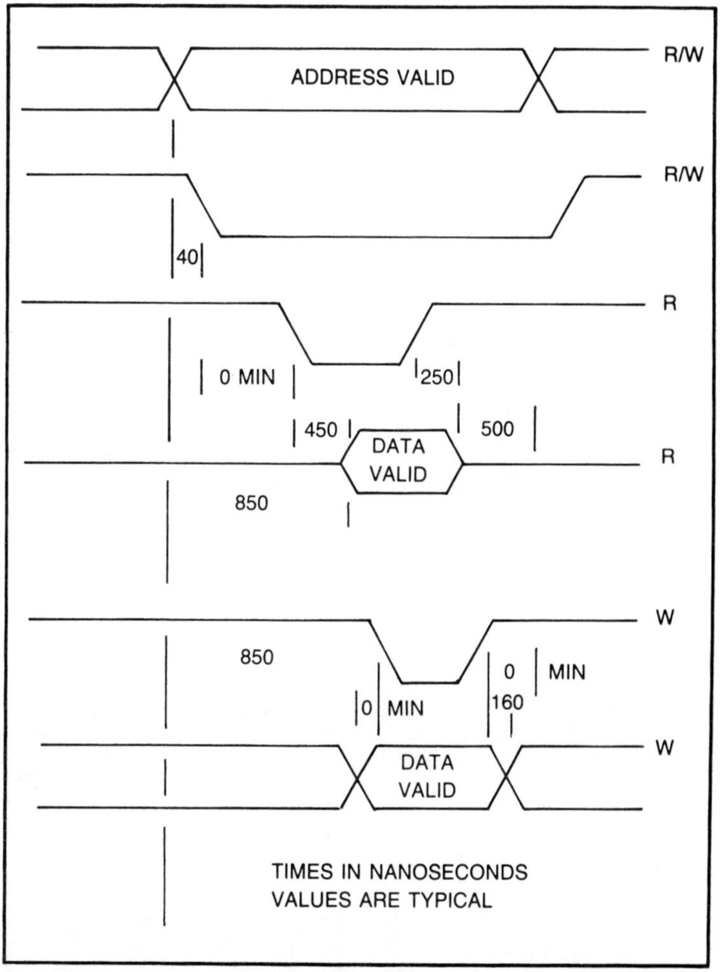


Fig. 11-7. Timing diagrams for the MM58174 chip, typical values. These are compatible with most microprocessor requirements.

**Table 11-2. Coding for the MM58174 Chip Address and Data Lines, for Setting, Reading and Interrupt Functions.**

A. Address Decoding

| Counter | | Address | | | | Mode |
|---|---|---|---|---|---|---|
| 0 | Test | 0 | 0 | 0 | 0 | W |
| 1 | Sec/10 | 0 | 0 | 0 | 1 | R |
| 2 | Sec | 0 | 0 | 1 | 0 | R |
| 3 | 10 Sec | 0 | 0 | 1 | 1 | R |
| 4 | Min | 0 | 1 | 0 | 0 | R/W |
| 5 | 10 Min | 0 | 1 | 0 | 1 | R/W |
| 6 | Hr | 0 | 1 | 1 | 0 | R/W |
| 7 | 10 Hr | 0 | 1 | 1 | 1 | R/W |
| 8 | Days | 1 | 0 | 0 | 0 | R/W |
| 9 | 10 Days | 1 | 0 | 0 | 1 | R/W |
| 10 | Day of Week | 1 | 0 | 1 | 0 | R/W |
| 11 | Month | 1 | 0 | 1 | 1 | R/W |
| 12 | 10 Months | 1 | 1 | 0 | 0 | R/W |
| 13 | Years | 1 | 1 | 0 | 1 | W |
| 14 | Start-Stop | 1 | 1 | 1 | 0 | W |
| 15 | Interrupt | 1 | 1 | 1 | 1 | R/W |

B. Interrupt Selection

| Function | Data, Single INT | | | | Data, Periodic INT | | | |
|---|---|---|---|---|---|---|---|---|
| No Interrupt | X | 0 | 0 | 0 | X | 0 | 0 | 0 |
| 60 Sec INT | 0 | 1 | 0 | 0 | 1 | 1 | 0 | 0 |
| 5 Sec INT | 0 | 0 | 1 | 0 | 1 | 0 | 1 | |
| 0.5 Sec INT | 0 | 0 | 0 | 1 | 1 | 0 | 0 | 1 |

C. Interrupt Read-Back

| | | | | |
|---|---|---|---|---|
| No Interrupt | X | 0 | 0 | 0 |
| 60 Sec INT | X | 1 | 0 | 0 |
| 5 Sec INT | X | 0 | 1 | 0 |
| 0.5 Sec INT | X | 0 | 0 | 1 |

D. Year Status Select

| | | | | |
|---|---|---|---|---|
| Leap Year | 1 | 0 | 0 | 0 |
| Leap Year +1 | 1 | 1 | 0 | 0 |
| Leap Year +2 | 1 | 0 | 1 | 0 |
| Leap Year +3 | 1 | 0 | 0 | 1 |

for example, in the command table of 24 commands of three different types, and in physical size and number of pins. A 28 pin package is used.

Because of its complexity, the logical block diagram is shown in three parts. Figure 11-9 shows the basic time bases, which offer a choice between a 50/60 Hz input, an internal crystal oscillator (4.096 MHz) or an external input at crystal frequency. The time base may be selected under computer control. The clock reset function also sets the seconds accumulation to a read precision of 0.1 second, and to an absolute precision of one clock cycle plus command signal delays. As for the previous chip, the basic time functions are driven at 0.1 Hz. The counters necessary for this are separate for the several inputs.

Figure 11-10 shows the time accumulation system. Two independent chains are provided. One gives time in seconds-minutes-hours format with each digit in BCD form, two digits per byte. The second gives a straight binary count of 0.1 second "ticks," starting at midnight. The binary range is through 863,999 decimal, held in three eight-bit registers with the most significant bit always 0.

There are separate counters for 12 and 24 hour format data. The last also drives the day of the week and the day-month-year counters, and in turn, counters for month length and leap year control. Digits are held in two per byte form, so that the major time divisions, seconds to years, can each be read from a single byte.

Figure 11-10 also shows the technique used for generating interrupts. Either binary or BCD setting may be used, and the interrupt may be one-shot or repetitive. The repeat period can be from

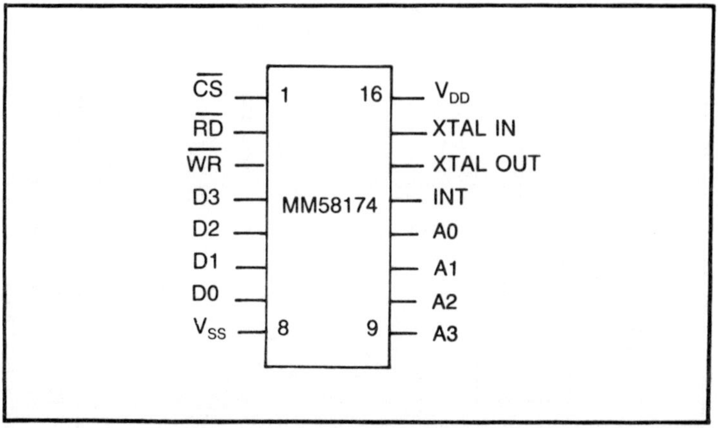


Fig. 11-8. Pin connections for the MM58174 chip. Designations are standard.

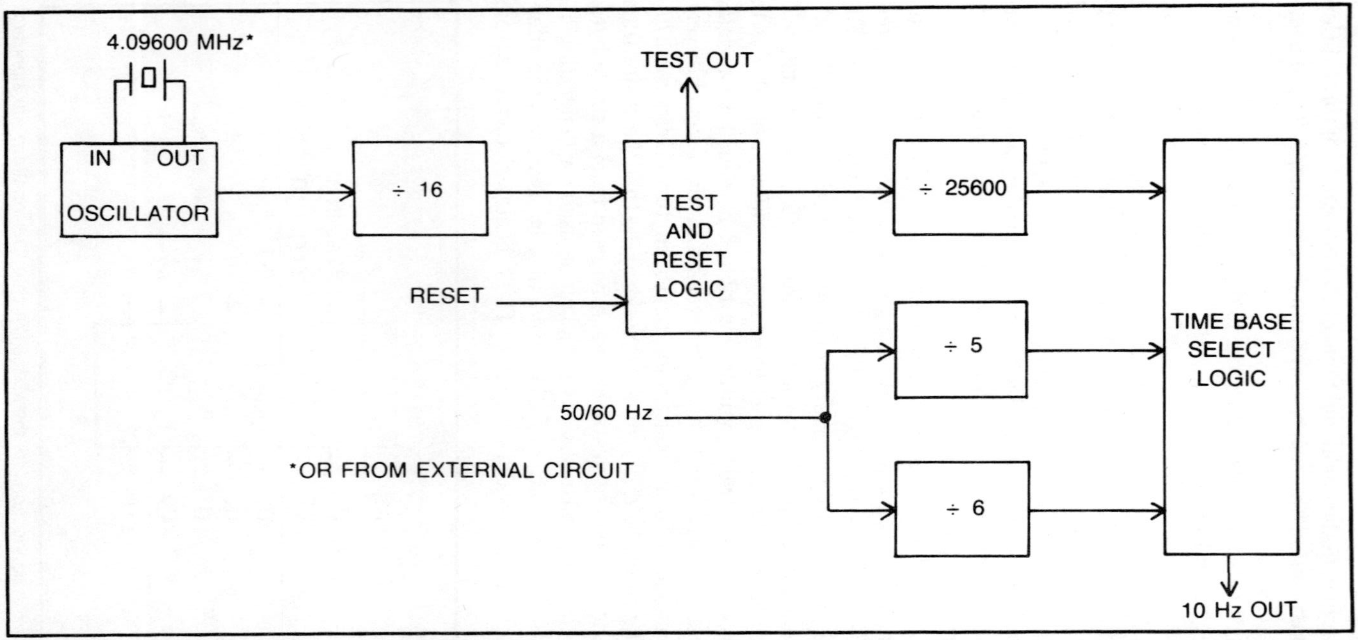


Fig. 11-9. The time base system of the WD2412 clock-computer chip. Either crystal or 50/60 Hz line operation can be selected. The test circuits appear to be intended for factory use, and are not fully discussed in the specification sheet. Note that this chip also uses 10 Hz as the common feed.

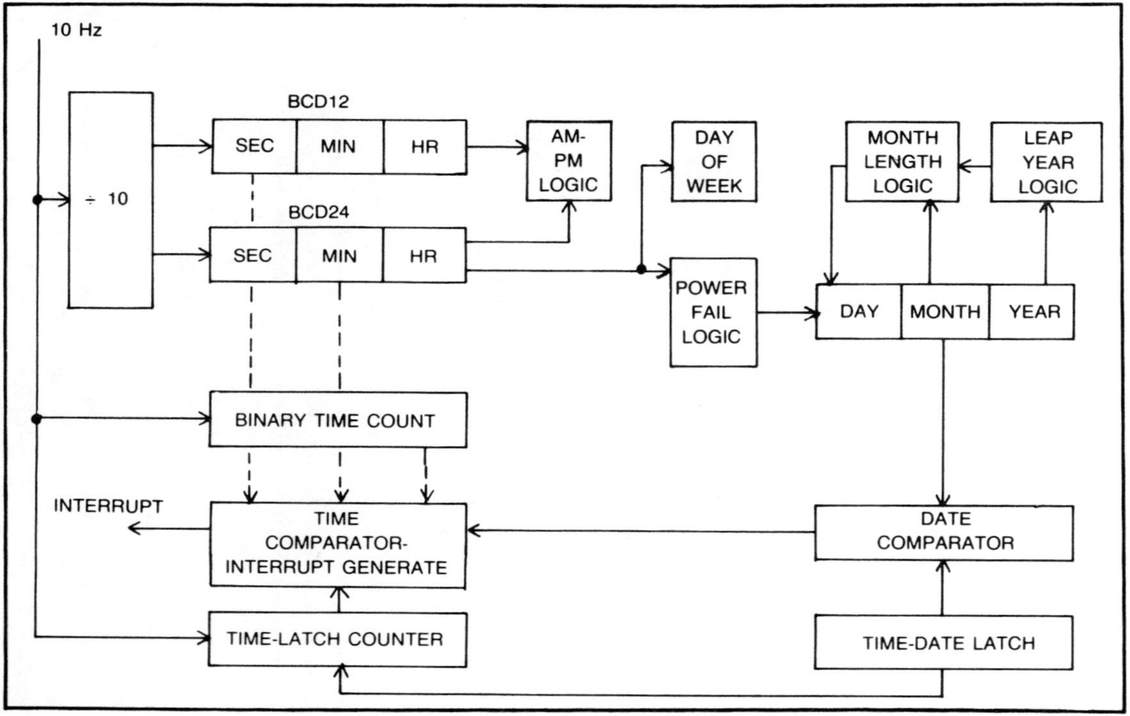


Fig. 11-10. Time accumulation and comparison circuits of the WD2412 chip. Note the separate twelve and twenty-four hour chains, and the separate BCD chain, which gives time in tenth seconds since midnight. Additional registers handle day, week and year accumulation. Comparators are arranged to allow periodic or "one shot" interrupts.

0.1 seconds to one day. The automatic repeat can be suspended by program control.

The last part of the system shown in Fig. 11-11 shows an unusual feature. The chip includes a display multiplexer which is capable of displaying six digits of time or date. Status signals show if the display is in 12 or 24 hour format, or if it is displaying the date.

A further unusual feature is associated with this display. A special register may be loaded from the computer, and its contents displayed. Valid characters are 0-9 and blank. The manufacturer's literature does not state whether this is a hardware limitation imposed by the chip, or by the decoder used in the typical connection shown, or whether full 16 BCD decoding is possible, even though some are invalid.

The command and read system uses only the data bus, together with a requisite chip select ($\overline{\text{cs}}$) and read or write ($\overline{\text{RE,WE}}$) signals. The commands and associated status signals are given by Tables 11-3, 4, and 5 for Type 1, Type 2 and Type 3 commands, respectively. Commands should not be transmitted until a chip status register read yields the byte EEH. Also, for either read or write, the receiving device should acknowledge with the status byte FFH.

There is an additional function, not explained by the manufacturer's literature. This is a pin marked TEST0. Presumably grounding this pin causes the registers to advance at a high rate (25,600:1?), to permit fast check of week to leap year registers.

Timing for the various signals is shown in Fig. 11-12. These are compatible with normal microprocessor signals, noting that the minimum width for the READ and WRITE strobes is 300 ns.

Despite its greater internal complexity, the chip is no more difficult to connect to a computer than the previous one. A block diagram showing some alternates is shown in Fig. 11-13. Chip select can be generated by one of the special chips or by comparators. In a simple system using the Z-80, it can even be an AND gate. The outputs are three state with the customary 1.6 mA drive capability, so normal bus loading rules can be followed.

In this figure, four types of time base source are shown. Only one need be used. Also shown are two other options, the $\overline{\text{MR}}$ lead which activates the crystal test circuit, and the crystal test lead which toggles at the crystal frequency divided by 16 when $\overline{\text{MR}}$ is active.

Figure 11-13 also shows the last option, the use of the display capability. Since the data output for this is in BCD form, a BCD-7 segment decoder is needed. This also serves as the segment driver.

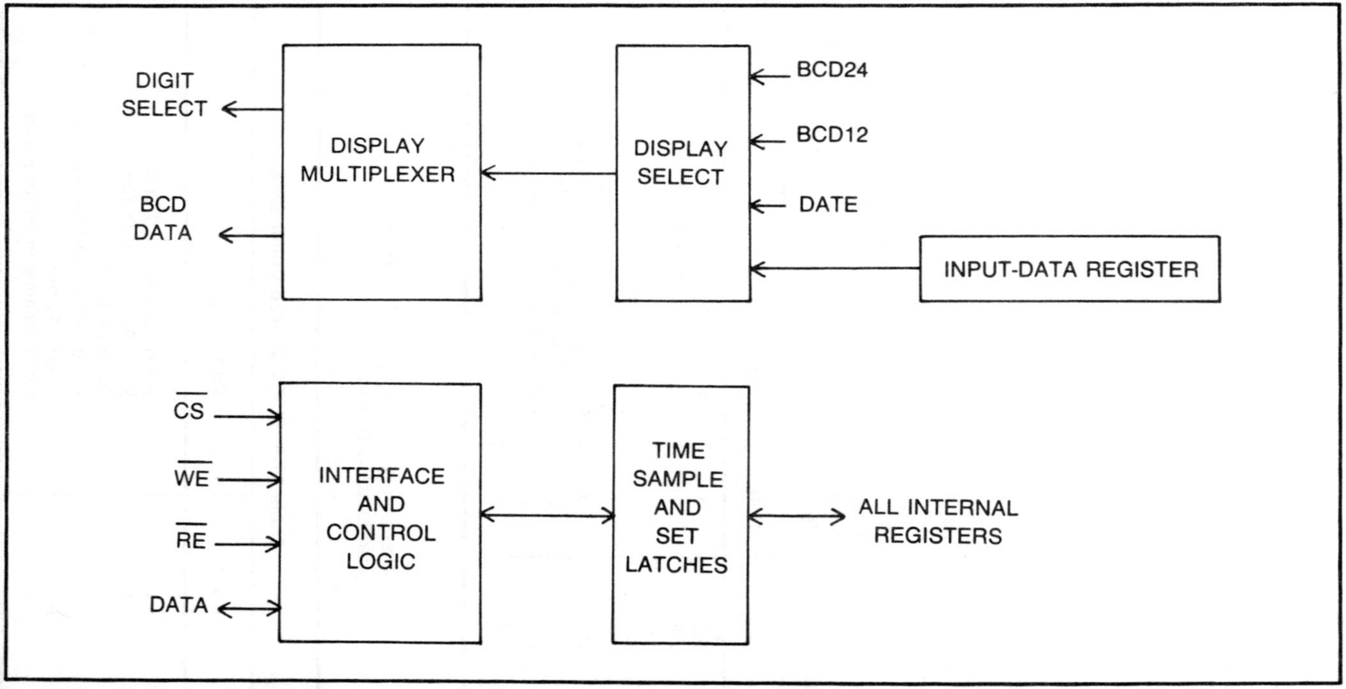


Fig. 11-11. Display and control elements of the WD2412 chip. The provision of a regular clock display is unusual in this class of chip.

| Code | Command |
|---|---|
| E0 | Display Off |
| E1 | Display 12 Hour Time |
| E2 | Display 24 Hour Time |
| E3 | Display Date |
| E4 | Display Input Data |
| E5 | Select Xtal Timebase |
| E6 | Select Ext. 50 Hz |
| E7 | Select Ext. 60 Hz |
| E8 | Disable Interrupts |

Note:
EE—Command Ready
FF—Data Transfer Acknowledge

**Table 11-3. WD2412 Commands for the Display and Time Base Selection, Plus Interrupt Disable.**

**Table 11-4. WD2412 Commands for Setting the Time Registers, and for Enabling Input and the Interrupts. Note That the Seconds Registers Cannot Be Set, and also That the Binary Register Resets When the Time Register Indicates Midnight.**

| Code | Notes | Command |
|---|---|---|
| E9 | 1 | Set 12 Hr BCD Time |
| EA | 1 | Set 24 Hr BCD Time |
| EB | | Set Date |
| EC | | Set Day of Week |
| ED | | Set Input Data |
| EF | 2 | Set Absolute 12 Hour Time Interrupt |
| F0 | 2 | Set Absolute 24 Hour Time Interrupt |
| F1 | 2 | Set Absolute Date Interrupt |
| F2 | 2 | Set Relative Interrupt |

Notes:

1. Also sets Binary and other BCD Register
2. Also Enables Interrupts

**Table 11-5. WD2412 Read Commands.**

| Code | Command |
|---|---|
| F3 | Read 12 Hour BCD Time |
| F4 | Read 24 Hour BCD Time |
| F5 | Read Binary Time |
| F6 | Read Date/Power Failure Flag |
| F7 | Read Day of Week |
| F8 | Read Relative Interrupt Status |

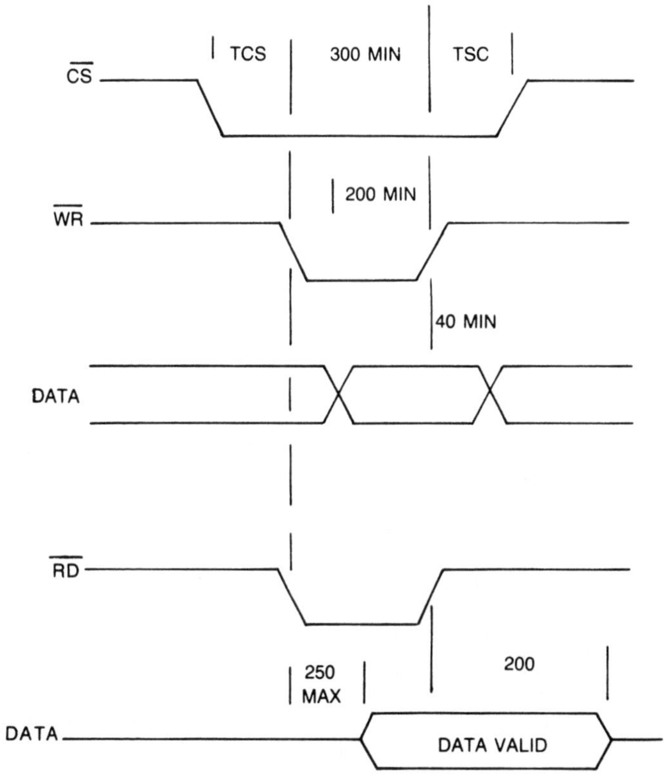


| | TCS,MIN | TSC,MIN |
|---|---|---|
| READ | 50 | 5 |
| WRITE | 20 | 20 |

Fig. 11-12. Timing of signals for the WD2412 chip. These are compatible with most microprocessors.

A digit driver is needed also. Depending on presentation mode chosen, up to four LED indicators are also needed.

This chip does not have a power-down mode. Since it requires up to 40 mA to operate, power interrupt circuitry will necessarily include more storage than for the previous chip. Beyond increasing battery voltage and size, the back up power does not need to be more complex, however.

While the computer interconnect is very simple, a considerable amount of programming will be required to take advantage of all of the features of this chip. Typical sequences in 8080 assembly language and in Microsoft BASIC are shown in the manufacturer's application note. These range from 10 bytes for the simplest machine code, to 25 Basic lines for setting and reading the clock. The preferred method, of course, is callable subroutines. Figure 11-14 gives the pin connections for this chip.

## OTHER CHIPS

The general characteristics of other chips have been tabulated in Table 11-1. Some additional features found in these are time coverage of up to 128 years, automatic Daylight Saving Time change, multiple interrupts, and provisions for microprocessor control (sleep and wake up modes). Study of these will show which is best for a particular application.

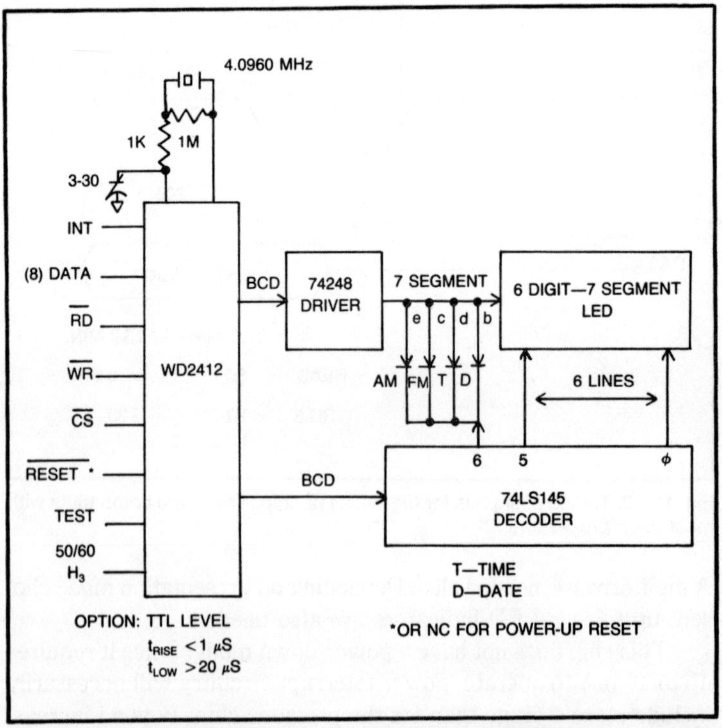


Fig. 11-13. Connection diagram for the WD2412 chip. Chip select must be developed by external components, and an interrupt recognition scheme established. The display elements are optional, but very useful, since computer information can be displayed as well as time.

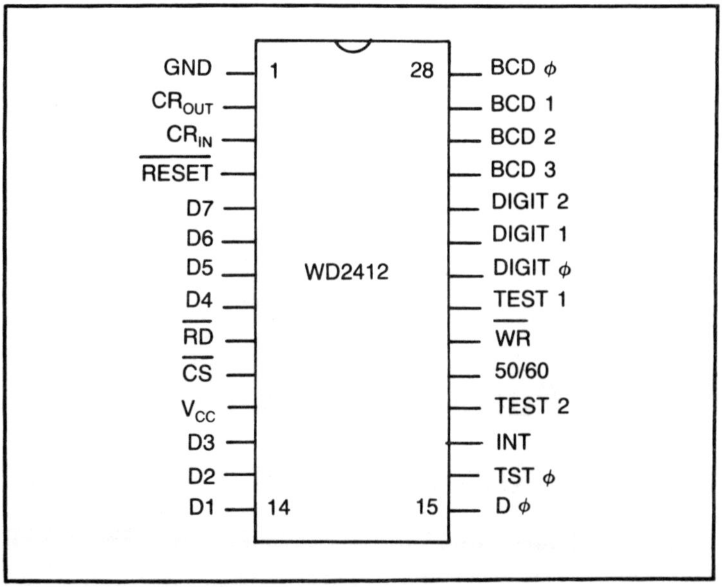


Fig. 11-14. Pin connections for the WD2412 computer-clock chip.

## A FINAL NOTE ON CLOCK CHIPS FOR COMPUTERS

The cost of these specialized microprocessor clock chips range from about the same as conventional clock chips, to perhaps twice as much. While not all are available through usual experimenter/hobby channels, a good selection is available.

Because of their greater capability, and their markedly simpler interconnections, these chips can yield both an overall cost and complexity advantage over conventional chips. The techniques of the first part of this chapter, while once necessary, are no longer recommended. They have been included for background only.

# Chapter 12

# Purchased Clocks, Clock Kits and Clock Repairs

One of the features of mass production, combined with mass distribution, is its effect on prices. As long as a standard design is needed, it is actually cheaper to buy the completely assembled standard unit than to purchase the parts which go into it.

This situation arises because of what is called "Original Equipment Manufacturer" pricing, or O.E.M. pricing for short. Basically, this is just a way of saying that there is a cost discount for the bulk buyer. It's actually for *any* bulk buyer, but only another manufacturer can buy complex chips in lots of 10,000 of a single type.

Manufacturers point out that the pricing is necessary. It costs essentially the same amount to process an order for one chip or for 10,000. Unit orders with any reasonable delivery time require a stock on hand, whereas manufacturers usually want regular shipments, which means that only a small inventory is necessary. Unit handling costs also go down with bulk packaging.

There is an important opposite facet to this. If you want a single unit of different design, you won't be able to buy it at any price. You must go to a prototype design organization and pay the entire engineering design and hand fabrication cost for your single unit. This is so unreasonable that usually the alternates are used—build the unit yourself, or do without.

Actually there is a very good financial reason for building the unit yourself—in addition to the reward of having something that's especially yours. For example, both Zenith and Collins Radio started as one man operations. Hewlett-Packard was different—two partners were involved. The opportunities still exist--look up the history of the Apple computer. Keep this in mind as a possibility if you come up with a design which you feel is both unique and marketable.

The kit sellers have managed to take advantage of these pricing policies by selling single unit kits at a little less than the cost of a completely assembled unit, or have offered some nonstandard features in designs at the same price. They also have an advantage in one respect—if you can build a unit from components, you can probably repair it. These days, that's a real advantage.

Repair problems exist because of two factors. One is the element of labor cost. Between direct wages, fringe benefits, taxes and the rest of the overhead items, labor charges are approaching one dollar a minute for any field involving training. And, the minutes start when the repairman picks up the work order.

The second factor is a recent tax decision on repair part inventory. Taxes must now be paid each year, instead of at time of sale. Parts are now far more expensive than production cost, or are no longer available at all. This is the result of shops trying to cut the tax load by reducing inventory.

In this complex situation, there are some guidelines which the experimenter can use to make his life at least less expensive. The important ones are:

—If a standard unit will do the job, it is probably the cheapest way to go. It is certainly the easiest and quickest.

—If a standard unit can be modified to do the job, it's probably cheaper than buying parts. It may be easier and no more time consuming.

—In either of the above cases, consider buying a second unit as a spare. If you are in the business, remember the WWII airplanes called Hangar Queens, which were the only source of spares.

—Keep an eye on the surplus and end-of-production sales. Look for bargains in assembled, or partly assembled, units or spare parts. You won't be taxed on your spares!

—Try to get a buying group together, say at a local Hobby Club, a Radio Club, or a Computer Group. There is usually a cost advantage in lots of ten or more.

—Try to set up a trading group in the same way. The cheapest part is the one you already have. The second cheapest is the one you can trade for.

With these techniques you have a good chance of having what you need, when you need it, at a reasonable price. You will need a little storage room, but some time spent on organization will do wonders here.

## PURCHASED CLOCKS

The state of electronic clock development has been so perfected that any clock on the market can be expected to be a good timekeeper. The ones which operate from the power line will be within a few seconds of exact time, subject to the occasional need for resetting due to power outages. Expect more of these in the future. Battery operated quartz clocks and watches approach chronometer accuracy, as discussed in an earlier chapter. They can be trusted to keep good time for about a year. However, despite this equality in timekeeping, all clocks are not equal. Some thought and some checks are in order.

In the chapter on clocks for the handicapped, it was mentioned that size and color influence readability. Perhaps these factors can be important for people whose eyesight is essentially normal, or who are only faced with the usual aging problems. Size effect would be expected, but the fact that color is important in readability is the oddity.

The phenomena involved are not well understood, and there seems to be little in the literature about it. Several factors seem to be involved. One is the ability to focus which reduces with aging. Eyeball shape which determines whether a person will be nearsighted, average, or farsighted when the eye muscles are relaxed is a factor. The final factor seems to be related to the matter of true color blindness, or even to the ability to distinguish between shades.

The net effect is that one person may see one type of clock display clearly across the room, and not see another. There is a marked difference between liquid crystal and the emitting displays. There is also a marked difference between colors of the emitting displays, especially between the fluorescent green and the LED red. Blue or amber displays are not common in clocks. Finally, there is apparently no experience with colors in the rear illumination of the liquid crystal displays in the night mode.

Because of these personal variations, it is suggested that you make a test for all members of the family, and record the results for later use. The test distance in feet, should be about 20 times the display height in inches. This would be about 15 feet for the large 0.7″ display. Try both normal room illumination and a darkened room simulating night conditions. Allow several minutes for eye adaptation to the dark. If a member of your family has been having vision problems, the results may be very surprising.

The other area where clocks are not equal is in the convenience of controls. For example, Fig. 12-1 shows a compact design, having a good size presentation. But notice that no controls are visible, except one that is obviously an add-on. For the regular controls it is necessary to go to the back of the clock, shown in Fig. 12-2. These controls succeed in violating a number of human engineering principles. The adaptation of controls to people is ignored.

The two buttons for slow and fast set are too close together, and cannot be told apart by shape. They are labeled, but it is necessary to pick up the clock, look at the back, and peer until the light finally picks out the tiny raised letters, which are the same color as the case. It is then necessary to remember the right one. The penalty for an error in use can be another several minutes required to go through the time or alarm set process.

Note also that there is a three position switch. This is recessed to keep it from "being bumped," but it also makes it hard to position. Also, the switch combines two radically different functions, that of turning on the alarm, and of placing the seconds indication on the face.

Fig. 12-1. A commercial clock. The unit is small, and easy to read. But the provided automatic brightness control did not have sufficient range, so was replaced by a manual control, the added knob.

Fig. 12-2. Back of the clock of Fig. 12-1. It is not possible to distinguish between slow and fast set, and two very different functions are combined on one switch.

This clock does have one feature which should be a help. It has an automatic-ambient-light-level control, intended to give a bright display during daytime and a subdued one at night. In this clock at least, the intensity change is barely noticeable and not a real help. The added manual adjustment works.

For almost complete contrast in ease of use, look at the clock of Fig. 12-3. It is larger and the extra size has been used to good advantage. All controls are now at the front and they are labeled with legible letters. There are two different sizes of buttons. The large one is isolated, since it will be used when senses are dulled—at bed time or on awakening. It is dual function but there is little chance of error.

However, there are still some relatively poor features. Both time and alarm setting require two button depression. It can be done with one hand, but it is almost a two-hand operation. And the time activate and the slow and fast buttons are the same size and shape and are grouped unnecessarily close together. These two examples should be sufficient to show the things to watch for.

Some people prefer the normal "hands" or clock face time presentation. These are now readily available in electronic designs, due to the development of low cost and low power stepper motors, which can be driven from a 1.5 volt digital circuit, and synced with a quartz crystal. At present, these designs are beyond experimenters capability. The complete drive motor assembly is avail-

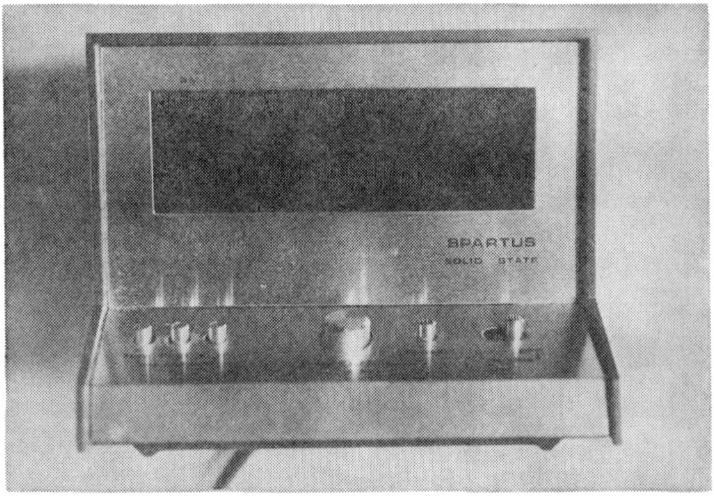

Fig. 12-3. Another commercial clock, for which some thought has been given to necessary manual operation. See text for residual deficiencies.

able, and can serve as a replacement, or as the basis for a new design or a reconstruction.

Figure 12-4 shows a typical application of one of these designs, a travel clock which is still large enough for general use. Time accuracy is very good, and the lack of ticking noise makes it a good bed-side clock. The hands are large enough to be visible, and are luminous at night.

Fig. 12-4. A "quartz" timepiece using a stepping motor drive to give a conventional clock face. Accuracy is very good.

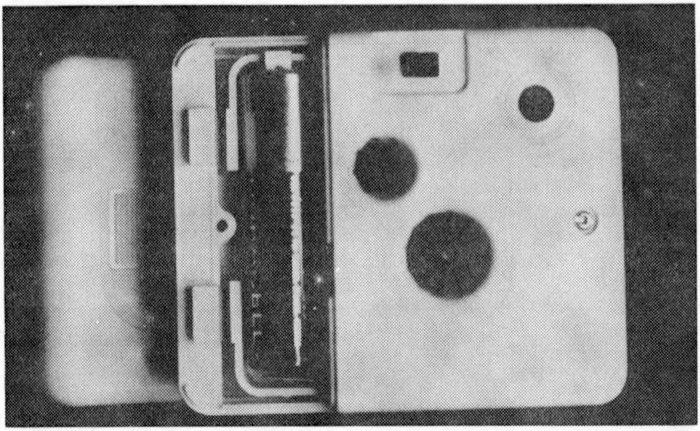

Fig. 12-5. The rear of the clock of Fig. 12-4. While not easy to set, battery operation plus good accuracy makes this necessary only once a year.

However, the controls leave a lot to be desired, as seen from Fig. 12-5. It is very difficult to see the small raised lettering which is the same color as the case. The circular time and alarm setting controls are set flush for antiknock protection, which makes setting difficult. This is probably acceptable for time setting, since it is needed only once a month or so. The alarm ON control is recessed also. While somewhat inconvenient, it serves its intended use but it would be impossible for many persons with physical or visual handicaps.

There are many other types of designs available, as clocks,

Fig. 12-6. A "stick-on" clock. Almost a watch, these units are available at very low prices. Accuracy of the units tested has been surprisingly high. Duplication is not a profitable experimental field, but the low cost and accuracy suggest that application experiments are in order.

clock radios, toys and even pieces of furniture. One that may be of interest to experimenters is the "stick-on" design of Fig. 12-6. Because of its low cost and small size, it can go almost anywhere.

## ABOUT WATCHES

You have probably noticed that there is very little information about watches in this volume. They are a specialized field. True, the basic principles of electronic watches and electronic clocks are the same, just as it was for the older mechanical types. But the constraints of size are just as severe in the electronic watches as in mechanical ones. Due to work involved in the cases, they require special tools for construction. Watches are beyond the capability of most experimenters.

The comments about legibility and ease of control are even more important in watches. This is true because of the inherent smaller size of the face and the area for controls. One thing to try to evaluate is the likelihood of control buttons jamming, especially those that are used rarely. "On the wrist" is really a severe environment because of the possibilities of shock, getting dipped in water, exposure to dusty areas, and so on.

## CLOCK KITS

Two types of clock kits are now available from several sources. One group includes complete clocks, including internal parts and cases. The second group includes modules. For example, a time module, a control module, and a power module in a modular kit which usually does not include any case or housing, is common.

Ads for these sources will be found in electronic hobby magazines, in computer magazines, in the catalogues of large kit sellers, and in catalogues of the parts distributors which cater to small customers. Kits and modules have been offered at hamfests and computer fairs, and have been observed for sale at flea markets.

The kits tend toward the upper end of clock prices, offering decorative features or exceptional displays. For example, the Heathkit GC-1720 Digital Wall Clock offers the following features:

—4 Digit Display, 1″ Height, Liquid Crystal
—12 or 24 hour format
—Battery operation
—One-two year battery life
—Crystal time base
—Wood grain case, 5″ × 7″

Since this is a wall clock, setting controls are not accessible.

The same company offers another design, with a 3 1/2″ display in an etagere, with an electronic version of Westminster chimes. This is a power line operated unit.

## CLOCK MODULES

A number of types of clock modules are available. These seem to have become standardized in format, in a four digit display. Some are 12-hour type only, others offer a choice. The modules are basically an assembly of a clock chip and a display on a single PC board. There is a choice of types for power line, automobile battery or dry cell. The latter uses a liquid crystal display, the others have LEDs. Digit size is usually reasonably large, 0.5 to 0.7 inches in height.

Kits of controls and power supply components are also offered. Some of these are designed to give a very compact arrangement of parts at the expense of control convenience. Cases do not seem to be offered.

Specifications for the module kits of one manufacturer are given in Table 12-1. Note particularly the wide range of supply voltages which are consistent with the design goal of power line, auto or battery use. Note also the optional capabilities. If you encounter

**Table 12-1. Characteristics of Selected Clock Modules That Are Readily Available from Hobby-Experimenter Supply Houses. Basically, These Modules Combine a Chip, Many Necessary Auxiliary Components and the Display. Supplemental Kits Are Available for Some, in the Way of Switches, Power Supply and Cabinets.**

| Type | Display | Voltage | 12/24h | Alarm | Sleep | Power Loss | Other |
|---|---|---|---|---|---|---|---|
| MA1003 | 0.3 Fluorescent | 9-24 | - | - | - | - | BC |
| MA1020 | 0.84 LED | 6.5-9 | L | L | L | L | BEL RS |
| MA1023 | 0.7 LED | 6.5-9 | L | L | L | L | BEL RS |
| MA1032 | 0.5 LCD | 1.4-1.6 | L | L | L | - | RS |
| MA1036 | 0.3 LED | 6.5-9 | L | L | L | L | BCS |

LEGEND:

B—Brightness Control
C—Crystal Timebase
E—50 Hz Option
L—Lamp Option
R—Radio Option
S—Seconds

one of these modules in a commercial clock, don't forget to look for these capabilities—they might be useful.

The last unit on the list is designed to a minimum cost objective. For example, the bare clock chip is mounted on the PC board, then a glob of epoxy placed over this, and cured. Board repair is impossible, but the high reliability of these designs has been surprising to me. The circuit of such a clock is shown in Fig. 12-7.

## SPECIALIZED CLOCK CHIPS, MODULES AND KITS

There are now a number of special clock products available. Generally, these combine the timekeeping of a clock chip with some other function or functions. Some use the clock display for the added function, some place the clock signal on another display, and others use multiple displays. The last may use separate components for each function.

An example of the clock display type is the National MA1026M Time-Temperature Module. As shown by the block diagram of Fig. 12-8, the unit is basically a clock with the added capability of reading temperatures remotely from a number of locations as determined by switching. There are also a selection of clock type output signals.

Temperature sensing is done by LM334 solid state temperature transducers, or the equivalent. Since these function as current sources, accuracy is essentially independent of the length of wire connecting the module and the sensor. However, one calibrating "set and forget" control is required for each sensor. The general characteristics of this module are shown in Table 12-2.

A unit designed to place clock data on another display is the General Instruments AY5-8320 chip. This is designed to take digit strobes and BCD encoded data from an AV-5-1203A clock chip, and encode it for presentation in color on a TV screen. It also accepts BCD coded channel selection data, for channels from 1 to 16. Several combinations of the display are possible.

The display is formed at the upper right of the screen. Characters are 1.7 inches high on a 25-inch screen, and are basically formed with a three dot by five line matrix. Each dot has a duration of 0.9 microseconds, so the character width is just over one inch.

The interconnections, shown in Fig. 12-9, are not excessively complex for the number of functions performed. As noted, digit and BCD data are taken from the clock chip. BCD channel data may come from a mechanical switch, or from a diode encoder if the tuner is varactor controlled.

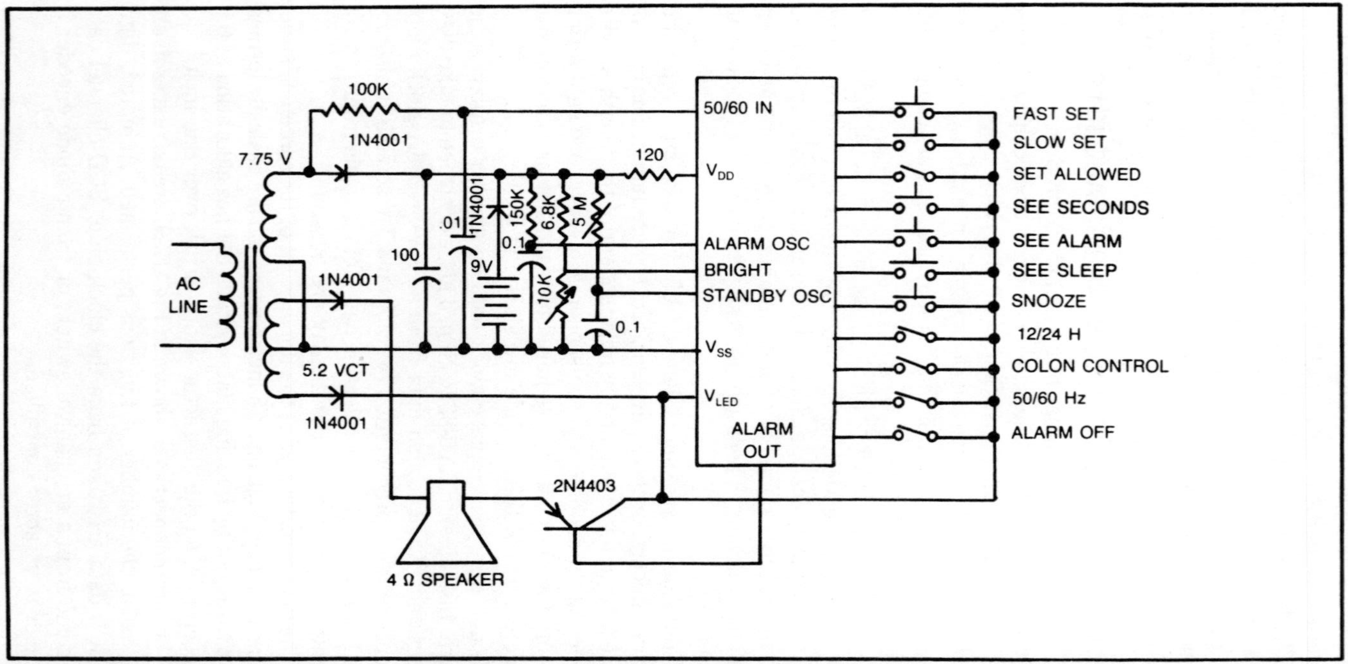


Fig. 12-7. Schematic of a clock based on the MA5036 low cost module. This has the basic clock chip cemented to the PC board by a drop of epoxy.

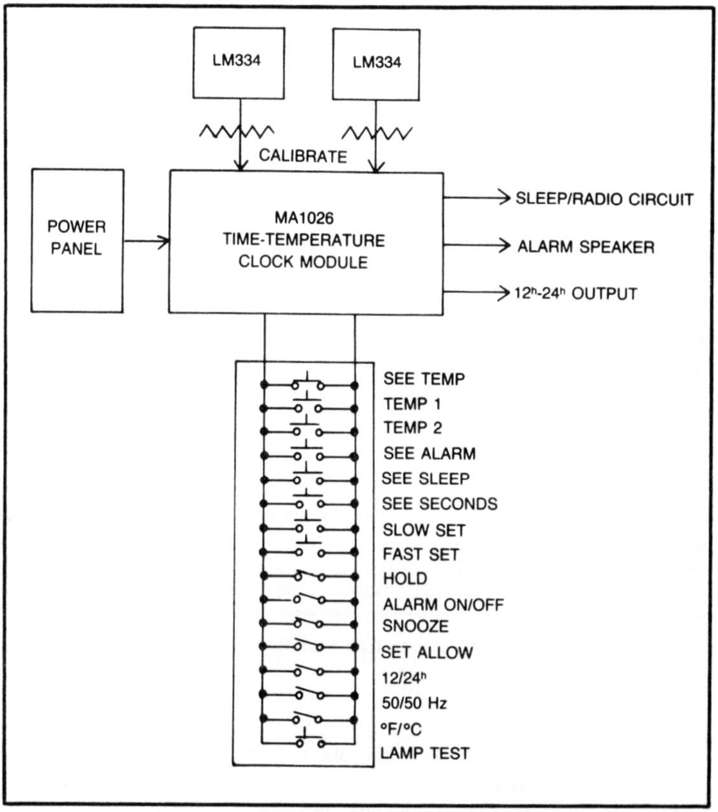


Fig. 12-8. Block diagram of the MA1026 Time-Temperature module. The number of temperature sensors can be increased by adding a bank selection switch. A calibrated potentiometer is required for each sensor. The unit can be built by combining several offered kits, and an equivalent complete kit is also available.

The TV circuits supply the Horizontal and Vertical sync pulses, which, through counters, position the display. The horizontal pulse also synchronizes the oscillator which generates the dots. There are two outputs to the TV, one for the character data, and one to set the background color by gating the chrominance channel or channels.

Since the presentation chip accepts BCD data, the presentation on the TV screen is not limited to just clock or channel numbers. However, there are limits to character length. The screen position can be controlled by additional circuitry, basically line and/or frame delay generators.

This class of chips is by no means the ultimate. The field seems to be converging toward the techniques employed to produce video displays for computers. At least two companies (Commodore and Sinclair) now produce single chip interfaces between the microprocessor and the TV receiver, and more can be expected. However, to date the companies have not released design or performance data on these chips. It remains to be seen whether the specialized chips of the type described above will be replaced by general purpose chips, such as microprocessors.

Last in this specialized chip review is one complex assembly, the Heathkit Time-Weather console. In addition to time, this presents temperature, barometric pressure, wind speed and direction. A number of presentation options are available, although the basic ones are always separate. You might want to look at this, or to consider a modular version, a start towards the "Home Electronic Center of Tomorrow."

**Table 12-2. Characteristics of a Time-Temperature System, Based on the MA1026 Module. Time Functions Are Complete. Temperature Accuracy Depends on the Sensor Used and on the Care in Calibration. See Application Note for the Sensor Used for Details.**

MA1026 Time-Temperature Module

No. Digits—4, LED, 0.7"

Operating Voltages
- MOS Elements −10.5 V, 10-15 mA
- LED Elements −3 V, 300 mA nominal

Display Modes
- Time, Hours—Minutes, 12/24 hr
- Time, Seconds—Minutes
- Alarm, Hours—Minutes
- Sleep, Minutes—10 minutes
- Temperature, °F or °C
- Lamp Test

Time Base—50/60 Hz, or 3.58 MHz with MM5369

Alarm—800 Hz nominal, 2 Hz Gating

Snooze—Repeatable 9 Minute Counter

Set Controls
- Slow
- Fast
- Hold

Power Failure—Flashing Lamp, Sets at 30% Voltage

Temperature Sensors—Any Number
- Type—LM334 +300 Ω Trimpotentiometer
- Range—40 to +193°F

Other—Radio Control via Sleep Circuit

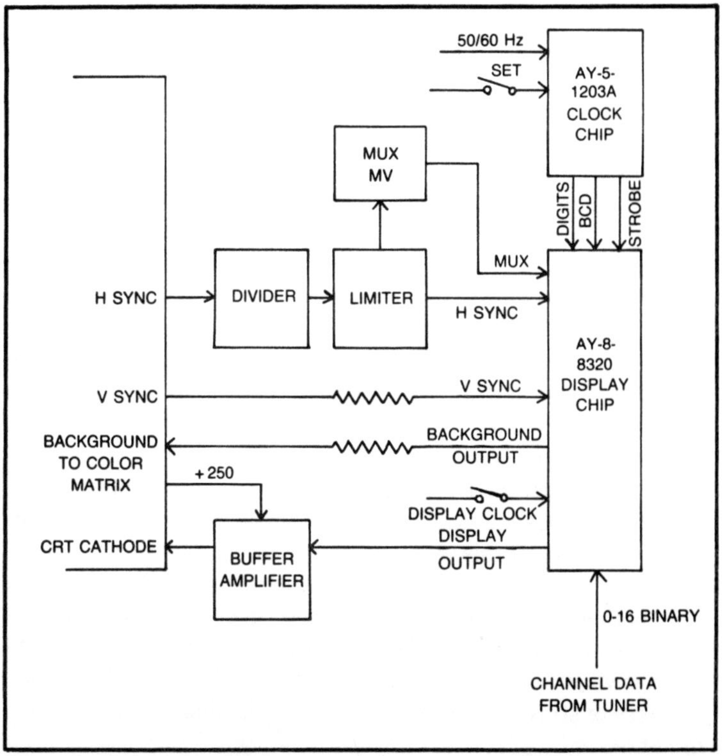


Fig. 12-9. Block diagram of Time-channel on-screen TV display based on the AY-8-8320 Display Chip. This is shown with an AY-5-1203A clock chip, but can be used with any chip with BCD output. Position on the screen is established by combining counts of the horizontal pulses (after the vertical sync pulse) with multiplexed clock data. The display chip can be used with other signals coded in the same format.

## CLOCK REPAIRS

The electronic clock has proven to be a reliable and long life piece of equipment. Because of the repair cost factor discussed at the start of this chapter and the steady design advances which have occurred, repair is often not attractive. It's easier, quicker and often no more expensive to buy a new unit.

However, because so many failures are really simple to repair, it is always worthwhile to look at your own unit when it fails, or to look at a friend's unit as a favor. After all, even though repair is too troublesome or expensive, you may have some more parts to play with. Or, if you have a repair business with some open time, think about incremental pricing—no guarantee on time of delivery and a no cure—no pay policy.

## THE COMMON FAILURE MODES

If a relatively new unit fails, there is one thing to look for first—a faulty solder joint. The number of these has decreased as solder techniques have improved, but they still occur. Probably they will show up after a marked change in temperature, or after a long period at high temperature with high humidity.

By far the easiest way to find these is visual inspection. A bright light and a good magnifying glass are the tools, the best being the special type with a circle fluorescent bulb surrounding a large glass. A binocular loupe is also good.

Most bad joints occur where leads go through the board. In a bad joint the lead will appear to be surrounded by a halo, usually due to flux which didn't do its job. Sometimes the halo will be dark, a sign of corrosion. Correcting the first type is easy—just a touch of the soldering iron. The second type may require removal of the component, polishing the lead with fine sandpaper, replacement and resoldering. If this is too much trouble, it may be possible to scrape the lead clean after the solder is removed from the joint.

Another way to check which is also good for other faults, is to use a dc-coupled scope. This gives signal and voltage levels, and also shows grounds and opens. The latter is indicated by the appearance of stray signals due to capacity coupling to the probe and circuit.

An ohmmeter can be used for power-off checking, but the oscilloscope attachment sometimes called an octopus is usually faster. The circuit for this is given in Fig. 12-10, which also shows a few scope patterns. An even faster method is to use an aural ohmmeter. This is easily built with a 555 oscillator by simply shunting part of the frequency setting resistor by a pair of test leads. A low resistance will cause the tone to decrease. Experiment with resistor values to get tones you find easy to detect. Note that it is usually possible to detect semiconductors by a change in tone quality as well as pitch.

If the unit under repair study is more than about ten years old, the problem is probably not soldered joints, but a component. A trick of speeding up work here is to first make a quick check for "brickbat" or catastrophic failures. If none are found, then replace any electrolytic capacitors—in simple clocks there is usually only one. It may still be good, but it probably has a high internal resistance and not much more expected life. Often this step will clear up several minor problems.

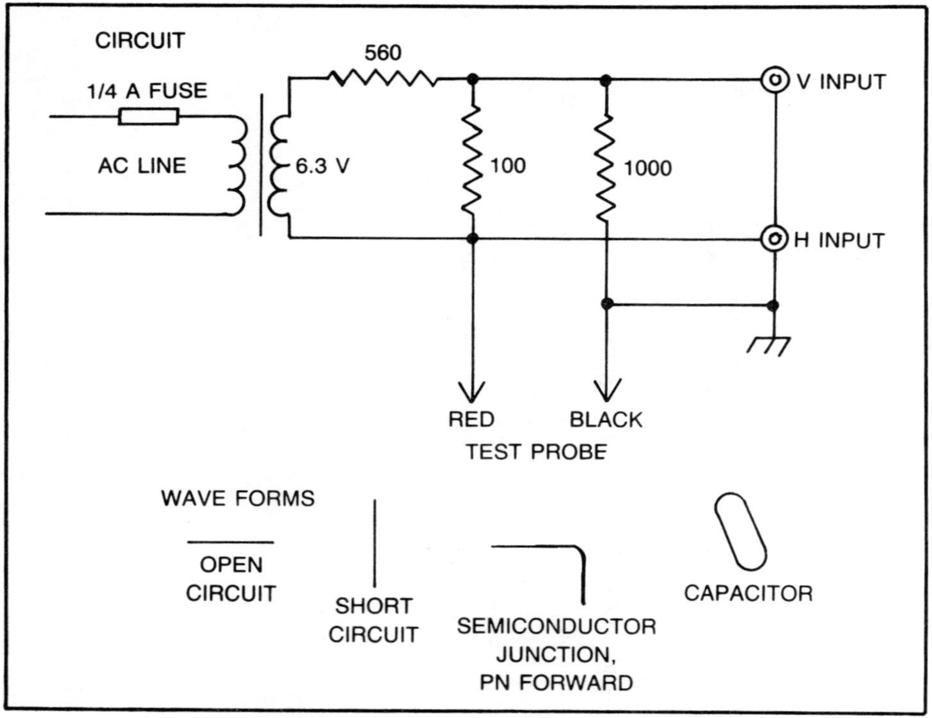


Fig. 12-10. Diagram of a simple and useful "in-circuit" tester, using oscilloscope presentation. This is often the only tester needed in troubleshooting. With a little experience, values of components can be estimated, and failed resistors, capacitors and semi-conductors found.

## OTHER FAILURES

Semiconductor devices do fail for a variety of reasons. In complex devices such as clock chips, this may show up as failure of a single function. Go to the affected circuit, then eliminate external components by test.

If nothing works, check the plus and minus supply, then go to the scan oscillator. Check the external components. If these are good, and there is no scan, the chip is bad.

Displays also fail. Failure may be complete, a digit or segment, or may just be a loss in output for an LED or contrast for a liquid crystal. For all of these, check the associated signal circuits for complete loss (if there is no display), or low level (if the display is dim).

## BATTERY OPERATED CLOCKS

For battery operated clocks, the most common problem is likely to be corrosion of the battery terminals. Once this happens, it is hard to keep it from occurring again. Try a thorough cleaning with a sand board (used for nail care) or a file. Then solder plate the contact with a hot iron. But don't use a soldering flux—rosin only, or there will be more trouble.

## FINAL COMMENT ON REPAIRS

Some units just don't respond to repair effort. Murphy's law was probably first stated by a repairman. Take the attitude that spending a reasonable amount of time is in order. But if there is no result, either set the unit aside for a later look, or put it in the salvage box.

# Chapter 13

# Additional Projects for the Experimenter

When this work was started, it appeared that the design of a clock chip for optimization of cost had resulted in a device of relatively limited usefulness. To some extent this was found to be true. However, as work progressed, it became apparent that there were a large number of applications of the chip beyond those selected for inclusion in the volume. In addition, there were a number of applications involving time which could either serve as interesting and useful applications of digital techniques, or required these techniques for realization.

A number of these additional projects are discussed briefly in the following material. Here, the purpose is not to show how to design to obtain a specific performance, but instead, to show the general end result of the device or experiment, and a single possible approach. The end desired, of course, can be modified to suit both the experimenter's need and his capabilities, and almost certainly, the approach outlined can be improved for the goal shown. The following material should be regarded as suggestions, rather than as complete projects.

## ADD DATE-TIME-INTERVAL CAPABILITY TO YOUR MICROCOMPUTER

Most large computers have a built-in clock for timekeeping. This is especially useful in such work as time sharing, where the charges are based on the amount of time the computer was used. It is necessary in some of the process-control computers, where events must be initiated at a specific time, or at a specific interval.

An appropriately selected clock chip plus a few interface devices will provide this same capability to the experimenter's microcomputer. Probably as easy a way to do this as any is to structure the chip-interface combination to provide signals which are identical to those of one of the input points of the computer. Considering the speed of events and the number of characters of data required, use of the teletype input point of the computer seems to be a good choice. The interface devices must provide from two to twelve digits of information, corresponding to seconds only, and up to full date-time information. Considering the method of clock functioning, it seems best to provide a synchronizing signal secured from the clock. Also, depending on the exact design of the clock chip, it may be necessary for the interface to include control signal provisions, as for requesting the date or minutes and seconds, rather than a normal hour-minute signal.

As an example of possibilities for such a device, consider Fig. 13-1. This is based on use of a 7002 clock chip, or similar, one chosen to provide both time and calendar information in BCD form. If the chip output is to be displayed, the interface must include a BCD-seven-segment converter. An alternate would be to use a 7001 chip to drive a display directly, either to provide a seven-segment to BCD decoder or to transmit data from the critical five segments directly to the computer, making the data transformation via software.

In this diagram, the scan rate has been set to give the character length corresponding to teletype data, 100 milliseconds for ASCII, or 163 milliseconds for the older five-level machines. External generation of the scan rate is by a 555 timer, chosen to assure scan stability. The exact structure of the interface buffers and the nature of the signals they must handle is determined by the microcomputer design.

The software for this clock addition should be straightforward. Date information would be available on request, as would minute-second information. If neither were requested, hour and minute information would be provided. These would be gated into the computer on request, and as determined by the synchronizing signal. Time interval would be obtained by storing the starting time of an event or process, and subtracting current time. The stop signal for process control would be obtained by storing the desired stop time,

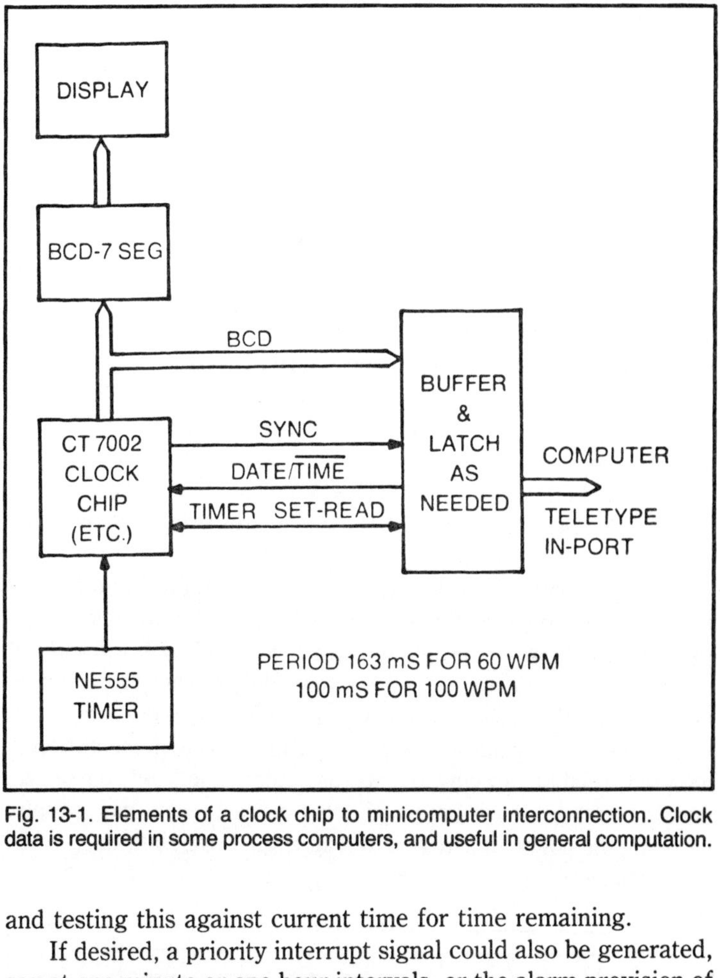


Fig. 13-1. Elements of a clock chip to minicomputer interconnection. Clock data is required in some process computers, and useful in general computation.

and testing this against current time for time remaining.

If desired, a priority interrupt signal could also be generated, say at one minute or one hour intervals, or the alarm provision of the chip could be used to generate such a signal.

## THE TIDE AND MOON CLOCK

A tidal clock is one which indicates the relative height of the tide at any time. Similarly, a moon clock is one which provides information about the moon, for example, the lunar phase. These two phenomena are associated, since the tide is determined primarily by the moon.

Of course, to a true seaman, the idea of a tidal clock is almost nonsense. The seaman almost automatically calculates the correction to clock or sun time and knows the state of the tide at any instant. However, this knack is very rare in weekend sailors. Since the tide state can be very important to the amateur sailor, this clock could make the basis of a worthwhile gift to a sailing friend.

The main feature of lunar time is that the moon is seemingly revolving around the earth in the same direction as the sun. Since this revolution takes just about 28 days, the moon passes a given point on earth in 1 + 1/28th of a day. Thus, to keep track of the tides, the clock used for this must run slower than a normal clock by about 50 minutes a day. This slow running of a clock is easily obtained by reducing the frequency of the driving ac voltage, the nominal change being from 60.00 Hz to 57.98 Hz.

The basic elements needed are shown in Fig. 13-2. The clock chip is driven by the frequency generated by a 555 timing oscillator. The clock is set to read twelve o'clock at the time of high tide. The clock should be on the twelve-hour scale, since the tides occur at double the frequency of the earth's rotation, a phenomena explained in most physics books.

The presentation of data from the clock can take a number of forms. For most chips, the simplest is to leave the presentation in the numerical form of a regular clock. Twelve o'clock will thus indicate high tide and six o'clock low tide. Half tide is approximately indicated by the three o'clock and nine o'clock time. Because of confusion with normal clock presentation, if this presentation is adopted, a label carrying the words "tidal time," or equivalent, should be placed immediately adjacent to the display. Incidentally, recoding the hours information to read High, Low, Half, etc., would be possible.

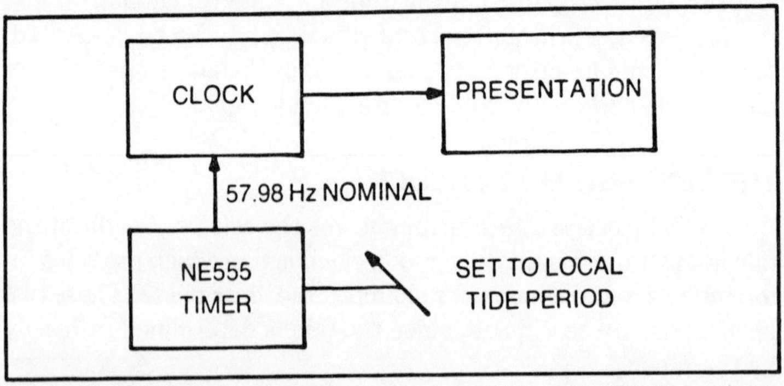


Fig. 13-2. Elements of a simple tidal period clock. The clock rate is set to give a 12:00 indication, or high tide, at the local tidal interval, about once each 12 hours 50 minutes.

An alternate method of presentation, especially useful if the decorator digital clock of Chapter 5 is used, is to arrange LEDs or other indicators in a circle labeled as in Fig. 13-3(A). This gives a better picture of tide state, and is less likely to be confused with normal clocks.

Actually, the motion of the tide is much more complicated than these simplified presentations implies. In many seaports, the two tides per day are of different height, and there may be a continual variation due to the interchange of effects between the sun and the moon. There is also the fact that the moon is in an inclined orbit, which changes the tidal force still further. Sample curves of this tidal phenomena are given in Bowditch and tabular values for most ports by the tide tables.

It is possible to secure a better approximation of the full tidal information by arranging the dot presentation into a graph rather than into a circle. A typical example of such a graph presentation is sketched in Fig. 13-3(B). Here, as successive LEDs are activated, the appropriate point on the graph is indicated. This presentation may be combined with a second graph showing the percent of full amplitude as a function of the lunar phase. This phase would be derived from the calendar information of the clock, changing each day. Because the lunar month and calendar months do not agree, it will be necessary to change the calendar setting at the end of each 28th day, or each lunar month.

True lunar information is difficult to derive. The single exception to this is the phase of the moon, which occurs in a reasonably simple cycle. The information for this is given by a clock which is keeping tide time. The date information from such a clock can be decoded to individual days, with these represented by LEDs, arranged in a circle as in Fig. 13-3(A). The labeling of the circle, however, would indicate full moon at the twelve o'clock position, new moon at the three and so on around the clock dial. The data can be derived by a divider chain, by dividers with a scale of 2, 2, and 7 fed from the A.M./P.M. indication of the clock.

Determining other lunar information, for example the time of moonrise, is much more difficult, due to the moon's inclination with respect to the equator. The information can be derived by varying the rate of a clock in a cyclic manner, as discussed in the next section in connection with sun time.

One design of a tidal clock was published in the magazine EDN, September 2, 1981. This design uses a crystal oscillator, a pair of precounters, plus a 12 stage shift register, and drivers feeding 12

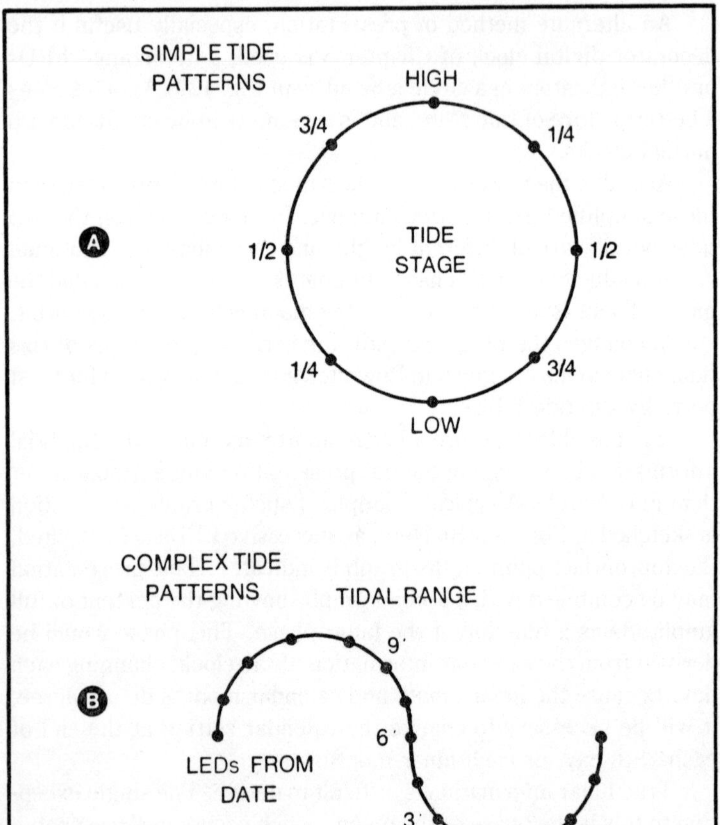


Fig. 13-3. Presentations for a tidal clock. The circular presentation is good for simple tide patterns, and often usable for complex ones. The separate presentation of stage and range can give a better approximation to actual state.

LEDs in a sine wave pattern. As designed, there is an error in timing amounting to 12 seconds per tidal day. Partial compensation for this can be secured by "pulling" the crystal by adjusting its capacitor, or complete compensation by ordering a special crystal (the design starts with a TV color crystal).

## A CLOCK FOR SUN TIME

In the first chapter, sun time, the time kept by a sundial, was discussed. It was noted that sun time can occur ahead of or behind the normal time, with the maximum difference between the two being about 16.4 minutes, this occurring in November. The difference between sun time and normal or mean time is primarily due to the fact that the earth's orbit is not a perfect circle.

Suppose it is desired to keep sun time by an electronic clock. The clock rate must vary, with the clock running fast as sun time moves ahead, and then running slow as the sun time moves behind. The amount of correction required can be developed from the nautical almanac, which shows the angular difference between the sun's position and the mean position for each hour of the day and each day of the year. The change from day to day is fairly small, the maximum being a change of 23 seconds of time in one day.

We can make an electronic clock give this change in time if we vary the frequency of its oscillator in proportion to the time change desired. By proportion, the normal 60 Hz oscillator must be changed to one which can vary between 59.948 Hz, the frequency which makes the clock run at the lowest slow rate and 60.016 Hz, the maximum fast rate frequency.

A clock chip which contains the calendar accumulators can provide the data to control the frequency change of the oscillator. A circuit for doing this is shown in block diagram form in Fig. 13-4. The calendar data from the clock is taken to a converter unit, which first converts it into a sawtooth with a period of one year, and then converts the sawtooth into the sinuous curve required to control the oscillator. The converter would use a set of diode nonlinear elements; this technique was used in one of the instruments in the previous book of this series to convert a sawtooth into a sine wave. The same technique would be used here, with constants being adjusted to give the required wave shape.

This output (which approximates a sine wave) with a period of one year is fed to a voltage-controlled oscillator, through a variable resistor provided to set the amplitude of frequency excursion. The voltage-controlled oscillator can be an NE555 with a varicap

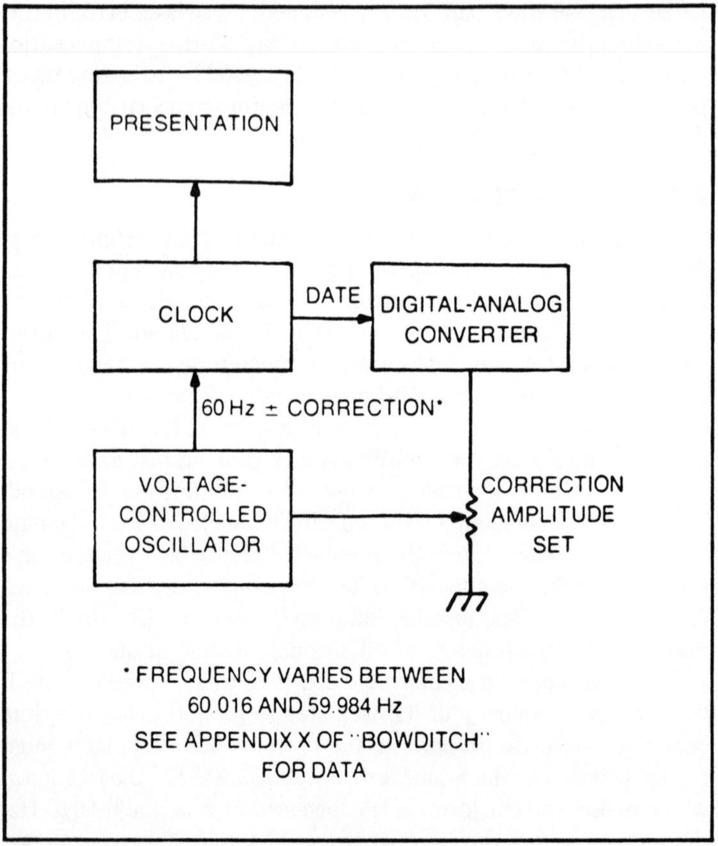


Fig. 13-4. A clock for keeping approximate sun time, the time given by a sundial. The correction is obtained by analog means.

connected in the frequency determining circuit, with values being selected to give the required control range. With the converter and oscillator adjusted, the clock would be set to local apparent sun time; this is easiest done by setting it to the twelve o'clock position, and then using data from the almanac to start the clock when the sun is on the meridian. The correction time is given in the American Ephemeris as the equation of time. Thereafter the clock will keep the time as seen by a sundial, but of course it will function day and night, which a sundial cannot do.

The same technique can be adapted to the problem of lunar timekeeping. The major difference would be that the frequency excursion required would be larger, and the period of the excursion would be faster, approximately 29.5 days per cycle.

An alternate possibility for keeping both sun and moon time is to add or gate out oscillator cycles at intervals. See Chapter 4 for a discussion. If a correction is made once an hour, the overall error would be very small.

## A JULIAN DAY CLOCK

In astronomy, the problem of keeping track of dates (which is due to the complexity of the calendar) is avoided by assigning a serial number to each day. The most common method of serializing is called the Julian Day. The serial count in this method assumes that Day One occurred at noon on 1 January, 4713 B.C. This strange date is the day on which a full moon occurred on Sunday, 1 January, which happens each 7980 years. Since this time, nearly two and one-half million days have passed. The Julian date corresponding to 4 July, 1976 is 2,442,962.

Any clock chip can be used to provide information to update a Julian day indication each day. All that is necessary is to provide a detector for the twelve-hour noon event, either twelve hour on a twenty-four hour scale, or twelve hours and the A.M. indication. The once-a-day pulse from this detector would be taken to a presettable counter, as shown in Fig. 13-5. Here a seven-digit counter is shown, but only three digits are needed if mechanical changing of display count every one hundred days is satisfactory. Four digits would allow for mechanical setting once every three years, and five digits for nearly an entire generation. The remaining digits could be hard-wired to remain on at all times.

In this figure the presettable counter is shown with a digit-setting switch, adjustable from 1 to 10. This digit-setting input can be switched to any digit of the counter, by means of a digit-select, five-pole switch, four poles being used for data, and one pole for the load signal.

The Julian Day addition to a regular clock might be used to provide a conversation item decorator clock, or it might be used for a gift to an astronomer friend.

For decorator purposes it might be worthwhile to consider using as an addition or alternate, the Mayan Calendar. This calendar is based on the interrelation between three cyclic counts, each having a name or number designation. Each also has an image, as shown on the calendar stone found in Mexico. A most unusual decorator clock could be developed by using these images and the date-time indications associated with them as an alternate presentation to a normal clock.

## A STAR TIME CLOCK

Normal time is based on the rotation of the earth, with the sun crossing the meridian at noon each day. Any star also crosses the meridian, but because the earth is revolving about the sun, the star will cross the meridian 366 times in a year, whereas the sun crosses only 365 times. Thus, star time differs from the sun time by the ratio of 366/365. The exact value of the ratio is 1.00273791.

If a clock is to keep star time rather than sun time or moon time, it must run faster than normal time by just this ratio. To build a star time clock, it is only necessary to change the oscillator frequency of a normal clock.

About the only people who use star time are astronomers. Their main use is not of the time itself, but of the rate, which is used to drive a telescope so that when placed on a given star, it will continue to point toward this star even though the earth is turning.

It is easily possible to provide this driving frequency and, at

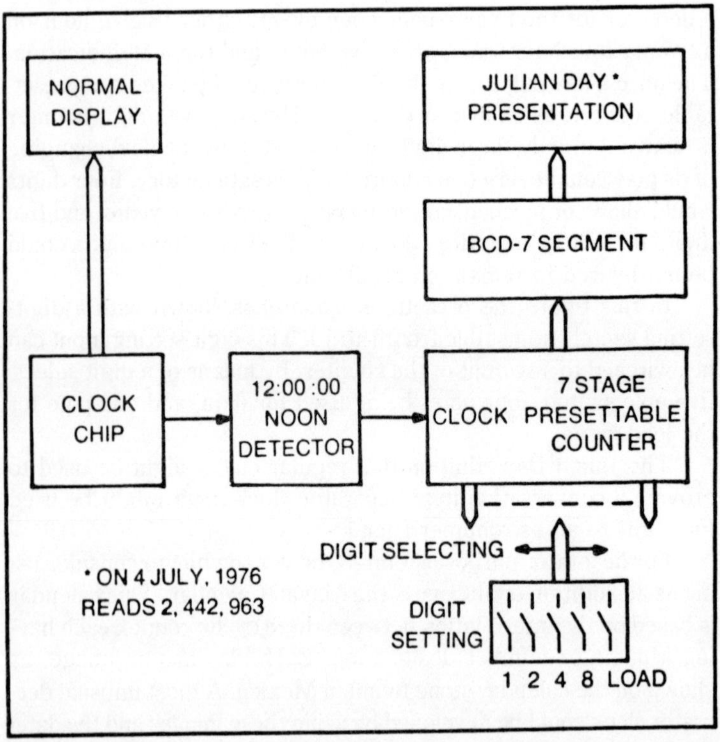


Fig. 13-5. A calendar clock keeping record of the Julian date, used primarily in astronomy.

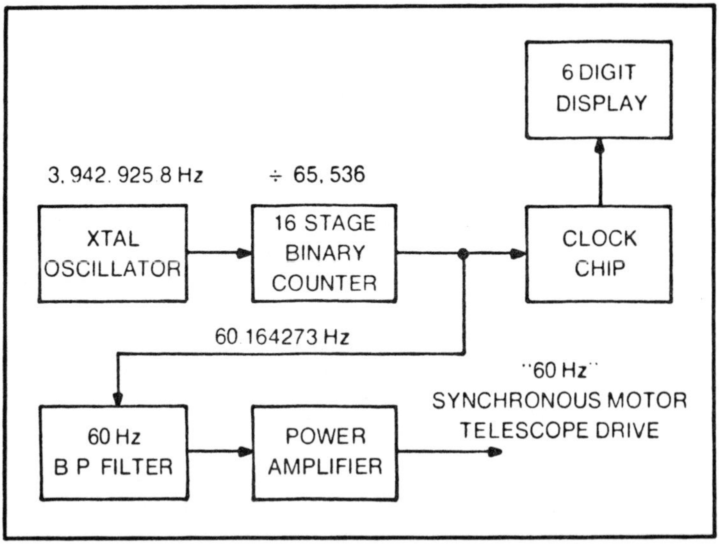


Fig. 13-6. A sidereal clock, which keeps track of star position. A power drive is included, to keep a telescope pointed to a given star.

the same time, the occasionally useful time indication. Since telescopes have a very small field of view, the time should be maintained with good accuracy. This means that it would be preferable to derive the clock driving frequency, as well as the frequency for driving the telescope drives, from a quartz crystal of good stability.

A suitable circuit for this is shown in Fig. 13-6. An AT-cut crystal oscillator is chosen because of its good stability. The divider chain which follows can be based on a scale of 2 counters with rest, or on a combination of a single scale of 8, plus 4 decade counters. The power amplifier could be similar to a dc 110-volt inverter.

## DIGITAL DERIVATION OF SUN TIME

Sun time, plus many other astronomical phenomena, such as sunrise and sunset, occur at a different time each day of the year, as a result of the relatively complex relations between such factors as orbital eccentricity, inclination of the earth's spin axis, inclination of the moon's orbit, and so on. Because of relations, the bodies involved appear to move at a nonuniform rate.

In contrast, clocks are designed to run at a constant rate to keep mean time. As a result, they are not well suited to directly showing the times at which astronomical phenomena will occur. In the older days, when clocks were gear operated, these were approximated by additional gearing, including gears which were not circular and so could produce nonuniform motion. Most large museums have clocks of this type on display.

Essentially, the same technique of adding nonuniform motion to a clock can be added to the output signals of digital clock chips. One suitable circuit is shown in block form in Fig. 13-7. Here, the difference between the clock time and the time of an astronomical event is contained in a lookup table, a programmable read-only memory (ROM). This would give the time of the event, based on the month and day information from the clock. The display would show, for example, time of sunrise.

An alternate is to record the change in the ROM. This correction would be added algebraically to clock date or time in a digital adder to obtain event time, which is then displayed. Figure 13-8 shows such a system. As the date changes, the lookup table shifts to another entry, updating the correction. This cycle continues throughout the year.

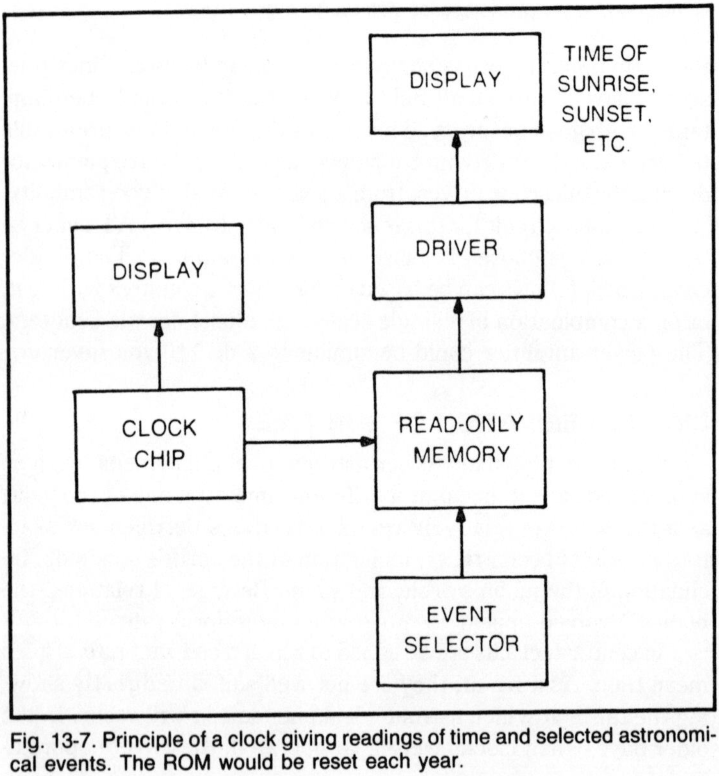


Fig. 13-7. Principle of a clock giving readings of time and selected astronomical events. The ROM would be reset each year.

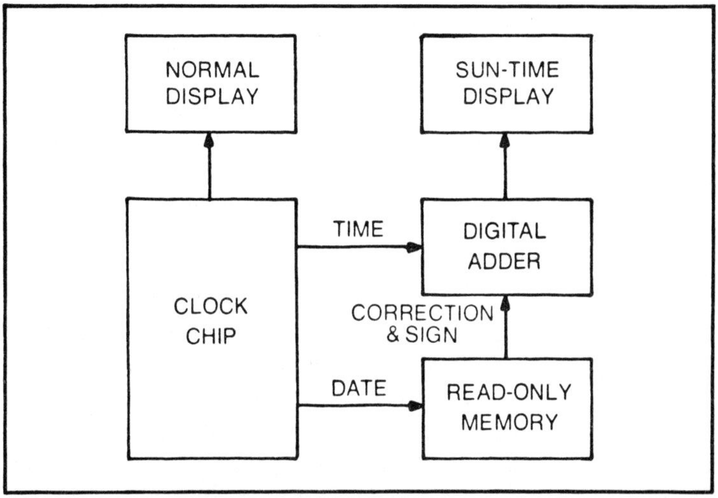


Fig. 13-8. Digital derivation of sun time. A ROM gives a correction to clock time.

The same generalized techniques can be used for such phenomena as sunrise, sunset, moonrise, moonset, and even for tides and phenomena on other planets. For these, however, the output of the lookup table would be the time of occurrence of the event, which is then displayed directly.

For most purposes, it would appear that the basic clock should remain in full operational condition, with a normal display. The additional display or displays should be separate and clearly labeled with the event occurring.

## GIANT DISPLAYS

As shown in Fig. 3-6, the normally encountered types of clock displays are only suitable for very close use, i.e., for distances up to perhaps fifteen feet. They are intended for home use.

At times, perhaps for public buildings, or perhaps for advertising purposes, giant displays are needed. These are not difficult to construct in a variety of sizes by making use of standard light bulbs. Figure 13-9 shows two possible constructions. The top illustration shows (in schematic form) the use of lumiline or fluorescent bulbs, each bulb being arranged to represent a segment of a normal seven-segment display. Since these types of bulbs are available in lengths from six inches to approximately 12 feet, displays sized for practically any purpose can be constructed.

The second part of the illustration shows the alternate technique of producing the bar segments from a series of discrete light sources arranged in a row. These may be small 110-volt light bulbs, or bulbs of other types, or even individual LEDs. In the figure, four such point sources are used to produce a bar; but in some applications, it might be better to increase the number to get a better representation of bar structure. Also, for some uses it may be desirable to introduce the additional segments discussed in Chapter 3, to allow formation of letters; however, special drivers will be needed.

These giant displays, since they require appreciable power, cannot be driven directly from the clock chip, so some interfacing will be necessary. For many purposes, use of SCRs (silicon-controlled rectifiers) would be advantageous. These are obtainable in a variety of sizes to suit lamp requirements, and additionally, are easily used in circuits which have sufficient memory to keep the bulbs on continuously during the scanning cycle, thereby eliminating need for overvoltage circuitry. Other types of interface are readily possible, of course, through cascaded transistors, or even by relays

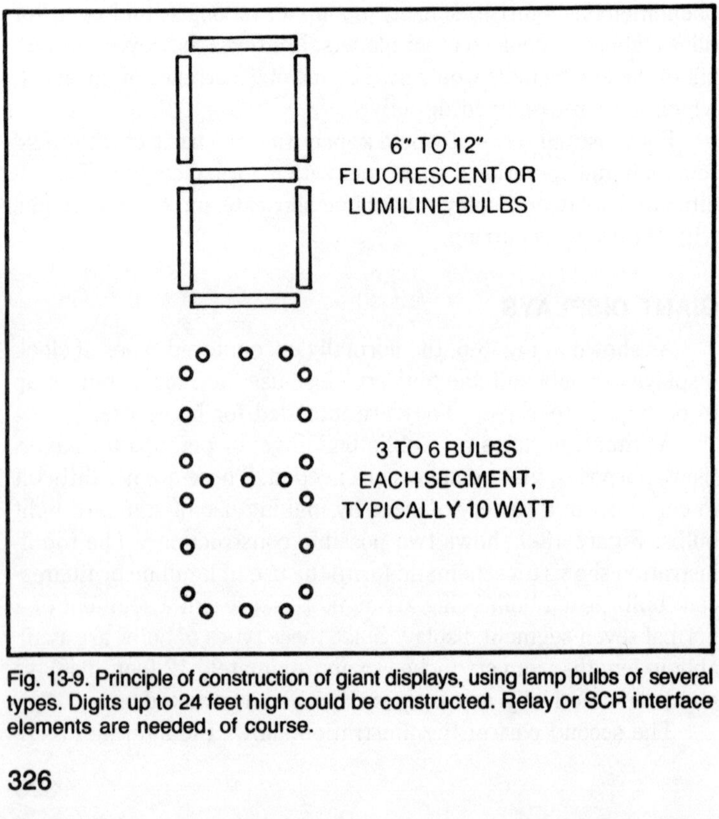


Fig. 13-9. Principle of construction of giant displays, using lamp bulbs of several types. Digits up to 24 feet high could be constructed. Relay or SCR interface elements are needed, of course.

driven by clock signals. In the smaller sizes using LEDs arranged in the bar form, it may be possible to secure sufficient power from the usual clock chip drivers to drive the display directly.

## MULTI-CITY CLOCK

It is sometimes desirable to show the times in various cities of the world, and it will be recalled that this is easily possible with a sundial. It can also be shown with normal clocks, by driving additional sets of hands. And, of course, it can be done for digital clocks by providing additional displays.

For most cities of the world, it is only necessary to make a correction in the hours figures from local time readings, since most cities keep Zone time, always one hour different from the next zone. However, there are a few places, some quite large, which keep other time, so inclusion of these cities will require more than a simple hours change. The most common other correction is in one-half hour steps, although there is an area in Africa which has a 44-minute correction.

Figure 13-10 shows a possible arrangement of a multiple-city clock. This is based on display of one city at a time, with the time display being determined from a lookup table whose entries are the cities selected and the local time reading. In this diagram, it is assumed that local time is kept by a 12-hour clock chip which has only a seven-segment readout. A simple decoder is used to take the hours and tens of hours information and convert it into unique readings as input to the lookup table contained in the read only memory.

The above, of course, is by no means the only way of securing the additional city data. For example, in the decorator digital clock of Chapter 5, the time in additional cities can be obtained by simple mechanical or electrical rearrangement of the hour or the hour and minute drive signals. Alternately, a movable face can be placed on the clock, just as is done with mechanical clocks. Another technique of developing additional city time would use presettable dividers, set to the city time by switches and BCD adders, with one input to the adder being switch controlled for the city selected. Of course, any of these can be designed to show a number of cities simultaneously, eliminating the switching operation.

## AN ADVANCED DESIGN WORLD TIME CLOCK

Because of geography and customs, there are many variations

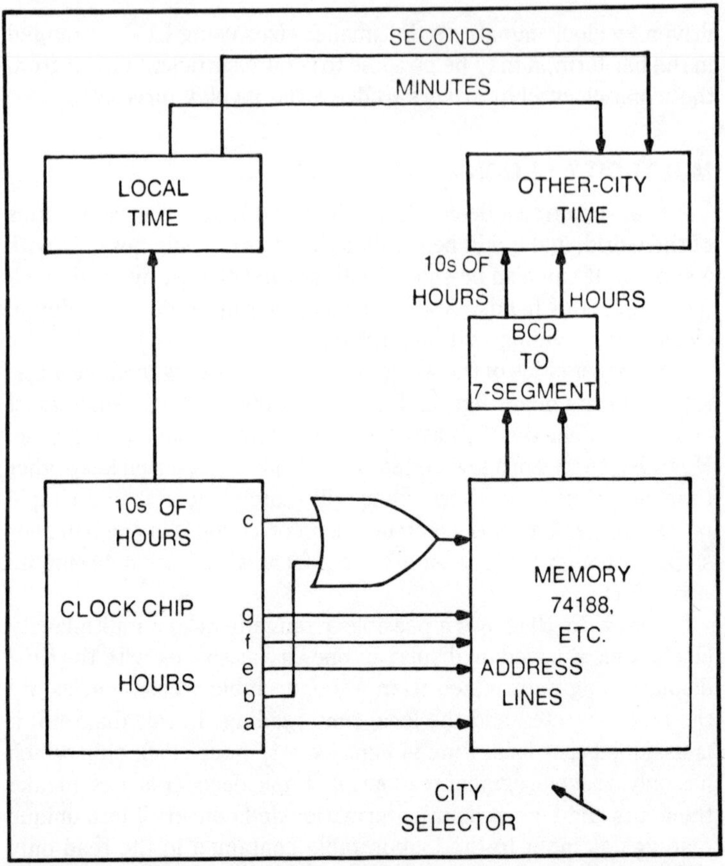


Fig. 13-10. Basic for a display which gives local time in two or more cities. Digital adders could also be used. The technique could be used for UT/local time clocks.

between nominal and adopted standard time. The extent of these can be seen from a Time Zone chart, for example, Hydrographic Office Chart 5192. There are so many, and the variations so arbitrary, that the only practical way of keeping track of all of them is by a table, keyed to cities and geographic areas.

Since table look-up and addition are simple tasks for a computer, the combination of a microprocessor, a processor-oriented clock chip, and a ROM would make a relatively simple and very powerful World Clock. With the WD2412 chip, for example, the chip driven display can be used for display of the computer derived time, as well as clock time. This makes presentation of time at any city in the table very easy.

Having gone thus far in combining techniques, it is easy to go far beyond mere time. Firmwave routines for sunrise and sunset, moonrise and moonset, and even events such as holidays and festivals can easily be added. Such a clock would be very useful in major airports.

## THE TIME-TO-GO CLOCK

A time-to-go clock measures the time remaining to some event, rather than regular time. It is often called a timer rather than a clock. Typical uses would be the time left before starting the oven, or the five minutes in a yacht race between the warning gun and the starting, or other such events. However, a time-to-go clock is really one that counts backward from normal time. So far we have not found a clock chip with down-count capability. If a time-to-go clock is desired it will have to be built.

The decorator digital clock of Chapter 4 is the closest to the time-to-go clock, since it can be converted to this mode by simple mechanical rearrangement of the leads to the LEDs. Instead of being connected to produce signals at their normal positions, the ONEs leads from the decoders would be connected to the LEDs at the eleven o'clock position, the TWOs leads to the LEDs at the ten o'clock position, and so on. A similar reconnection of the minutes display would also be needed.

The decorator digital display is easily extended to read 0 to 60 seconds as well as 0 to 60 minutes. This could be done by using 60 LEDs in a third circle. Time remaining in seconds and up to 12 minutes can be shown with two circles. Several other time arrangements are easily possible.

If a true numerical reading of time-to-go is desired, it will be necessary to use a counter chain to obtain this. A suitable chain is shown schematically in Fig. 13-11. Here, type 74190 up-down counters are shown, since they can be used in either a normal up or down counting or in a preset mode. The normal counting mode would be used if the clock is always to start at a particular time point, say the five minute interval of a yacht timer. The preset characteristic should be used if the clock is to be arbitrarily preset to a given time remaining before starting. While a universal arrangement probably could be worked out, the range of requirements is sufficiently great that a design to suit needs is probably better.

One of these types of time-to-go clocks in portable form would make an excellent gift to a yachting friend, and would be useful in other sports.

## AN AUTOMATIC ALARM SETTER

If it is necessary to reset the alarm very often, the operation becomes a considerable nuisance. This is due to the fact that even the most flexible clocks only have provisions for presetting the alarm to a single time, usually midnight. To reach any other setting, the alarm must run through a count sequence. True, the count sequence usually has slow and fast setting provisions, but it is very easy to miss the correct setting and have to repeat the entire operation.

The addition of a few components externally to the alarm chip will permit automatic setting of the alarm, or of the clock itself. This can be done by providing dials which are set to the desired time, and determining the difference between the reading coming from the chip and the dialed-in reading. When these are equal, a stop-set signal is generated, turning off the process. A suitable circuit for this is shown schematically in Fig. 13-12. Here, type 7485 magnitude comparers are used to make the hours, minutes and se-

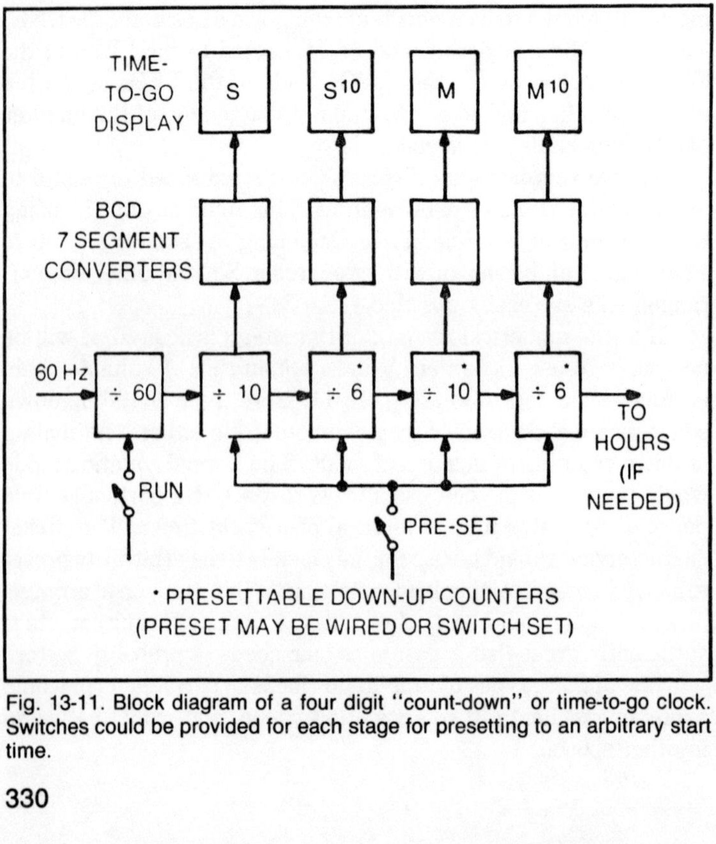


Fig. 13-11. Block diagram of a four digit "count-down" or time-to-go clock. Switches could be provided for each stage for presetting to an arbitrary start time.

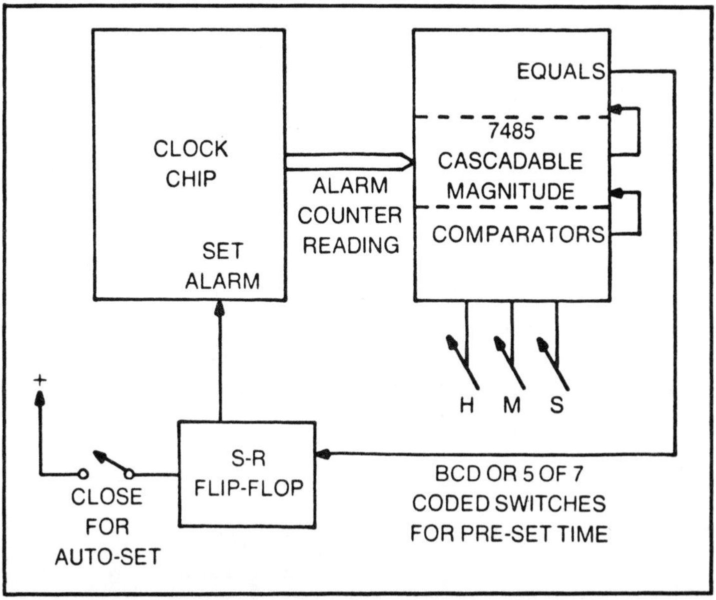


**Fig. 13-12. Principle of an automatic alarm setter. The BCD switches are set to the desired alarm time, and the reset initiated. Reset is stopped automatically when the alarm counter reaches the preset count.**

conds comparison. One input to the comparator comes from the clock. The second input comes from a set of dials, which are preset to the desired time setting. The process of setting is initiated by a set-reset flip-flop, which generates the signal to the clock chip. This flip-flop remains in the set position until the output of the comparator indicates that the clock signal and the dial signals are equal. The flip-flop then goes to the reset state, stopping the setting action. In this simplified diagram no multiplexing is assumed. If the chip signals are multiplex, various timing signals derived from the digit drives will be needed in addition to the ones shown.

Switching could be provided to allow the same circuitry to set the alarm, the clock reading, or the auxiliary circuits, where these are provided. There are various other possibilities by the use of comparators; for example, the generation of several successive alarm signals by comparison of the clock reading with a bank of comparators, each having its own dial-in time setting. Opportunities for process control and such events are large. Various combinations can be worked out, for example, the use of one of the comparator signals during the A.M. hours and the second during the P.M. See Fig. 13-13.

## DIGITAL CLOCKS WITH ANALOG PRESENTATION

Because the standard clock face is so familiar and because it merges easily with decorator motifs, there are many uses for clocks which appear conventional, but which are based on electronic digital techniques.

This has not been an easy field for the experimenter, because of the precision needed for some components and the lack of suitable components in the market place. The lack of components is changing, however, and some ideas have also been developed which makes the analog-digital "marriage" easier. Here are some starting points for further experimental work.

## DRIVING STANDARD ELECTRIC CLOCKS

A large fraction of older electric clocks, and many current ones,

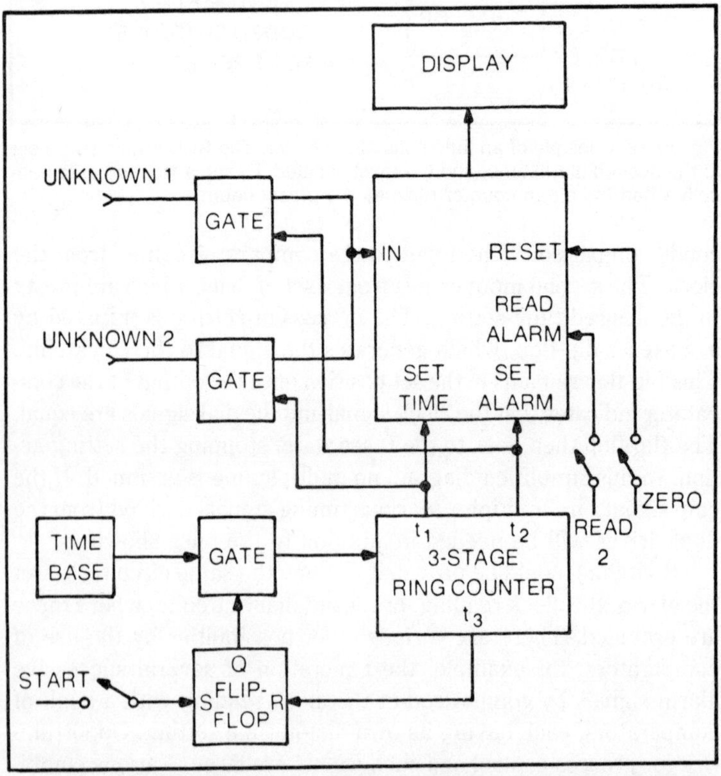


Fig. 13-13. Use of an alarm clock chip to compare two unknown numbers of events. A signal is provided when they are equal. The magnitude of each can be read, subject to some limitations inherent in chip design.

use a small synchronous electric motor to drive a gear train which turns the hands. These motors are of the hysteresis type, with many poles, so that they turn at relatively low speed. There are often internal as well as external gears. Many types are found in surplus.

Because of the multiplying effect of the gear train, the driver motor does not need to provide much torque, and therefore is very low power. It is rare to see these "telechron" or similar motors rated at more than three watts, even in heavy duty timers, and clock ratings are often only 0.8 watts.

These motors are not very sensitive to wave form, and most will also work properly even if there is a small dc component present. The low voltage types can often be directly driven by IC circuit output. However, standard electric clocks require 100-120 volts to work. This is beyond the rating of most semiconductors.

These standard clocks can be driven by adding a small driver amplifier and a voltage step-up transformer, as shown in Fig. 13-14. The amplifier can be a package unit, with a rating of about twice that of the clock motor. For example, an LM380 would be ample for almost any clock. The output transformer should be selected to give 100-120 volts at the clock. An old filament transformer would do.

The drive power required can probably be reduced if some wave shaping is done. A simple form is an RC integrator at the input to the amplifier, which converts the digital square wave to a peaked one. Alternately, or in addition, a capacitor can be placed across the output of the transformer to tune it to 60 Hz. A series capacitor can also be used.

If you can find a low power clock, a single IC circuit can be built from standard parts. As shown in Fig. 13-15, the Intersil ICM7038A is intended for just this service, being the interface between a 4.194 MHz crystal and a 64 Hz, 3 mA clock motor. However, since the chip is basically a 16-stage divider plus a 1—16 MHz oscillator and output drivers, it is equally easy to secure 60 Hz, by changing the crystal to $60 \times 2^{16}$, or 3,932,160 Hz. This circuit can be used to drive a stepper motor with a capacitor in series to block any dc component which may be present.

Another chip for analog drive is the National MM5822. This chip uses a 13-stage divider, giving 4 Hz output for a 32.768 Hz crystal, which can be excited by an internal circuit. The output is intended to drive a stepping motor directly. This chip can be used with standard clocks by changing the crystal frequency to $60 \times 2^{13}$, or 491,520 Hz. An output amplifier would be necessary.

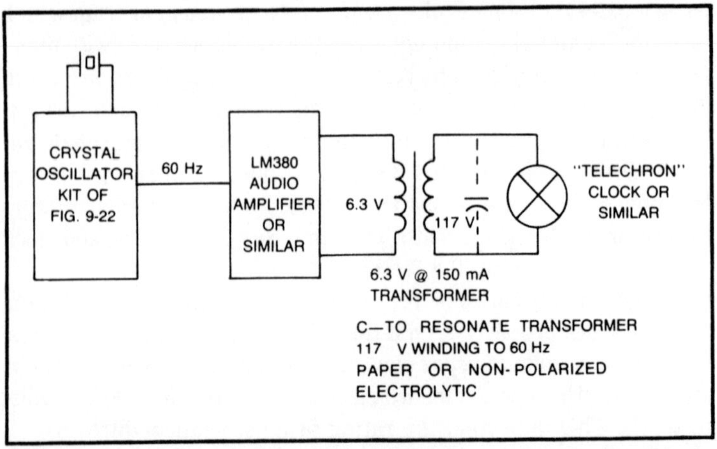


Fig. 13-14. A driver circuit for standard "Telechron" or other motor driven clocks designed for power line use. The crystal circuit will improve the accuracy markedly. If a battery back-up supply is provided, reliability will also be increased.

## SYNCHRONIZATION OF CLOCKS

Rather than driving the clock by digital means, you might wish to try synchronizing the clock to a digital circuit. The major advantage of this is that the digital circuit doesn't have to supply drive power, but only the very small signal required to change the clock rate slightly.

One method of synchronizing is to do this manually. Suppose we are dealing with a grandfather clock. We could flash an LED each second, and note the position of the pendulum. If it changes with time, the clock rate device would be adjusted. Very close control would be possible.

School laboratories often have a seconds pendulum, in which a fine point on the pendulum sweeps through a pool of mercury to give an electrical pulse each second. Digital comparators can be used to get synchronism with great precision.

Another method of synchronizing a clock is to place a small iron core coil near the pendulum or balance wheel. It could be fed with a pulse for a pendulum, or a sine wave or approximation of one for a balance wheel. The clock rate will adjust until there is minimum interaction between the electrical and mechanical circuits, if energy interchange relations make this possible. Over a long term, this gives the synchronized condition. Some experimenting may be needed to secure the proper combination of coil size, distance and current.

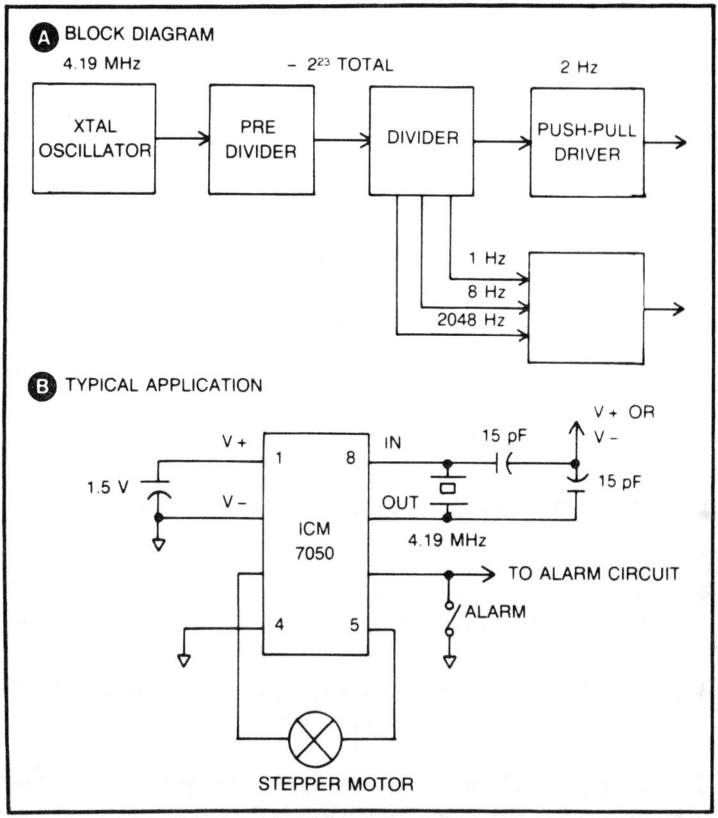


Fig. 13-15. Block diagram and typical application circuit for an ICM 7050 "stepping motor" type of digital electric clock. Other designs are available for use with 32 kHz crystals, and are commonly found in the smaller type of "quartz" clock with a conventional clock face.

While these steps are not suggested for general use, they would allow a family heirloom to be converted from a dustcatcher to an accurate timepiece without affecting heirloom or antique value.

## CONCLUDING NOTES

Clock chip manufacturers have to think about sales potential when they establish their product line. As a result, they must concentrate on high volume items—standard clock chips.

For the experimenter this has a special meaning. Special types of clocks, or even whole classes of decorator clocks, are not going to be on the market. If the experimenter wants one, he must build it. The field for experimenters is going to continue for a long time.

# Bibliography

Chapter 1:

See the *Encylopedia Britannica* for articles on Sun Dials, Clocks and Watches, and Concepts of Time

Bowditch, *The American Practical Navigator*, Oceanographic Office, Washington D.C.: Current edition

C. L. Strong, *The Scientific American Book of Projects for the Amateur Scientist*, Simon and Schuster, N.Y., 1960

*The American Ephemeris and Nautical Almanack*, Oceanographic Office, Washington, D.C.: Current edition

Chapter 2:

Fred Blechman, "Digital Clock Roundup," *Radio Electronics*, Aug.-Sept., 1976

P. S. Duryee, "Counter Designs Without Gates," *Electronic Design*, Dec. 6, 1967

V. Rao, "Two LEDs Blend and Blink to Indicate Six States," *Electronic Design*, Aug. 5, 1982

Original data on line voltage is from an article by Hall and Watts, QST, April, 1976

Chapter 3:
D. J. Renn, "Consider LCDs For Your Next Design," *Electronic Designs*, et al, Sept. 13, 1975

Jules H. Giler, "Focus on Displays," *Electronic Design*, Sept. 13, 1975

Chapter 4:
National, MM5311 series data sheets

Fairchild, FND70 Data Sheet

Chapter 5:
*The TTL Data Book*, Texas Instruments, Dallas, Texas, 1973
"Solid State (Column)," *Popular Electronics*, April, 1976

G. D. Gaugler, *EDN*, "Construct a Circular Clock with LEDs," May 12, 1982

Chapter 7:
*Popular Electronics*, "Update Your Digital Clock with Add-ons," February, 1975

Chapter 9:
R. Silberstein, "An Experimental Frequency Standard Using ICs," *QST*, Sept., 1974

R. L. Harrison, "Crystal-controlled Oscillators," *Ham Radio*, Mar., 1976

P. H. Mathieson, "Simple Crystal Oven," *Ham Radio*, April, 1976

M. S. Robbins, "NBS Frequency Calibration," *Radio Electronics*, Sept., 1976

H. D. Olsen, "Transformerless Mecham Bridge (Oscillator)," *Electronic Design*, Sept. 27, 1969

Chapter 10:
G. W. Horn, "Braille Tactile Transducer-New Freedom for the Sightless," *QST*, December, 1981

M. A. Rigsby, "Dissecting the TI Speak and Spell," *Byte*, September, 1980

Chapter 11:
J. C. Mein, "Real-time LSI Clock Chips Ease $\mu$P-control tasks," *EDN*, January 6, 1983

S. Ciarcia, "Everyone Can Know the Real Time," *Byte*, May, 1982

# Index

**D**

# Other Bestsellers From TAB

☐ **ENCYCLOPEDIA OF ELECTRONICS**

Here are more than 3,000 complete articles covering thousands of electronics terms and applications. A must-have resource for anyone involved in any area of electronics or communications practice. From basic electronics terms to state-of-the-art digital electronics theory and applications . . . from microcomputers and laser technology to amateur radio and satellite TV, you can count on finding the information you need! 1,024 pp., 1,300 illus. 8 1/2 ″ × 11″.
**Hard $60.00 Book No. 2000**

☐ **THE ILLUSTRATED DICTIONARY OF ELECTRONICS—3rd Edition—Turner & Gibilisco**

Identifies and defines over 27,000 vital electronics terms—more than any other electronics reference published! More than *2,000 new topics* have been added to this state-of-the-art 3rd Edition! *Every* term has been revised and updated to reflect the most current trends, technologies, and usage. Covers basic electronics, electricity, communications, computers, and emerging technologies! 720 pp., 395 illus. 7″ × 10″.
**Paper $21.95 Hard $34.95**
**Book No. 1866**

☐ **ELECTRONIC AND MICROPROCESSOR-CONTROLLED SECURITY PROJECTS**

Have the latest security systems protecting your family, home and property for a fraction of the price of installing a commercially-available system. With only a basic understanding of electronic principles, you can design, build and install devices ranging from an electronic light controller to a complete 6502-microcomputer based security system. 240 pp., 119 illus. 7″ × 10″.
**Paper $14.95 Hard $21.95**
**Book No. 1957**

☐ **A BEGINNER'S GUIDE TO MAKING ELECTRONIC GADGETS—2nd Edition**

You can build a metal detector, light-operated control or alarm devices, a light beam radio, an inductive loop radio control . . . the list of practical, fun and easy-to-build projects goes on and on! If you're interested in electronics, but unsure of how to get started building projects that actually accomplish something, this is the ideal guide for you. It includes step-by-step how-to's and illustrations. 182 pp., 152 illus.
**Paper $8.95 Hard $14.95**
**Book No. 1793**

# Other Bestsellers From TAB

☐ **BASIC ELECTRONICS THEORY—WITH PROJECTS AND EXPERIMENTS—2nd Edition—Horn**

If you're looking for an introduction to modern electronics . . . or if you're an experienced hobbyist or technician in need of a quick-reference on the many facets of today's analog or digital electronics practice . . . there's simply no better sourcebook than this new second edition of *Basic Electronics Theory—with Projects and Experiments*! It includes all the basics plus the most recent digital electronics developments! 672 pp., 645 illus.

**Paper $18.95 Hard $29.95**
**Book No. 1775**

☐ **CMOS/TTL—A USER'S GUIDE WITH PROJECTS—Carr**

If you're an electronics enthusiast or a computer hobbyist, this is the perfect guide to get you up-to-speed in digital electronics and to show you how CMOS/TTL devices work and how they are used in specific applications. It's a book that can add greatly to your enjoyment of an electronics hobby, and provide the basis for new directions in spare-time or on-the-job accomplishment. 368 pp., 327 illus.

**Paper $13.50 Hard $19.95**
**Book No. 1650**

☐ **THE AUTOMOTIVE SECURITY SYSTEM DESIGN HANDBOOK—Gifford**

With the expert guidance provided by this state-of-the-art security handbook, you cannot only protect your vehicles, but do it at a fraction of the cost of commercially made and installed security units! Absolutely nothing is left out from selecting the type of device needed to step-by-step construction and installation procedures. You can rest assured that your vehicles are safe from thieves and vandals! 240 pp., 106 illus.

**Paper $12.95 Hard $18.95**
**Book No. 1734**

☐ **THE GIANT BOOK OF ELECTRONIC PROJECTS**

From amateur radio to computers, from just-for-fun experiments to practical, problem-solving equipment, from simple do-it-in-an-evening gadgets to highly sophisticated electronics gear, this huge compendium of do-it-yourself electronics projects includes step-by-step instructions for building and using everything from an EPROM eraser to an automated CW for a level I TRS-80. The easy-to-follow directions make it possible for you to have all kinds of sophisticated electronics gear at far less than off-the-shelf prices. 504 pp., 343 illus.

**Paper $17.95 Book No. 1367**

*Prices subject to change without notice

**Look for these and other TAB books at your local bookstore.**

**TAB BOOKS Inc.**
**P.O. Box 40**
**Blue Ridge Summit, PA 17214**

**Send for FREE TAB catalog describing over 1200 current titles in print.**